Lecture Notes in Artificial Intelligence

Subseries of Lecture Notes in Computer Science
Edited by J. Siekmann

Lecture Notes in Computer Science
Edited by G. Goos and J. Hartmanis

Ph. Jorrand   J. Kelemen (Eds.)

Fundamentals of Artificial Intelligence Research

International Workshop FAIR '91
Smolenice, Czechoslovakia, September 8-13, 1991
Proceedings

Springer-Verlag
Berlin Heidelberg New York
London Paris Tokyo
Hong Kong Barcelona
Budapest

Series Editor

Jörg Siekmann
Institut für Informatik, Universität Kaiserslautern
Postfach 3049, W-6750 Kaiserslautern, FRG

Volume Editors

Philippe Jorrand
Institut IMAG-LIFIA
46, Avenue Felix Viallet
F-38031 Grenoble Cedex, France

Jozef Kelemen
Department of Artificial Intelligence
Faculty of Mathematics and Physics, Comenius University
Mlynskádolina, 842 15 Bratislava, Czechoslovakia

CR Subject Classification (1991): I.2.3-4, F.4.1-2, I.2.6

ISBN 3-540-54507-7 Springer-Verlag Berlin Heidelberg New York
ISBN 0-387-54507-7 Springer-Verlag New York Berlin Heidelberg

Typesetting: Camera ready by author
Printing and binding: Druckhaus Beltz, Hemsbach/Bergstr.
2145/3140-543210 - Printed on acid-free paper

Foreword

This volume contains the text of 7 invited lectures and 13 submitted contributions to the scientific programme of the international workshop *Fundamentals of Artificial Intelligence Research, FAIR '91*, held at Smolenice Castle, Czechoslovakia, September 8-12, 1991.

The international workshop *Fundamentals of Artificial Intelligence Research* was jointly organized by the Association of Slovak Mathematicians and Physicists, by the Laboratory of Fundamental Informatics and Artificial Intelligence of the Institute of Informatics and Applied Mathematics of Grenoble, France, and by the Department of Artificial Intelligence of the Comenius University, Bratislava, Czechoslovakia, under the sponsorship of the European Coordinating Committee for Artificial Intelligence, ECCAI.

FAIR '91, the first one of an intended series of international workshops, adresses issues which belong to the theoretical foundations of Artificial Intelligence (AI) considered as a discipline focused on concise theoretical description of some aspects of intelligence by tools and methods adopted from mathematics, logics, and theoretical computer science. The intended goal of the FAIR workshops is to provide a forum for the exchange of ideas and results in a domain where theoretical models play an essential role. Such theoretical studies, their development and their relations to AI experiments and applications have to be promoted in the AI research community.

The scientific programme of FAIR '91 was prepared by the Programme Committee consisting of Steffen Hölldobler (Technical University Darmstadt, Germany), Klaus - Peter Jantke (Technical University Leipzig, Germany), Philippe Jorrand (chairman, Institute IMAG - LIFIA, Grenoble, France), Jozef Kelemen (Comenius University, Bratislava, Czechoslovakia), Augustin Lux (Institute IMAG - LIFIA, Grenoble, France), Alberto Martelli (University of Torino, Italy), and Henri Prade (Institute IRIT - LSI, Toulouse, France). The editors wish to thank all of them for their meritorious work in consultating the invited speakers and evaluating the papers submitted to response to the call for papers as well as to all of subreferees who assisted the Programme Committee members.

On behalf of all the participants of FAIR '91 we express our gratitude to the members of the organizational staff of the workshop, especially to Peter Mikulecký for chairing the Organizing Committee, to Juraj Waczulík, Richard Nemec, and Michal Winczer for technical assistance, and to Alica Kelemenová for her substantial help in finishing the editorial work.

We are highly indebted to all the contributors for preparing their texts carefully and on time. We would also like to gratefully acknowledge the support of all the above mentioned cooperating institutions. Last but not least we wish to thank Springer-Verlag for excellent cooperation in publication of this volume.

July 1991

Philippe Jorrand
Jozef Kelemen

Contents

Part III: Appendix

Part I
Invited Lectures

User-oriented theorem proving with the ATINF graphic proof editor*

Ricardo Caferra, Michel Herment & Nicolas Zabel

LIFIA-IMAG
46 Av. Félix Viallet, 38031 Grenoble Cedex, FRANCE
{caferra@neptune.imag.fr; michel@lifia.imag.fr; zabel@nereide.imag.fr}

Abstract

An inference laboratory called ATINF, developed at LIFIA since 1985 is presented. Its main design principles and some of its components are described. The presentation gives the greatest place to the graphic proof editor of ATINF and an example shows how proof edition can be profitably used in user-oriented theorem proving. Particularly interesting in this concern are the hierarchical presentation of proofs and the possibility of handling natural tricks such as renaming and taking advantage of symmetries.

1. Introduction

In 1965 J.A. Robinson published his seminal paper on theorem proving "A *machine-oriented* logic based on the resolution principle". The resolution principle, born in 1963, was 25 years old in 1988. In this occasion, during CADE-9, J.A. Robinson gave a very nice -unfortunately unpublished- talk entitled (if we recall correctly) "How *human-oriented* can a mechanical proof be?". In his conference Robinson emphasized the key concept of *proof outline*, which is a very natural notion, especially if one really wants to use a mechanical theorem prover as an assistant mathematician (in the sense for example of [Wos88]). The philosophy underlying this talk illustrates very well the evolution of theorem proving in the last 25 years. It is worthwhile to notice a convergence with ideas coming from Mathematics, particularly with a deep reflection about a broad sense of the notion of "mathematical proof" and on the impossibility to understand or to communicate "big proofs" (see for example [DLP79], [DH86]).

These ideas lead naturally to the conclusion that *it is necessary to have in a theorem prover different levels of presentation of proofs.*

This necessity has been one of the guiding principles of our project of Inference Laboratory, named ATINF and developed at LIFIA since 1985.

In this paper we give in section 2 an overview of the present state of the ATINF project. In section 3, one of its major components, a graphic proof editor, is

---
* This work was partially supported by MEDLAR (Mechanizing Deduction in the Logic of Practical Reasoning) ESPRIT-BRA 3125 project.

described and some of its original features are evidenced. Finally in section 4 a "long proof" of a nontrivial theorem is presented. The capabilities of the proof editor and a unified approach theorem proving/proof checking allows us to take advantage — in a "mathematician-like" style — of the particularities of this theorem. Only the most relevant parts of the proof are displayed.

2. Brief description of the Inference Laboratory ATINF

ATINF is an acronym for ATelier d'INFérence which stands in French for Inference Laboratory. Its design was founded mainly on the following bases:

1- Unified approach of theorem proving and proof checking (as in [McC61]).

2- Unified approach of search for proofs (or refutations) and counterexamples (or models).

3- Cooperation (integration) of different theorem provers for classical and non classical logics. One of the characteristics of the integration is that it is possible to complete different parts of a single proof using different proof systems.

4- User-oriented, principally through graphic proof edition. Particularly, we clearly distinguish between the *way of proving* from the *way of presenting* a proof.

Usual requirements such as efficiency and ease of maintenance and portability have not been neglected.

Point 3 corresponds to a presently important trend of research: design of *hybrid reasoning sytems* (see for ex. [FC91] and also [B+90]).

Point 4 is capital in our approach. The idea behind it can be summarized in distinguishing two levels in the handling of proofs. To do so we have borrowed and adapted two notions and the correponding terminology from Martin-Löf's philosophy of mathematics (as we understand it). The two notions are those of *derivation* (i.e. an object able to convince us of the truth of a judgement) and *proof* (i.e. an object containing all the computational information needed by a mechanical agent in order to verify a proposition).

What we find in mathematical books and papers and in everyday mathematics are *derivations, not proofs*.

It is therefore natural to reserve *derivations for communication purposes* between a prover (human or machine) and a human being and *proofs for mechanical proofs (or checking)*. To convince the skeptical reader about this choice it is enough to consider the well kown examples of proofs with length of hundreds of pages and the obvious impossibility of humans to be able to understand or even to check such proofs (see for example [DLP79], [DH86]).

Theoretical and practical results on ATINF have been published elsewhere ([BC87a], [BC87b], [BdIT90], [BC88], [BCC88a], [BCC88b], [BC90], [CZ90a], [CZ90b], [CDH91a], [CDH91b], [Cha89], [Cha91], [Del91]); others are in progress ([CDH91b], [Cha91]).

The present state of the ATINF project is the following:

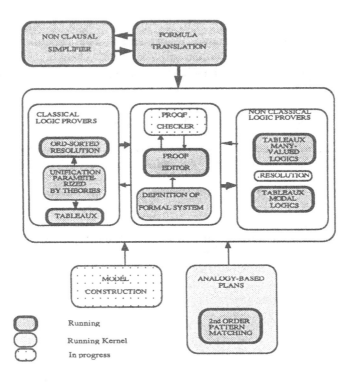

We mention here only the possibilities of the two theorem provers for nonclassical logics and of the parameterized unification module.

The tableaux theorem prover for non classical logics handles linear temporal logic and several first order modal logics (normal as well as non normal ones).

The (parameterized) many-valued theorem prover handles: logics with arbitrary finitely valued propositional logics, in particular some paraconsistent logics and in the near future some propositional infinitely valued logics. Truth values can be only partially ordered. It is also possible to deal with first-order many valued logics with equality.

The general e-unification (i.e. unification modulo equational theories) module has been designed in order to be used in a general purpose theorem prover: it balances efficiency with completeness. It can be profitably used in mechanizing non classical logics as in the translation method (see for example [Ohl88]).

3. Graphic proof edition in ATINF. Its philosophy and possibilities

It should be clear that in ATINF we are not looking for a logic or formalism as general as possible in order to be able to *program any theorem prover* or to *formalize all the mathematics* (excellent works exist in this field, see for example [DeB80], [C+86], [HHP87], [CH88], [Pau89]) but for a formalism able to *describe* proof systems corresponding to as much logics as possible. These proof systems will correspond in general to *already existing* theorem provers or proof checkers.

The difference is not a shade of meaning but a deep one: we do not need to care about difficult matters such as soundness, completeness,...which are supposed to be already solved for each theorem prover (proof checker).

The task of the proof editor is to present proofs in different ways and to communicate with the theorem provers and with the user. We need only be interested in abstract proof representation (internal) and proof presentation (external). Here "abstract proof representation" means, obviously, unique and independent of the machine representation for a given theorem prover.

It is natural to ask about the prize to pay for the facility of editing proofs coming from different theorem provers (including those not in ATINF). It is very low: it requires, for each theorem prover, to write a program that is the equivalent of a (very primitive) pretty printer (say 2 hours programming); for details see [CDH91b].

Incidentally, the boxes discipline of the editor can typically be used for presentation of natural deduction proofs as, for example, in [Fit90] and fits very well (and very intuitively) with possible worlds semantics: change of boxes corresponds to change the state of a world or change of world.

Considering the practical importance (in classical and nonclassical logics) of a structured presentation of proofs it is astonishing to notice how proof edition has been neglected in the field of theorem proving.

The most important features of our proof editor are the following:

- Based on a graphic approach (boxes discipline).
- Multi-windowing system.
- Possibility of hierarchical representation of proofs.
- Possibility of memorize parts of proofs and use them again
- Ability to define a rather large class of formal systems.
- Possibility of verify that some parts of a proof can (may be after a renaming) be reused.
- Possibility of "putting in a box" an arbitrary number of proof steps. It is therefore possible to handle inference steps of arbitrary "size" — or equivalently, arbitrary derived inference rules.

One of the few systems that incorporates proof edition as one of its major components is Nuprl [C+86]. It is one of the most important and interesting existing systems for mechanizing mathematics. We select 3 characteristics of the Nuprl editor in order to illustrate some of the originalities of our approach.

In [C+86] we can read:

p.67: "...using the refinement editor, or red for short, to navigate through the proof tree".

It is possible to navigate, but not to see or to handle arbitrary parts of the proof.

p. 206: "If in a middle of a proof one discovers that the current subgoal is similar to one proved previously, the pair of tactics mark and copy can be used..."

The approach in the ATINF proof editor fits much more the usual mathematical practice: parts of a proof which the user thinks to be essentially the same as needed in other parts of the proof can be "pasted" elsewhere (if allowed by the editor-unifier — see section 4) without using any programming or command language.

p. 125: "The proof editor window always displays the *current node* of the proof...".

That is to say, proof edition in Nuprl is *local*. In this respect our approach, essentially *global*, is definitely more useful because if one wants to be able to recognize different patterns it is necessary *"to see"* the concrete realizations of these patterns.

4. An example

We have chosen to treat a problem due to J. Los because it belongs to "...a type of problems whose statement can be quite simple but whose proofs are nevertheless quite difficult for ATP (and people) to find" [PR86].

This problem has been considered as a challenge by the automated deduction community as evidenced by the series of papers published in the AAR Newsletters and the diversity of techniques employed to solve it.

We give here a refutation using the ATINF tableau prover for classical logics. Of course similar work can be done using the ATINF resolution theorem prover. Automatically generated proofs are in general not easy to read. The hierarchical and modular presentation done by the ATINF proof editor enables the user to grasp at a glance a long proof (length of approximately 10 pages), to handle the symmetries and to reuse parts of the proof in the way a mathematician would do it.

We give a proof for the original statement of the problem (see [Rud 87]). We have chosen it because it allows us to show how the facilities of the editor enables the user to see different levels of the proof and to use natural tricks. Then we indicate how to modify this proof to get a proof of the simplified version of the theorem, which is the version considered in the different publications.

$$
\begin{array}{ll}
\text{(tp)} & \forall x\, \forall y\, \forall z\, (Pxy \wedge Pyz \Rightarrow Pxz) \\
\text{(tq)} & \forall x\, \forall y\, \forall z\, (Qxy \wedge Qyz \Rightarrow Qxz) \\
\text{(sq)} & \forall x\, \forall y\, (Qxy \Rightarrow Qyx) \\
\text{(sp)} & \forall x\, \forall y\, (Pxy \Rightarrow Pyx) \\
\text{(sr)} & \forall x\, \forall y\, (Pxy \vee Qxy) \\
\hline
\text{(c)} & \forall x\, \forall y\, Pxy \vee \forall x\, \forall y\, Qxy
\end{array}
$$

A natural (and trivial) remark is the symmetry of P and Q. The user will normally look for a proof exploiting this symmetry.

The following series of screen-copies presents the proof using the ATINF editor. For the sake of simplicity, we give the skolemized version of the problem, that is to say, we consider the negation of (c) and introduce the skolem constants A, B, C and D.

The first screen copy displays a compact representation of the proof.

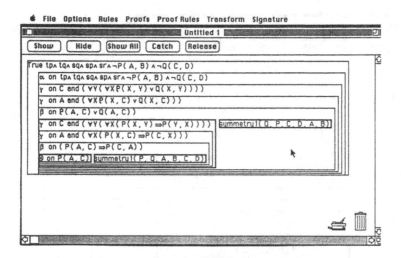

The rationale for this proof structuring is the following:

When the user considers the case analysis upon an instance of (sr): P(A,C) ∨ Q(A,C) — topmost β in the screen — (s)he will very likely notice that using the substitution P ↦ Q, Q ↦ P, A ↦ C, C ↦ A, B ↦ D, D ↦ B, the right hand side of the screen copy (the Q(A,C) case) becomes a set of formulae containing the axioms and the literal P(C,A). In order to get a left-hand subtableau containing this set of formulae, (s)he adds to the P(A,C) case the deduction of P(C,A) from (sp) and P(A,C), and considers, this subtableau is closed as the right one is. This appears in the screen as the box containing the last four lines in the left-hand side of the screen.

The renaming explained above is mirrored by the two boxes "**symmetry1**...".

symmetry1 has been memorized as a subproof

If the user wants to see the fully expanded tableau, (s)he must "look inside" the two boxes "**symmetry1**...". If (s)he "looks" for example "inside" the box "**symmetry1 (Q,P,C,D,A,B)**", (s)he will see the subproof corresponding to the case P(B,D) ∨ Q(B,D). The case P (B,D) can also be handled using the symmetry of the problem ("**symmetry2**..." boxes).

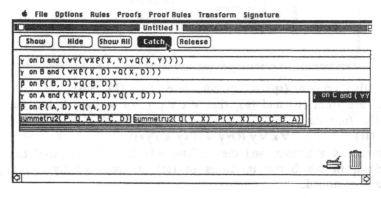

Now, if the user decides to develop the darkened part of the proof (right-hand side of the screen), (s)he can do so clicking on "**catch**".

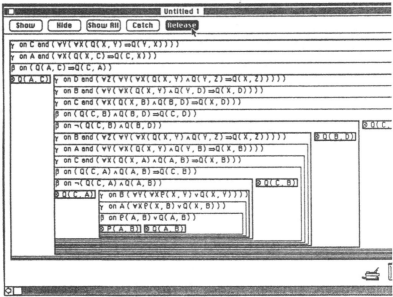

By clicking on **"release"**, the user comes back to the previous screen. So the user can examine for example **"symmetry2(P,Q,A,B,C,D)"**, (s)he gets:

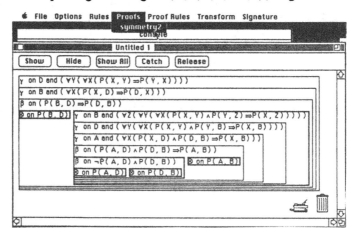

In the simplified version, (sp) is not considered (see [PR 86]).

It is easy to see from (sr) and (sq) that $\forall x \, \forall y \, (Qxy \vee Pyx)$ holds also. By using (sr) again the formula

$$(lr) \qquad \forall x \, \forall y \, (Qxy \vee (Pxy \wedge Pyx)),$$

can be deduced as a lemma, and the skeleton of the previous proof can be reused, as a guideline, in which the instances of (sr) are replaced by (lr) and the (sp) instances are removed.

These four screens show essential parts of the proof and should give a clear idea how the ATINF editor works.

5. Bibliography

[B+90] F. Baader, H-J. Bürckert, B. Hollunder, W. Nutt, J.H. Siekmann: *Concept logics* in "Computational logic" (J.W. Lloyd ed.) Springer-Verlag 1990, pp. 177-201.

[BC87a] T. Boy de la Tour, R. Caferra: *Proof analogy in interactive theorem proving: a method to use and express it via second order pattern matching.* Proc. AAAI-87, Morgan and Kaufmann 1987, pp. 95-99.

[BC87b] T. Boy de la Tour, R. Caferra: *L'analogie en démonstration automatique: une approche de la généralisation utilisant le filtrage du second ordre.* Actes 6ème RFIA. Tome 2. Dunod Informatique, 1987, pp. 809-818.

[BdlT90] T. Boy de la Tour: *Minimizing the number of clauses by renaming.* Proc. of CADE-10, LNAI 449, Springer-Verlag 1990, pp. 558-572.

[BC88] T. Boy de la Tour, R. Caferra: *A formal approach to some usually informal techniques used in mathematical reasoning.* Proc. of ISSAC-88. LNCS 358, Springer-Verlag 1988, pp. 402-406.

[BCC88a] T. Boy de la Tour, R. Caferra, G. Chaminade : *Some tools for an Inference Laboratory (ATINF).* Proc. of CADE-9, LNCS 310, Springer-Verlag 1988, pp. 744-745.

[BCC88b] T. Boy de la Tour, R. Caferra, G. Chaminade: *Towards an inference laboratory.* Proc. of COLOG-88 (International Conference in Computer Logic) P. Lorents, G. Mints and E. Tyugu eds. , part II, pp. 5-13.

[BC90] T. Boy de la Tour, G. Chaminade: *The use of renaming to improve the efficiency of clausal theorem proving.* Proc. AIMSA '90. North-Holland 1990, pp. 3-12.

[CZ90a] R. Caferra, N. Zabel: *Extending resolution for model construction.* Proc. of Logics in AI - JELIA '90, LNAI 478, Springer-Verlag 1990, pp. 153-169.

[CZ90b] R. Caferra, N. Zabel: *An application of many-valued logic to decide propositional S5 formulae: a startegy designed for a parameterized tableaux-based theorem prover.* Proc. AIMSA '90, North-Holland 1990. pp. 23-32.

[CDH91a] R. Caferra , S. Demri , M. Herment : *Logic morphisms as a framework for backward transfer of lemmas and strategies in some modal and epistemic logics.* To appear in Proc. of AAAI-91.

[CDH91b] R. Caferra, S. Demri, M. Herment: *Logic-independent graphic proof edition : an application in structuring and transforming proofs.* In preparation.

[Cha89] G. Chaminade: *An implementation oriented view of order- sorted resolution.* Revue d'Intelligence Artificielle Vol. 3 (1), August 1989, pp. 7-30.

[Cha91] G.Chaminade: *Intégration et implémentation de mécanismes de déduction naturelle dans les démonstrateurs utilisant la résolution.* Ph.D thesis. Institut National Polytechnique de Grenoble. Forthcoming (October 1991).

[CH88] T. Coquand, G. Huet: *The Calculus of Constructions,* Information and Computation 76 (1988), pp. 95-120.

[C+86] R. Constable et al.: *Implementing mathematics with the Nuprl development system.* Prentice-Hall 1986.

[DH86] P.J. Davis, R. Hersh: *Descarte's dream.* Harcourt Brace Jovanovich 1986.

[deB80] N.G. de Bruijn: *A survey of the project AUTOMATH* in "To H.B Curry: Essays on Combinatory Logic, Lambda calculus and formalism" (J.P. Seldin and J.R. Hindley eds.) Academic Press 1980, pp. 579-606.

[Del91] B. Delsart: *General e-unification: balancing efficiency and completeness.* To be presented in Unif'91.

[DLP79] R.A. De Millo, R.J. Lipton, A.J. Perlis: *Social processes and proofs of theorems and programs.* Comm. of the ACM, Vol. 22, Number 5 (1979), pp. 271-280.

[Fit90] M. Fitting: *First-order logic and automated theorem proving.* Springer-Verlag 1990.

[FC91] A.M. Frisch, A.G. Cohn: *Thoughts and Afterthoughts of the 1988 Workshop on principles of hybrid reasoning.* AI Magazine, Vol. 11, Number 5, January 1991, pp. 77-83.

[HHP87] R. Harper, F. Honsell, G. Plotkin: *A framework for defining logics,* Proc. of the 2nd Annual Logic in Computer Science, Ithaca, NY, 1987.

[KF90] T. Käufl, N. Zabel: *Cooperation of decision procedures in a tableau-based theorem prover.* Revue d'Intelligence Artificielle. Special issue on Automated Deduction. Vol. 4 (3), November 1990, pp. 99-125.

[McC61] J. Mc Carthy: *Computer programs for checking mathematical proofs.* Proc. Amer. Math. Soc. Recursive Function Theory, 1961, pp. 219-227.

[Ohl88] H-J. Ohlbach: *A resolution calculus for modal logics.* Proc. of CADE-9, LNCS 310, Springer-Verlag 1988, pp. 500-516.

[Pau89] L.C. Paulson: *The foundations of a generic theorem prover.* Journal of Automated Reasoning 5 (1989), pp. 363-397.

[PR86] F.J. Pelletier, P. Rudnicki: *Non-Obviousness.* AAR Newsletter N° 6, September 1986, pp. 4-5.

[Rob65] Robinson J.A.: *A machine-oriented logic based on the resolution principle.* Journal of the ACM, Vol. 12, N° 1, January 1965, pp. 23-41.

[Rud87] P. Rudnicki: *Obvious inferences.* Journal of Automated Reasoning. Vol. 3, N° 4, December 1987, pp. 383-393.

[Wos88] Wos L.: *Automated Reasoning: 33 basic research problems.* Prentice-Hall 1988.

A MODAL ANALYSIS OF POSSIBILITY THEORY

Luis Fariñas del Cerro, Andreas Herzig

IRIT
Université Paul Sabatier
118 Route de Narbonne, F-31062 Toulouse Cédex, France
email: {farinas,herzig}@irit.fr, fax: (33) 61 55 62 58

Abstract

In this paper we study possibility theory from the point of view of modal logic. Our first and main result is that the logic of qualitative possibility is nothing else than Lewis's conditional logic VN. Second, we propose a multi-modal logic able to support possibility theory. Some connexions between these formalisms are stressed.

1. Introduction

To find models able to support reasoning under uncertainty is an important problem for Artificial Intelligence as well as Logic and Formal Philosophy. Several models have been proposed, like probability theory, belief structures, decomposable confidence measures, qualitative neccessity or possibility theory. This last theory (and dually necessity theory) has been introduced by D. Dubois and H. Prade in 1986 [Dubois & Prade 88] and seems to be a "nice" point of view from which several other models can be observed. Given a set of events this theory allows to associate an uncertainty degree to each event, thanks to a function P mapping events (being sets of elementary events) into the real interval [0,1] which satisfies the following axioms:

P1. $P(\text{True}) = 1$

P2. $P(\text{False}) = 0$

P3. $P(A \cup B) = \max(P(A), P(B))$

The function P is called a *possibility measure*. It induces a relation "$\geq$" between events defined by $A \geq B$ if and only if $P(A) \geq P(B)$. We call $\geq$ a relation *agreeing strictly* with P. $A \geq B$ means that A is at least possible as B. This relation is called *qualitative possibility relation* and satisfies the following conditions:

QP1. (tautology) $\text{True} \geq F$

QP2. (compatibility) $A \geq B$ or $B \geq A$

QP3. (transitivity)   if $A \geq B$ and $B \geq C$ then $A \geq C$

QP4. (non triviality) $\text{True} > \text{False}$

QP5. (disjunctive stability) if $B \geq C$ then for each A, $A \cup B \geq A \cup C$

where A, B and C represent sets of events; False stands for the empty set and True for the total set of events.

Just as Kraft, Pratt, and Seidenberg did for probabilities [Scott 64], D. Dubois [Dubois 86] established the formal relation between possibility theory and qualitative possibility relations by means of the following theorem:

Dubois's Theorem. The only functions mapping events into [0,1] which strictly agree with qualitative possibility relations are possibility measures. Moreover a strictly agreeing possibility measure always exist.

In a recent work P. Gärdenfors and D. Makinson have established a similar relation between quantitative and qualitatives plausibility measures. Such measures allow them to characterize an interesting class of nonmonotonic reasoning [Gärdenfors, Makinson 91].

2. From Qualitative Possibility to Lewis's System VN

In this section we present a conditional logic obtained from qualitative possibility relations, and we prove that it is exactly Lewis's VN logic. Thus we proceed in the same manner as Segerberg did for probability measures [Segerberg 71].

2.1 Qualitative Possibility Logic (QPL)

Considering that events are represented not by sets of elementary events, but by propositional formulas, we can represent the qualitative possibility relation as a particular dyadic connective. Formally we define a propositional language with the classical connectives : $\land, \lor, \neg$, etc. plus $\geq$. Formulas of QPL (called qualitative formulas) are defined as usual, in particular $F \geq G$ is a qualitative formula if F and G are qualitative formulas. The axiomatics of QPL is the following:

QPL0.   Classical axioms
QPL1. $(F \geq G \land G \geq H) \rightarrow F \geq H$
QPL2. $F \geq G \lor G \geq F$
QPL3. $\neg(\text{False} \geq \text{True})$
QPL4. $\text{True} \geq F$
QPL5. $F \geq G \rightarrow F \lor H \geq G \lor H$

with the inference rules

QPL6.   modus ponens
QPL7.   If $F \leftrightarrow G$ then $F \geq G$

Since the axioms of this conditional logic and the axioms of qualitative possibility are exactly the same we can assume that this logic is the logic of qualitative possibility. We stress that the axioms Q1,..., Q5 for qualitative possibility relations correspond to formulas without nested $\geq$-operators. On the contrary in logic QPL, nested $\geq$-operators are allowed. Remark also that switching from sets of events to propositional formulas forces the introduction of QP7. It will be used in the sequel that QPL4 can be deduced from QPL0, QPL2, QPL3 and the inference rule QPL6: As $\neg(\text{False} \geq \text{True}) \to \text{True} \geq \text{False}$ is an instance of QPL2, modus ponens on this formula and QPL3 produces $\text{True} \geq \text{False}$.

QPL can be considered a conditional logic: As indicated above, the formula $F \geq G$ expresses that the possibility degree of G is less or equal than the possibility degree of F. In other words, it is at least as possible that F as it is that G.

2.2   Lewis's System VN

D. K. Lewis [Lewis 73] defined a conditional logic called VN. Its language is that of classical logic augmented by a conditional dyadic connective. Let us to denote this conditional connective by $\geq$ as previously. Hence the language of VN is the same language of qualitative formulas as for QPL. One reading Lewis gives for a formula of the form $F \geq G$ is that "it is at least as possible that F as it is that G". This gives already a hint about the proximity of VN and QPL. The axioms of VN are

VN0.	Classical axioms
VN1.	$(F \geq G \land G \geq H) \to F \geq H$
VN2.	$F \geq G \lor G \geq F$
VN3.	$\neg(\text{True} \geq \text{False})$

and the inference rules are

VN4.	modus ponens
VN5.	If $F \to (G_1 \lor \ldots \lor G_n)$ then $(G_1 \geq F) \lor \ldots \lor (G_n \geq F)$

2.3   The Equivalence Theorem

Theorem (Equivalence of QPL and VN). QPL0, ..., QPL7 is an alternative axiomatization of the conditional logic VN.

Proof. The axioms VN0,..., VN3 are the same as QPL0,..., QPL3, respectively, and VN4 the same as QPL6. Moreover, as we have noted in 2.1., QPL4 can be deduced from the other axioms and inference rules of QPL. It remains to establish that VN5 corresponds to QPL5 and QPL7. First of all, it can be shown that if we replace QPL5 by the two axioms

QPL5'.	$(F \geq (F \lor G)) \lor (G \geq F \lor G)$
QPL5".	$(F \lor G) \geq F$

then the resulting axiomatics is an alternative axiomatization for QPL. Second, it can be shown that if we replace the inference rule VN5 by the axiom

VN5'.	$(F \geq (F \lor G)) \lor (G \geq (F \lor G))$

plus the inference rule

VN5".	If $F \to G$ then $G \geq F$

then the resulting axiomatics is an alternative axiomatization for VN. Then immmediately, QPL5' corresponds to VN5'. Now on the one hand QPL7 can be deduced with VN5", and QPL5" is obtained from VN5" (by inferring $(F \lor G) \geq F$ from $F \to (F \lor G)$). Finally, it suffices to show that VN5" can be deduced with QPL5" and QPL7.

3. Semantics for Qualitative Possibility

In this section we shall present two different semantics for qualitative possibility theory, one based in the notion of a sphere, and another in traditional relational Kripke style.

3.1 Sphere Semantics

D. K. Lewis proposed for his logic VN a semantics in terms of spheres. Due to its completeness and to the above equivalence theorem, sphere semantics will be a semantics for possibility theory.

Let W be a nonempty set of *possible worlds* and S be a mapping which assigns to each world w a nonempty set $S(w)$ of sets of worlds. S is called a *system of spheres*, and each $S(w)$ is called a sphere around w, if and only if, for each world w in W, the following conditions hold:

- $U \subseteq V$ or $V \subseteq U$ for every $U, V \in S(w)$
- for every subset $s \subseteq S(w)$, $\cup s, \cap s \in S(w)$

An *sphere model* $M = (W, S, m)$ for VN is defined by a set of worlds W, a system of spheres S, and a meaning function m mapping propositional variables into sets of worlds. *Satisfaction* of a formula in a world w of a model M is defined as usual, in particular for the case of the conditional operator we have

$$M, w \models F \geq G \text{ if for every sphere V in } S(w),$$
$$\text{if there is a w' in V such that } M, w' \models G$$
$$\text{then there is a w" in V such that } M, w" \models F.$$

A formula F is *true in a sphere model* M if and only if for every world w in M we have that $M, w \models F$. A formula is *valid in sphere models* if and only if it is true in every model. D.K. Lewis proved the following completeness and soundness theorem for VN.

Theorem (Soundness and completeness of VN). Let F be a qualitative formula. F is a theorem of VN if and only if F is valid in sphere models.

3.2 Multi-Relational Semantics

In the spirit of the usual semantics for multi-modal logics we define possible world models with a set of accessibility relations. A set P is called a *parameter set* if the constant 1 is in P. By a *multi-relational model indexed by P* we mean a structure of the form

$$M = (W, \{R_p : p \in P\}, m)$$

where W is a nonempty set of worlds, and each R_p is a binary relation on W. The set $\{R_p : p \in P\}$, m) is a totally ordered set w.r.t. relation inclusion whose supremum is R_1, and R_1 is serial (i.e. for every world w there is a world w' such that $w R_p w'$). m is a meaning function mapping propositional variables into sets of worlds. *Satisfaction* of formulas in a world w of a model M is defined as usual, in particular for the case of the conditional operator we have

$$M,w \models F \geq G \text{ if for every p in P, if } M,w' \models G \text{ for some w' such that } w R_p w'$$
$$\text{then } M,w' \models F \text{ for some w' such that } w R_p w'$$

A formula F is *true in a multi-relational model M indexed by P* if and only if for every world w in M we have that $M,w \models F$. A formula is *valid in multi-relational models* if and only if it is true in every model indexed by any P.

Theorem (Soundness of QPL). Every axiom of QPL is valid in multi-relational models, and every inference rule of QPL preserves validity.

According to Lewis we can view - at least in the finite case - every sphere model as a multi-relational model. Therefore and in consequence of the above lemma, multi-relational models are a candidate for qualitative possibility semantics.

4. Possibility logic (PL)

Originally, possibility theory has been introduced in terms of a measure on events, and in the literature one can find many applications of possibility theory where these measures are manipulated directly. That motivates search of logics which model such numerical representations explicitly in the language. In this section we present a tentative of such a formalization by introducing a multi-modal logic which also axiomatizes the multi-relational models of the previous section.

Up to now we have used the concept of possibility instead of the dual notion of necessity. As the latter concept is traditionally used as a primitive in modal logics, we have preferred to state our multi-modal logic in terms of necessity[1].

---

[1] We can switch easily from one concept to the other thanks to their duality: To every possibility measure P there is associated a function N mapping events into [0,1] by N(A) = 1 - P(-A). N is called a *necessity measure*, and it satisfies

N1.   N(True) = 1
N2.   N(False) = 0
N3. $N(A \cap B) = \min(N(A),N(B))$

As before, we can associate to N a relation "$\geq$" between events defined by $A \geq B$ if and only if $N(A) \geq N(B)$ such that the necessity counterpart of Dubois's theorem holds. This relation is called *qualitative necessity relation*. (We use the same symbol for the relation than before.) Dually to the axiomatization of qualitative possibility logic QPL in section 2, we can define qualitative necessity logic (QNL) by axioms QNL0, ..., QNL7, all of which coincide with that of QPL except QPL5 which is replaced by

QNL5. $F \geq G \rightarrow F \wedge H \geq G \wedge H$

4.1 Axiomatics

The language of a possibility logic PL$_P$ is a multi-modal extension of the classical propositional language with a parameter set P. To each parameter $p \in P$ we associate a modal operator [p]. Formulas of PL$_P$ (called quantitative formulas) are defined as usual, in particular [p] F is a formula if F is a formula. We may read [p] F as "the necessity of F is at least p". As usual, <p> F is an abreviation for $\neg$ [p] $\neg$F.

The axiomatics of possibility logic PL$_P$ is an extension of classical logic:

PL0.   Classical axioms
PL1.   [p] True
PL2. $\neg$ [1] False
PL3.   ([p] F $\wedge$ [p] G) $\leftrightarrow$ [p] (F $\wedge$ G)
PL4.   ([p] F $\rightarrow$ [p] G) $\vee$ ([q]G$\rightarrow$ [q]F)

with the inference rules

PL5.   modus ponens
PL6.   If F $\leftrightarrow$ G then [p] F $\leftrightarrow$ [p] G

Thus, PL$_P$ is a multi-modal logic where the operator [1] is axiomatized as in modal logic KD, and every other operator [p] as in modal logic K. Moreover, PL$_P$ states a particular interaction between modal operators: The axiom PL4 is classically equivalent to

PL4'.   ([p] F $\wedge$ [q] G) $\rightarrow$ [q]F $\vee$ [p]G

This formulation makes it clearer that PL4 axiomatizes that P is ordered linearly.

4.2 Semantics

The semantics of PL$_P$ is defined on multi-relational models indexed by the parameter set P. *Satisfaction* of formulas in a world w of a model M is defined as before, and the only thing new is the case of the modal operator:

$$M,w \models [p] F \text{ if } M,w' \models F \text{ for every } w' \text{ such that } w \ R_p \ w'$$

Using standard techniques in modal logic as in [Hughes & Cresswell 68], we have the following completeness theorem.

Theorem (Completeness of PL). Let F be a formula of PL$_P$. F is a theorem if and only if F is true in every in multi-relational model indexed by P.

4.3   From Qualitative Possibility Logic to Possibility Logic

We define a translation mapping qualitative formulas of QPL into formulas of the possibility logic PL$_P$ indexed by a *finite* parameter set P = {p$_1$,... ,p$_n$}. This will allow to embed qualitative possibility logic QPL into possibility logic.

Definition. Let P a parameter set. The translation T$_P$ from the language of QPL into that of PL$_P$ is such that

- T$_P$(F) = F if F is an atomic formula
- T$_P$(F ≥ G) = (<p$_1$>T$_P$(G) → <p$_1$>T$_P$(F)) ∧ ... ∧ (<p$_n$>T$_P$(G) → <p$_n$>T$_P$(F))
- and homomorphic elsewhere

Then we have the following theorem.

Theorem (Translation from QPL to PL). Let F be a qualitative possibility formula and P a finite set of parameters. F ↔ T$_P$(F) is true in every multi-relational model indexed by P.

As a corollary of the translation theorem we get that *every expression of possibility theory representable by a qualitative possibility formula is captured exactly by a quantitative possibility formula*.

5.   Discussion

There are quantitative formulas of possibility logic which cannot be represented in the language of qualitative possibility logic, for example [1] F and [0] False. So what about the status of such formulas in possibility theory? Fortunately, interpreting [p] F as P(¬F) ≤ 1-p we can see that every possibility logic axiom corresponds to a valid principle of possibility theory, and every inference rule of possibility logic preserves validity, too.

The other way round, not every valid expression of possibility theory corresponds to a theorem of possibility logic. For example ¬ [p] False for some p ≠ 1 is not a theorem of possibility logic, whereas the corresponding expression P(True) > 1-p is valid in possibility theory. Informally, we can say that possibility logic axiomatizes qualitative possibility more economically than possibility theory does.

Naturally, an interesting research direction is to enrich possibility logic in order to capture exactly possibility theory. Then the major difficulty seems to know exactly what is the minimal underlying numerical structure which possibility theory needs. We think as has been suggested by D. Makinson [Makinson 91] what is basically needed from the real interval [0,1] is that its elements are ordered linearly. This is illustrated by P. Gärdenfors and D. Makinson in [Gärdenfors

Makinson 91] where they develop a plausibility theory, which is at the base of possibility theory and in which the axiom ¬(True ≥ False) is omitted. In this case the corresponding conditional logic is Lewis's logic V which is axiomatized by VN0, VN1, VN2, VN4 and VN5. The counterpart of this logic in terms of multi-modal logic is the logic obtained from possibility logic after removing the axioms PL2 and PL6. Note that the constant "1" does not occur any longer in the axioms. Such a correspondence gives a first answer to the question about the relation between modal logic and uncertainty measures.

6. Conclusion

We can summarize the analysis presented in this note by the following three points:

1. The logic of qualitative possibility is exactly Lewis's logic VN without nested ≥ -operators.
2. The sphere semantics of VN as well as multi-relational semantics are good candidates for furnishing meaning to possibility logic.
3. A natural and economic logical counterpart of qualitative possibility theory is possibility logic.

7. Acknowledgements

We are indebted to our neighbours and friends D. Dubois and H. Prade for useful suggestions concerning possibility theory, and to D. Makinson for his comments on a first version of this note.

8. References

D. Dubois (1986), Belief Structures, Possibility Theory, Decomposables Confidence Measures on finite sets. Computer and Arti. Intell. vol 5, N° 5 pp 403-417.

D. Dubois, H. Prade (1988), Possibility Theory: An Approach to Computerized Processing of Uncertainty. Plenum Press, New York.

P. Gärdenfors, D. Makinson (1991), Nonmonotonic Inference based on Expectation Ordering. Manuscript.

D. Makinson (1991), letter, march 1991.

G. E. Hughes , M. J. Cresswell (1968), An Introduction to Modal Logic. Methuen & Co.

D. K. Lewis (1973), Counterfactuals. Harvard University Press.

D. Scott (1964), Measurement structures and linear inequalities. Jounal of Mathematical Psychology, Vol. 1, 233-247.

K.Segerberg (1971), Qualitative probability in a modal setting. In: Proc. of the Second Scandinavian Logic Symposium (ed. J. E. Fenstad), North Holland.

Making Inconsistency Respectable:
A Logical Framework for Inconsistency in Reasoning, Part I - A Position Paper

Dov Gabbay and Anthony Hunter

Department of Computing, Imperial College,
180 Queen's Gate, London SW7 2BZ, UK

Email: { dg , abh } @doc.ic.ac.uk

Abstract

We claim there is a fundamental difference between the way humans handle inconsistency and the way it is currently handled in formal logical systems: To a human, resolving inconsistencies is *not necessarily* done by "restoring" consistency but by supplying rules telling one how to act when the inconsistency arises. For artificial intelligence there is an urgent need to revise the view that inconsistency is a 'bad' thing, and instead view it as mostly a 'good' thing. Inconsistencies can be read as signals to take external action, such as 'ask the user,' or invoke a 'truth maintenance system', or as signals for internal actions that activate some rules and deactivate other rules. There is a need to develop a framework in which inconsistency can be viewed according to context, as a vital trigger for actions, for learning, and as an important source of direction in argumentation.

1.   Our position

Unfortunately the consensus of opinion in the logic community is that inconsistency is undesirable. Many believe that databases should be completely free of inconsistency, and try to eradicate inconsistency from databases by any means possible. Others, address inconsistency by isolating it, and perhaps resolving it locally. All seem to agree, however, that data of the form q and ¬q, for any proposition q, can not exist together, and that the conflict must be resolved somehow.

This view is too simplistic for capturing common-sense reasoning, and furthermore, fails to use the benefits of inconsistency in modelling cognition. Inconsistency in information is the norm, and we should feel happy to be able to formalize it. There are cases where q and ¬q can be perfectly acceptable together and hence need not be resolved. In other cases, q and ¬q, serve as a welcome trigger for various logical actions. We see inconsistency as useful in directing reasoning, and instigating the natural processes of learning. We must put forward a framework for handling inconsistencies. We need to classify the various aspects of inconsistency and present logical systems which capture various strategies for reasoning with inconsistency

To summarize, inconsistency in a logical system should be dealt with in a fashion that is more akin to that of human reasoning, namely:

$$INCONSISTENCY \Rightarrow ACTION$$

Having identified an inconsistency, we need to know whether to act on it, and if so, what kind of action to take.

In this series of papers, we attempt to clarify these points, and discuss some relevant work that has been undertaken to address aspects of these issues. We feel that by taking a more appropriate view on inconsistency, we can develop an intuitive framework for diverse logics for common-sense reasoning. In this, part I of the series, we present our position.

2.   Introduction

Classical logic, and intuitionistic logic, take the view that anything follows from an inconsistency. Effectively, when inconsistency occurs in a database, it explodes. The semantics reinforces the nature of inconsistency - there is no classical or intuitionistic model for inconsistent data.

Yet, from a proof theoretic perspective, inconsistency is a powerful theorem proving technique for classical and intuitionistic logics - for any formula, if we show inconsistency holds in the database, then the formula holds. Indeed for classical logic it is even stronger - if we assume the negation of a formula, and show inconsistency holds, then the formula holds.

It is well known, that for practical reasoning, such proof theoretical power is not compatible with human reasoning. The following example will illustrate the point:

```
The Americans are very much worried these days about computer
information and technology falling into the wrong hands. As part
```

of their security measures, they have instructed the CIA and the FBI to keep detailed files on computer scientists. The CIA and FBI co-operate in amalgamating their databases.

In the case of Professor Nobody, an unfortuante discrepency had occurred. The FBI had his address as Stanford University, and the CIA had his address as Imperial College. When the two databases were joined together, the union contained contradictory information. The computer using classical logic, inferred that Professor Nobody was none other than Public Enemy No 1.

There are several points of principle here. First classical and intuitionistic logics do not deal with contradictions correctly. Although it is logically correct to infer from the above contradiction that Professor Nobody is Public Enemy No 1, it is clearly an incorrect step in terms of human reasoning. More importantly, Professor Nobody's address is completely irrelevant to the question of whether he is Public Enemy No. 1 or not.

Examples of this nature prompted the logic community to study such logics as relevant, and paraconsistent logics, which do indeed isolate inconsistency by various means. But these logics do not offer strategies for dealing with the inconsistency. There remains the question of what do we do when we have two contradictory items of information in the database. Do we choose one of them? How do we make the choice? Do we leave them in and find a way "around" them? Other logics, such as certain non-monotonic logics, resolve some forms of inconsistency, but do not allow the representation of certain forms of inconsistent data, or give no answer when present.

A second point is more practical. The CIA agent may investigate Professor Nobody and find the charge the computer made to be ridiculous. They may suspect that there is a contradiction in the database but they may not know how to locate it. Generally the contradiction may involve several steps of reasoning and may not be as blatant as in the case of Professor Nobody. We may have several simple and innocent looking data items and some very reasonable rules which together give the wrong answers, but no single item is to blame. How do we debug our system in this case? Systems have been proposed that address aspects of these questions, including Truth Maintenance Systems, but their underlying philosophy is still to eradicate inconsistency at all costs.

From this second point we can see that inconsistency is actually a spur to act on the basis of an inconsistency. If we are making a decision and we find an inconsistency in our reasoning, we seek to identify further information to resolve it. The following is an example (based on an example from Nute 1988):

> Professor Nobody had purchased a tropical fish from the pet shop,
> together with a guide on tropical fish. The book stated that if
> the fish is an etropline, it comes from sea-water, and if the fish
> is a cichlid, it comes from fresh-water.When he bought the fish he
> was told by the shop-owner that the fish was a cichlid, but
> unfortunately, when he got home he realized that the bag
> containing the fish stated it was an etropline. Immediately, he
> rang the shop-owner who clarified the situation by informing him
> that etroplines are a form of cichlid. Relieved, he put the fish
> in sea-water.

This example also shows how strategies can be used to resolve inconsistency. In this case, it became apparent to Professor Nobody that since etroplines are a form of cichlid, the contradiction arising from the two results from the fact with regard to water, etroplines are one exception to the general rule for cichlid. This is an example of a strategy based on using the more specialized rules when making decisions.

So far in our discussion, we have seen the need to isolate, remove, or over-ride inconsistencies in data in some way, or alternatively use them as a spur to acquire more information. Yet, for some kinds of reasoning, it seems desirable to maintain the inconsistency:

> Professor Nobody was 55 years old and wanted an early retirement.
> He could in fact retire with a full pension if he were ill. So
> Professor Nobody presented his Head of Department, Professor
> Somebody, with a letter certifying he had a heart condition. He
> was thus able to retire. His wife, Mrs Faith Nobody, however,
> heard of this letter and was worried. She asked Professor Nobody
> about this letter and Professor Nobody told her that he was
> actually quite healthy and the letter was a trick to get an early
> retirement. Mrs Nobody was relieved. Unfortunately, Professor
> Somebody overheard the conversation, and very angrily confronted
> Professor Nobody. Professor Nobody was undisturbed, he explained
> to Professor Somebody that he had to tell his wife what he had
> told her, in order to stop her worrying. This may have been the
> end of the story except that, unfortunately, Mrs Nobody overheard
> the conversation with Professor Somebody and was worried again.
> Professor Nobody assured his wife that he was quite healthy and
> that he had to tell Professor Somebody what he had told him in
> order not to get his pension revoked.

There is a basic inconsistency here, but there is no need to "restore" consistency. In fact, to restore consistency in this case, is to cause disaster. If Professor Somebody meets Mrs Nobody in a party, he will have to pretend that her husband is healthy in order not to worry her, or at least avoid the subject. Mrs Nobody will have to pretend that her husband is not healthy, in order to keep the pension. Professor Nobody himself will pretend one way or the other depending on the occasion. The database as we described it *does not care* for the inconsistency. There are no means to resolve it and it makes no difference what "the truth" is.

There are many situations where inconsistency can be used to best advantage in arguments. Take the following example:

> Professor Nobody was attending an ESPRIT project meeting in Brussels until Saturday. He had booked a Saturday afternoon flight back to London and had arranged for his wife to pick him up at the airport. However, on the Friday afternoon, he had a row with his project partners and had decided to fly back to London that evening.
>
> Without telling his wife of his altered travel plans, he returned to his home in the early hours of Saturday morning intending to give her a pleasant suprise.He tiptoed to his bedroom intending to quietly slip into bed. But to his suprise – there was *another man* in his bed with his wife.Both were fast asleep. He was shocked and angry, but being a logician, he exercised self-constraint and paused to think. Then he left the house quietly, and went back to the airport. He caught a morning flight back to Brussels, and then returned to London on the flight his wife was expecting him to be on.

We know that Professor Nobody is a logician, but his behaviour does seem to be inconsistent. Most would have expected that perhaps he should have hit the other man, or perhaps he should have made a row. Indeed we have inconsistencies at the object-level and the meta-level. The object-level inconsistencies include all expectations about his wife, Faith, being in the bed on her own, etc, and his new knowledge about his wife being unfaithful.

At the meta-level, if we view Professor Nobody as a classical database management system, we should expect, by comparison with the database system, that he resolve the conflict by adopting some strategy, such as killing the other man, (i.e. databse deletion to maintain consistency), or alternatively he should update the database by say, leaving his wife. However, his action seems to be contrary to the rules of a classical database management system - he seems to be refusing to accept the new input. By going back to Brussels, and following his original plans, he is acting as if the new input was never provided. The refusal of input is therefore inconsistent with the meta-level rules of a classical database management system.

However, instead of making a decision of what action to take when he saw his wife in bed with another man, he chose to pretend that he did not even know. By not taking action at the time, he could choose to raise the issue with his wife when it suited him best. Essentially, he chose to adopt a strategy of 'keeping his options open'. Such strategy therefore requires a more sophisticated formalization of the meta-level rules of the 'database management systems' which define for capturing aspects of human reasoning. Thus the database management system, which is more human-like, can delay or refuse input.

Indeed the above example shows how argumentation is based on more sophisticated meta-level theories for using inconsistentcies. To highlight this we show the same use in another example:

```
Professor  Nobody  had  a  research  assistant  called  Dr
Incompetent.One Monday Professor Nobody asked why Dr Incompetent
hadn't come to work the previous week. Dr Incompetent said that he
had been ill in bed all of the last week. Professor Nobody
expressed  sympathy  at  Dr  Incompetent's  ill  health,  but
unfortunately for Dr Incompetent, Professor Nobody knew otherwise.
For Professor Nobody had seen Dr Incompetent sunbathing in the
Park every day last week. However, he didn't tell Dr Incompetent
that he knew.Over the course of the summer this scenario was
repeated four times. Upon the fifth occurence, Professor Nobody
sacked Dr Incompetent.
```

As in the previous example, Professor Nobody, knew information to which he didn't admit. Indeed he acted as if he knew the contrary. For the above example, he talked with Dr Incompetent, as if he believed him, but he later he expressed his lack of belief in him. In order to support such reasoning, he uses essentially inconsistent information in way that benefits his overall argumentation strategies. Similarly in the earlier example, he pretended that he didn't know his wife had been unfaithful, and then later used the contrary information to his advantage.

When reasoning with these examples we are happy with the inconsistencies because we read them as signals to take action or signals which trigger some rules and deactivate other rules. We do not perceive ourselves as living in some platonic world contemplating for eternity which of two contradictory items to disregard. Rather we constantly face inconsistencies in the real-world, most of which we deal with without any problems.

We believe a solution for better handling of contradictions can be found by looking closely at the ways humans deal with them. In developing a new framework, the following points need to be considered:

(1) We do not share the view that contradictions in a database are necessarily a "bad" thing, and that they should be avoided at all costs. Contradictory information seems to be part of our lives, and sometimes we even prefer ambiguities and irreconcilable views. We must therefore seek logical principles which will allow for contradictions in the same way that we humans allow for them, and even make contradictions useful.

(2) Humans seem to intuitively grasp that some information is more relevant than other information. The notion of relevance should be developed and used.

(3) There seems to be a hierarchy of rules involved in rationalizing inconsistency. Rules of the form "When contradictory information is recieved about A, then do B" seemed to be used constantly. These are meta-rules, i.e. rules about rules. Full exploitation of these rules require our database language to be able to talk about itself.

(4) There is a need to be careful about throwing things out of the database. Always keep an open mind, that although A is rejected now (because of new information contradicting it), it may become useful again. Perhaps some uncertainty values may be attached to all data items.

(5) Learning is a process that is directed by inconsistency. For example, if we have a hypothesis to explain a system, we check our hypothesis by making a series of observations of the system: If the hypothesis predicts the observations, then we have increased confidence in the hypothesis, but if the hypothesis negation of the observation, then we have inconsistency. This inconsistency should initiate further learning about the system.

(6) Argumentation proceeds via inconsistency. If two people, say Jack and Joe, undertake a dialogue, the inconsistencies between Jack's and Joe's views constitute foci for the dialogue. If the objective of the dialogue is for one of the participants, say Jack, to exert a particular view upon Joe, then Jack will use inconsistencies in Joe's argument as support for Jack's point of view.

(7) Despite everything we do, although we may give various heuristic rules dealing with

contradictory items of equal reliability and relevance, there will be no way of deciding which of the two items is the correct one. In this case, we can only wait and be suspicious of any computation involving them.

These points indicate that a wide range of cognitive activities are involved in reasoning with inconsistent information including uncertainty reasoning, meta-reasoning, learning, and argumentation. Yet central to our position is that we should suspend the axiom of absurdity (*ex contradictione quodlibet*) for many kinds of reasoning. Furthermore, the cognitive activities involved in reasoning with inconsistent information seem to be directly related to the kind of inconsistency. In the next section we consider issues formalizations for inconsistent information, and of categories of inconsistent information.

3. A Case Study

It is valuable to illustrate our ideas through a simple case study. We take, a database with a few simple clauses:

t_1: $\forall x$ (bird(x) → fly(x))

t_2: $\forall x$ (bird(x) ∧ big(x) → ¬fly(x))

t_3: bird(a)

t_4: big(a)

t_5: hungry(b)

t_6: hungry(a)

Our query is ?bird(b).

Classical and intuitionistic logic will derive a contradiction from (t_1) - (t_4) and then deduce the goal. Relevance logic will notice that the contradiction involves another part of the database, namely the part dealing with "a" and will not infer the goal. A paraconsistent logic will allow the inference of both consequents, though some such as IDL (Pequeno 1991) will annotate the consequents to show they are defeasible.

A truth maintenance system will recognize the inconsistency, and possibly take out some of the data and arrive at a new consistent database. For example, it could take out (t_2) or even modify (t_1) to (t_1^*):

t_1^*: $\forall x\ (\ bird(x) \land \neg big(x) \rightarrow fly(x)\ )$

A defeasible logic will resolve the inconsistency by some means. For example, in this case the rules are ordered according to "specificity". The rule (t_2) has a more specific antecedent than (t_1). In a logic such as LDR (Nute 1988), this ordering allows for a resolution of the conflict by prefering the conclusion from the more specific rule. However, a defeasible system cannot always resolve inconsistency. If we add the following rule:

t_7: $\forall x\ (\ hungry(x)\ \rightarrow \neg fly(x)\ )$

then (t_7) and (t_1) are in conflict, but neither is stronger than the other. Thus the query ?fly(a) will get no answer.

There are some other approaches which use multiple truth values, such as the values { T, F, {}, {T,F} }, where the value {T,F} would be assigned to fly(a). Such an assignment captures the situation of both fly(a) and ¬fly(a) following from the database (Belnap 1977).

Our framework generalizes on all of the above solutions, both technically and conceptually. First we conceptually allow for both fly(a) and ¬fly(a) to be derivable, or even explicitly reside in the database. We regard this favourably. Second, we need to know know what *action* to take when facing any inconsistency, and that this action leads to desirable results.

What do we mean by action? The notions and notation for actions is external to the database. We can invoke some form of truth maintenance system to restore consistency, or we can refer to a user, or we can attempt to obtain more data. In each case the 'action' and the 'action results' are expressed in a special langauge, supplementary to the langauge of the database.

Such a special language is not entirely new. Planning systems exhibit aspects of this behaviour. Consider the blocks world - the predicates on(x,y) and free(y) are database predicates, and the predicate move(x,y) is an action predicate. A mixed language can state what happens when we execute move(x,y). For example the following is a mixed language statement:

$$move(x,y) \rightarrow [\ one(x,y) \land \neg free(y) \ ]$$

We are thus proposing that inconsistencies should be dealt with in a framework whose data and actions are mixed. The paradigm, we require, can be captured by the following. This is a similar idea to planning, but is different in the way it is developed:

$$\text{INCONSISTENCY} \Rightarrow \text{ACTION}$$

Action statements are also used elsewhere in real practice to augment declarative information such as in relational database technology and logic programming. For example in a warehouse database, if certain stock level parameters fall below a predetermined level, then the database program can automatically generate an order for more supplies for the warehouse. Such an example indicates how declarative information in the database can be linked to actions.

In our framework, the new concept of a database is a pair, the ordinary data (corresponding to the old notion of a database), and the supplementary database, describing the actions. Given the same ordinary data, we get different databases if we vary the actions of the supplementary part, and different logics result from different diciplines for supplementary databases. The supplementary actions capture how we can use and change an object-level database when inconsistencies arise. For example, according to our framework, the database $(t_1) \ldots (t_6)$, given above, is the object-level data. We need to supplement this with the action data of the form below, where Γ is the inconsistent subset of the database, and QUERY_USER is an action predicate that queries the user about the inconsistency:

$$\text{INCONSISTENCY}(\Gamma) \rightarrow \text{QUERY\_USER}(\Gamma)$$

Alternatively, we could supplement $(t_1) \ldots (t_6)$, with the following action data, where INVOKE_TMS is an action predicate that invokes some form of truth maintenance system:

INCONSISTENCY(Γ) → INVOKE_TMS(Γ)

In both cases, the action predicate can cause a new object-level database to be formed, and that the inconsistency has been resolved in the new database.

As another example, we consider the above example of Professor Nobody seeking early retirement. For this each agent has a supplementary database that provides information from the object-level database according to who that agent is dealing with. For Professor Nobody we could represent it as follows, where Γ is the object-level database, QUERY(x) is a query x of the object-level database, WHO_IS_ASKING is an action to ascertain who is making the query, ASKING represents the reply to the action, and USE_FOR_QUERY is the subset of the database that can be used to answer the query:

Γ = { ¬professor_nobody_ill , professor_nobody_ill }

QUERY(Γ,x) → WHO_IS_ASKING(x)

ASKING(mrs_nobody) → USE_FOR_QUERY(({¬professor_nobody_ill}))

ASKING(professor_somebody) → USE_FOR_QUERY(({professor_nobody_ill}))

In this way, any query of the database causes the supplementary database to check the identitity of the querier, and then use the appropriate subset of the database to answer the original query. This not resolving the inconsistency, but is instead uses the action language as a way of dealing with the inconsistency.

As a final example, we consider the above example of Professor Nobody and his unfaithful wife Faith. In the supplementary database for our model of Professor Nobody, we represent the following information, where UPDATE is a request to update the object-level database Δ with the data x, INCONSISTENT holds if $\Delta \cup \{x\}$ is inconsistent, SERIOUS_IMPLICATIONS captures the situations where $\Delta \cup \{x\}$ has particular consequences, PRETEND_NO_UPDATE supports the dual reasoning that allows the agent to act on an update by x, and also act as if there has been no update by x:

$$\text{UPDATE}(\Delta, x) \wedge \text{INCONSISTENT}(\Delta \cup \{x\}) \wedge \text{SERIOUS\_IMPLICATIONS}(\Delta \cup \{x\})$$
$$\rightarrow \text{PRETEND\_NO\_UPDATE}(\Delta, x)$$

This inconsistency causes the supplementary database to invoke an updating of the database that would allow 'keeping options open'.

These examples are intended to illustrate how a supplementary database can be used to support a more human-oriented approach to reasoning with inconsistent information. Even though, some of the likely examples of inconsistent information can be dealt with by existing aproaches, such as truth-maintenance systems, paraconsistent logics, etc, the intention of the framework is to identify general principles of reasoning with inconsistent information. Some strategies maybe quite specific to certain examples, others may be quite generic, however, we believe that it is necessary to develop a framework in which the variety of approaches can be formalized and developed.

As we have seen from these examples, how we view inconsistency depends on various factors. We need to be able to differentiate between trivial inconsistencies that we can ignore, and significant inconsistencies that we need to act upon. Furthermore, we need to classify inconsistency according to the appropriate action that is required. We list possible categories of action that might be adopted in a supplementary database:

(2)   **Learning action:** This should result in some form of revision of information. We consider inconsistencies that cause such actions as useful, since they constitute part of the natural learning processes.

(3)   **Information acquisition action:** This type of action involves seeking further information to reconcile an inconsistency. We consider inconsistencies that cause such actions as useful, since they direct the process of acquiring knowledge for decision making.

(3)   **Inconsistency removal action:** This type of action adopts strategies for resolving an inconsistency such as localizing the inconsistency, or adopting some belief revision, or truth maintenance, heuristic to resolve the inconsistency. We consider inconsistencies that cause such actions as an integral part of non-monotonic reasoning.

(5)   **Inference preference action:** Some inconsistencies can be resolved if we have a strategy for preferring some inferences over others. Examples include preferring inferences from more specialized rules. We consider inconsistencies that cause such actions as an integral part of non-monotonic reasoning.

(6)   **Argumentation action:** Inconsistencies that occur during the course of a dialogue between two agents can serve to direct further dialogue. An inconsistency can be a focus for a dialogue. We consider such inconsistencies as useful, since they constitute the basis of natural argumentation processes.

4.   Discussion

It is inconsistency that, at least in part, makes us human. we couldn't conduct our normal ("consistent") lives if we did not ignore the majority of inconsistencies around. There are too many inconsistencies in the real-world for us to be able to resolve them all. Our view on inconsistencies depends on the kind of inconsistency. However, we do stress that inconsistency is important for cognitive activities and as such should considered to be a desirable notion in logics for artificial intelligence. Our view is different from the current view that contradictions are "bad". We think that they are an essential part of life and some can even be made useful. In this paper we have outlined a series of actions. We now require a clearer logic formalization coupling inconsistency with action or meta-action.

We propose that reasoning with inconsistency can be studied as a Labelled Deductive System (Gabbay 1989), where deduction is done both on labels and on formulae. The framework of the Labelled Deductive System (LDS) is that a basic unit of information is a labelled formula, and that a logic can then be defined in terms of allowed operations on the labelled formulae. Hence, for example, logical consequence can be defined on labelled formulae, where l_i are labels and A_i are formulae:

$$l_1{:}A_1, \ldots, l_n{:}A_n \vdash l{:}B$$

The actions of the supplementary database will be linked to sets of labels. Thus very briefly, if we can prove $l_1{:}A, \ldots, l_n{:}\neg A$ with different labels, then the set $(\{\ l_1, \ldots, l_n\ \}, A)$ will trigger an action. The action could be internal, namely some further non-monotonic mechanism such as choose $l_1:A$, or some external action, such as query the user. The labels give us the tight control over a derivation that is necessary in deciding which action to take. In this framework it is possible to trigger action even when there is no inconsistency, such as for formalizing the warehouse database system which generate actions according to certain stock level parameters. We will pursue this issues in subsequent work.

Acknowledgements

This work is supported by SERC project *Rule-Based Systems*, GR/G 29861. Special thanks are due to Lydia Rivlin, Ruth Kempson, and Mark Reynolds for reading earlier drafts of this paper at short notice.

References

Belnap N (1977) A useful four-valued logic, in Dunn J & Epstein G, Modern Uses of Multiple-Valued Logic, 5 - 37

Gabbay D (1989) Labelled Deductive Systems, Technical Report, Department of Computing, Imperial College, London, also to be expanded in Gabbay D (in preparation) Labelled Deductive Systems, Oxford University Press

Nute D (1988) Defeasible reasoning and decision support systems, Decision Support Systems, 4, 97-110

Pequeno T & Buchsbaum A (1991) The logic of epistemic inconsistency, Proceedings of the Second International Conference on the Principles of Knowledge Representation and Reasoning, 453 - 460, Morgan Kaufmann

RELATIONAL PROOF SYSTEMS FOR SOME AI LOGICS

Eva Orlowska*
Institute of Theoretical and Applied Computer Science
Polish Academy of Sciences

Abstract
Relational methodology of defining automated proof systems has been applied to a modal logic for reasoning with incomplete information and to an epistemic logic for reasoning about partial knowlege of groups of agents.

1. Relational logic and its proof system

We define a formal language with expressions of the two kinds: relational expressions, interpreted as binary relations, and formulas, interpreted as schemes of sentences saying that a pair of objects stands in a relation. Expressions of the language are built with symbols taken from the following pairwise disjoint sets:

VAROB   a set of object variables

CONREL   a set of relational constants such that $1, I \in$ CONREL interpreted as the universal relation and identity, respectively

OPREL=$\{\cup, \cap, -, ;, ^{-1}, /\}$ the set of relational operations of union, intersection, complement, composition, converse, and right residuation $((y,z) \in R/P$ iff $(x,y) \in P$ implies $(x,z) \in R$ for all $x)$, respectively

Set EREL of relational expressions is the smallest set including set CONREL of relational constants and closed with respect to the relational operations. Set FORREL of formulas is the set of expressions of the form xPy for $x, y \in$ VAROB and $P \in$ EREL.

We define semantics of the language by means of notions of model and satisfiability of formulas in a model. By model we mean a system of the form M=(OB, $\{r_P\}_{P \in CONREL}$, m), where OB is a nonempty set of objects, for each $P \in$ CONREL r_P is a binary relation in set OB, and m is a meaning function providing interpretation of relational expressions. We assume that function m satisfies the following conditions:

$m(P)=r_P$ for any $P \in$ CONREL,

----

*Mailing address: Azaliowa 29, 04-539 Warsaw, Poland

$m(1)=OB\times OB=1_M$, $m(I)=\{(s,s):s\in OB\}=I_M$,

$m(-P)=-m(P)$, $m(P^{-1})=(m(P))^{-1}$,

$m(P\cup Q)=m(P)\cup m(Q)$, $m(P\cap Q)=m(P)\cap m(Q)$, $m(P;Q)=m(P);m(Q)$,

$m(R/P)=m(R)/m(P)$.

By a valuation we mean a function $v:VAROB\rightarrow OB$ which assigns objects to object variables. We say that in model M valuation v satisfies formula xPy whenever objects $v(x)$ and $v(y)$ stand in relation $m(P)$:

M,v sat xPy iff $(v(x),v(y))\in m(P)$.

The logic defined above is called LREL. We say that a formula F is true in a model M ($\vDash_M F$) iff M,v sat F for all valuations v. A formula F is valid in LREL iff it is true in all models. We say that formulas $F_1,\ldots,F_n$ imply a formula F iff for any model M if $F_1,\ldots,F_n$ are true in M, then F is true in M.

Below are listed examples of properties expressible in the language of LREL.

Proposition 1.1

(a) $\vDash_M xPy$ iff $m(P)=1_M$

(b) $\vDash_M x(1;-P;1)y$ iff $m(P)\neq 1_M$

(c) $\vDash_M x(-P\cup R)y$ iff $m(P)\subseteq m(R)$

(d) $\vDash_M x((-P\cup R)\cap(-R\cup P))y$ iff $m(P)=m(R)$

(e) $\vDash_M x(1;(-(R_1\cap\ldots\cap R_n));1\cup R)y$ iff $m(R_1)=1_M,\ldots,m(R_n)=1_M$ imply $m(R)=1_M$.

Now we recall the deduction rules for the presented language (Orlowska 1991a,b). The rules apply to finite sequences of formulas, and they enable us to decompose formulas in a sequence into some simpler formulas. As a result of decomposition we obtain a single sequence or a pair of sequences. We admit the following rules:

$$(\cup)\quad \frac{K,\ x(P\cup Q)y,\ H}{K,\ xPy,\ xQy,\ H}$$

$$(-\cup)\quad \frac{K,\ x(-(P\cup Q))y,\ H}{K,\ x(-P)y,\ H \qquad K,\ x(-Q)y,\ H}$$

$$(\cap)\quad \frac{K,\ x(P\cap Q)y,\ H}{K,\ xPy,\ H \qquad K,\ xQy,\ H}$$

$$(-\cap)\quad \frac{K,\ x(-(P\cap Q))y,\ H}{K,\ x(-P)y,\ x(-Q)y,\ H}$$

$$(--)\quad \frac{K,\ x(--P)y,\ H}{K,\ xPy,\ H}$$

$$(^{-1})\quad \frac{K,\ xP^{-1}y,\ H}{K,\ yPx,\ H}$$

$$(-^{-1})\quad \frac{K,\ x(-(P^{-1}))y,\ H}{K,\ y(-P)x,\ H}$$

(;)
$$\frac{K, \ x(P;Q)y, \ H}{K, \ xPz, \ H, \ x(P;Q)y \qquad K, \ zQy, \ H, \ x(P;Q)y}$$
where z is a variable

(-;)
$$\frac{K, \ x(-(P;Q))y, \ H}{K, \ x(-P)z, \ z(-Q)y, \ H}$$
where z is a variable which does not appear in any formula
above the line

(/)
$$\frac{K, \ xR/Py, \ H}{K, \ z(-P)x, \ zRy, \ H}$$
where z is a variable which does not appear in any formula
above the line

(-/)
$$\frac{K, \ x(-(R/P))y, \ H}{K, \ zPx, \ H, \ x(-(R/P))y \qquad K, \ z(-R)y, \ H, \ x(-(R/P))y}$$
where z is a variable

(symI)
$$\frac{K, \ xIy, \ H}{K, \ yIx, \ H}$$

(I1)
$$\frac{K, \ xRy, \ H}{K, \ xIz, \ xRy, \ H \qquad K, \ zRy, \ xRy, \ H} \qquad z \text{ is a variable}$$

(I2)
$$\frac{K, \ xRy, \ H}{K, \ xRz, \ xIy, \ H \qquad K, \ zIy, \ xRy, \ H} \qquad z \text{ is a variable}$$

In all the above rules K and H denote finite (possibly empty)
sequences of formulas.

A sequence of formulas is said to be fundamental if it contains formulas of one of the following forms:
(f1) xPy, x(-P)y for some P∈EREL and some x,y∈VAROB
(f2) xIy for some x,y∈VAROB
(f3) xIx for some x∈VAROB.

A sequence K of formulas is said to be valid if for every
model M and for every valuation v there is a formula F occurring
in K such that M,v sat F. A rule of the form K/H is said to be
admissible if sequence K is valid iff sequence H is valid. A rule
of the form K/H, H' is admissible iff sequence K is valid iff
both H and H' are valid.

Proposition 1.2
(a) Every fundamental sequence is valid
(b) The rules are admissible.

Proofs of the lemmas can be easily obtained from definition of validity, properties of relational operations, symmetry of I, and the law R;I=R=I;R.

Given a formula of the form xPy, where P is a compound relational expression, we can decompose it by successive application of the rules. In the process of decomposition we form a tree whose vertices consist of finite sequences of formulas. Each vertex of the decomposition tree has at most two successors. We stop decomposing formulas in a vertex after obtaining a fundamental sequence, or if none of the rules can be applied to the formulas in the vertex. A branch of a decomposition tree is said to be fundamental if it contains a fundamental sequence. A decompositon tree is fundamental if all of its branches are fundamental. The following completeness theorem holds in the relational logic (Orlowska 1991a,b).

Proposition 1.3
The following conditions are equivalent:
(a) A formula F is valid in LREL
(b) There is a fundamental decomposition tree of F.

2. Interpretation of multimodal logics in the relational logic

We consider a multimodal propositional language with a family of modal operators determined by accessibility relations. Expressions of the language are constructed with symbols from the following pairwise disjoint sets:

VARPROP   a set of propositional variables

CONAC   a set of accessibility relation constants

OPAC={-,∪,∩,;,$^{-1}$}   the set of relational operations

OPPROP={¬,∨,∧,→,[],⟨⟩}   the set of propositional operations of negation, disjunction, conjunction, implication, necessity, and possibility, respectively.

Set EAC of relational expressions representing accessibility relations is the smallest set including set CONAC and closed with respect to relational operations -,∪,∩,;,$^{-1}$.

Set FORMOD of modal formulas is the smallest set satisfying the following conditions:

VARPROP⊆FORMOD

F,G∈FORMOD imply ¬F,F∨G,F∧G,F→G∈FORMOD

F∈FORMOD, R∈EAC imply [R]F,⟨R⟩F∈FORMOD.

By model of the modal language we mean a system:
M=(ST, {r_R}$_{R∈CONAC}$, m), where ST is a nonempty set whose elements are called states, for each constant R∈CONAC r_R is a binary

relation in set ST, and m is a meaning function satisfying the following conditions:

$m(p) \subseteq ST$ for $p \in VARPROP$, $m(R) = r_R$ for $R \in CONAC$,

$m(-R) = -m(R)$, $m(R^{-1}) = (m(R))^{-1}$,

$m(R \cup S) = m(R) \cup m(S)$, $m(R \cap S) = m(R) \cap m(S)$, $m(R;S) = m(R);m(S)$.

We say that in model M a state s satisfies a formula F (M,s sat F) iff the following conditions are satisfied:

M,s sat p   iff $s \in m(p)$ for $p \in VARPROP$

M,s sat ¬F   iff not M s sat F

M,s sat F∨G   iff M s sat F or M s sat G

M,s sat F∧G   iff M s sat F and M s sat G

M,s sat F→G   iff M s sat (¬F∨G)

M,s sat [R]F iff for all $s' \in ST$ $(s,s') \in m(R)$ implies M,s' sat F.

M,s sat ⟨R⟩F iff there is $s' \in ST$ such that $(s,s') \in m(R)$ and M,s' sat F

The logic defined above will be called LMOD. We say that a formula F is true in a model M iff M,s sat F for all $s \in ST$. A formula F is valid in LMOD iff F is true in all models. A formula F is satisfiable iff there are M and s such that M,s sat F. A formula F is unsatisfiable iff it is not satisfiable.

Now we define a translation of the modal language into relational language. Assume, that a one-to-one mapping t':CONAC∪ VARPROP→CONREL is given, assigning relational constants to accessibility relation constants and propositional variables, and preserving 1 and I. We define translation function t:EAC∪FORMOD→EREL assigning relational expressions to accessibility relation expressions and modal formulas:

$t(R) = t'(R)$ for $R \in CONAC$,

$t(-R) = -t(R)$, $t(R^{-1}) = t(R)^{-1}$,

$t(R \cup S) = t(R) \cup t(S)$, $t(R \cap S) = t(R) \cap t(S)$, $t(R;S) = t(R);t(S)$,

$t(p) = t'(p);1$ for $p \in VARPROP$,

$t(\neg F) = -t(F)$,

$t(F \vee G) = t(F) \cup t(G)$, $t(F \wedge G) = t(F) \cap t(G)$, $t(F \rightarrow G) = t(\neg F \vee G)$,

$t([R]F) = t(F)/t(R)^{-1}$, $t(\langle R \rangle F) = t(R);t(F)$.

Usually, we assume that mapping t' is an identity on the set of relational constants, that is $t'(R) = R$ for $R \in CONAC$.

Proposition 2.1

For any formula of logic LMOD we have $t(F);1 = t(F)$.

The theorem says that modal formulas become binary relations with a special property, namely the domain of relation t(F) is the set of states in which F is true, and the range of t(F) is the whole set of states. Any relation R such that R;1=R is called a right ideal relation.

The logic of relations can be considered to be a theorem prover for modal logics, namely we have

Proposition 2.2

(a) A formula F is valid in LMOD iff xt(F)y is valid in LREL

(b) F is unsatisfiable in LMOD iff x(-t(F))y is valid in LREL.

In logic LREL we can also express derivability of modal formulas from some other modal formulas.

Proposition 2.3

The following conditions are equivalent:

(a) $F_1,...,F_n$ imply F in LMOD

(b) $x(1;-(t(F_1)\cap...\cap t(F_n));1\cup t(F))y$ is valid in LREL.

Let PROP be a property of relations, and let EPROP(R) be a relational expression which reflects possession of PROP by R. Namely, EPROP(R) represents universal relation iff R possesses PROP. For example, if PROP=reflexivity, then EPROP(R)=-I∪R, since we have R reflexive iff I⊆R iff -I∪R=1. Let R be an accessibility relation expression, and let F(R) be a formula of LMOD in which R occurs. By proposition 2.1(e) we easily obtain the following theorem.

Proposition 2.4

The following conditions are equivalent:

(a) Relation represented by R possesses PROP implies F(R) is valid in LMOD

(b) EPROP(t(R))=1 implies t(F(R))=1

(c) x(1;-EPROP(t(R));1∪t(F(R)))y is valid in LREL.

Example 2.1

We prove in LREL that if accessibility relation is reflexive, than F(R)=[R]p→p is valid in LMOD. We have EPROP(t(R))=-I∪t(R), and $t(F(R))=-(t'(p);1/t(R)^{-1})\cup t'(p);1$. For the sake of simplicity we assume that t'(p)=p, and t(R)=R. We show that formula $x(1;-(-I\cup R);1\cup-(p;1/R^{-1})\cup(p;1))y$ is valid in LREL.

$$x(1;-(-I\cup R);1\cup-(p;1/R^{-1})\cup(p;1))y$$
$$\Big|(\cup)$$
$$x(1;-(-I\cup R);1)y,\ x-(p;1/R^{-1})y,\ x(p;1)y$$

(;) to F=x(1;-(-I∪R);1)y / ... \ new variable:=x

x1x   x(-(-I∪R);1)y
x-(p;1/R⁻¹)y   x-(p;1/R⁻¹)y
x(p;1)y, F   x(p;1)y, F
fundamental   (;) to F1=x(-(-I∪R);1)y
                                       new variable:=x

        x-(-I∪R)x   x1x
        x-(p;1/R⁻¹)y   x-(p;1/R⁻¹)y

$$x(p;1)y, F, F1 \qquad\qquad x(p;1)y, F, F1$$
$$(-\cup) (--) \qquad\qquad\qquad \text{fundamental}$$

xIx	x-Rx
x-(p;1/R⁻¹)y	x-(p;1/R⁻¹)y
x(p;1)y, F, F1	x(p;1)y, F, F1
fundamental	

$$(-/) (^{-1})$$
$$\text{new variable}:=x$$

x-Rx	x-Rx
xRx	x-(p;1)y
x(p;1)y, F, F1	x(p;1)y, F, F1
x-(p;1/R⁻¹)y	x-(p;1/R⁻¹)y
fundamental	fundamental

3. Relational proof system for a knowledge representation logic

A logic considered in the present section is a variant of logics investigated in Farinas del Cerro and Orlowska (1985), Orlowska (1985, 1988), Vakarelov (1987, 1989). The logic provides a tool to represent information in a knowledge base, taking into account uncertainty and incompleteness of that information.

A universe of each model of the logic is a collection of atomic facts which are pieces of information of the form: an object, a list of properties of the object. Properties of objects are articulated in terms of attributes and values of those attributes. For example, the property 'to be green' is expressed as the pair (colour, green), the property 'to be large' is represented as the pair (size, large). In those examples 'colour' and 'size' are attributes of objects, and 'green', 'large' are values of those attributes, respectively. Thus we consider information systems of the form $(OB, AT, \{VAL_a\}_{a \in AT}, f)$ where OB is a nonempty set of objects, AT is a nonempty set of attributes, for each $a \in AT$ set VAL_a consists of the values of attribute a, and f is an information function which assigns the properties to objects, and satisfies $f(o,a) \in VAL_a$ (Pawlak 1981).

The information function can be deterministic or nondeterministic. A deterministic information function $f: OB \times AT \rightarrow VAL = \bigcup\{VAL_a : a \in AT\}$ assigns a single value of an attribute to an object. For example, if the colour of object o is green, then we have f(o, colour)=green. A nondeterminstic informatin function $f: OB \times AT \rightarrow P(VAL)$ assigns a set of values of an attribute to an object. It covers the case when we are not in a position to state with certainty the value taken by a given object for a given attribute. For example, we may not know the exact age of a person p, the only information we have is that p is between 20 and 25. In that case f(p, age)={20,...,25}.

To make an intelligent synthesis of the atomic facts we look for a deep structure of the set of objects manifested by relationships between its elements. The following three classes of relations are of special importance. Let a set A⊆AT of attributes be given. We define an indiscernibility relation ind(A)⊆OBxOB determined by set A:

(o1,o2)∈ind(A) iff f(o1,a)=f(o2,a) for all a∈A

Proposition 3.1

(a) Indiscernibility relations are equivlence relations

(b) ind(A∪B)=ind(A)∩ind(B)

(c) ind(∅)=OBxOB

(d) A⊆B implies ind(B)⊆ind(A).

Indiscernibility relations reflect the discrimination ability of the properties assumed in an information system. If two distinct objects have the same characterization in terms of the given properties, then we conclude that the knowledge provided by the information system is not complete, it does not enable us to recognize the objects as distinct entities. The detailed discussion of the role of indiscernibility in knowledge representation can be found in Orlowska (1983, 1988, 1989).

Let f be a nondeterministic information function. For a set A⊆AT of attributes we define similarity relation sim(A)⊆OBxOB and relation of informational inclusion in(A)⊆OBxOB:

(o1,o2)∈sim(A) iff f(o1,a)∩f(o2,a)≠∅ for all a∈A

(o1,o2)∈in(A) iff f(o1,a)⊆f(o2,a) for all a∈A.

For example, if the age of a person p is between 20 and 25, and the age of a person q is between 23 and 27, then we may say that p and q are of the similar age. Simlarly, assume that person p knows the three languages French, German, and Swedish, and person q knows French, German, Swedish, and Spanish. In that case we say that p and q stand in relation in(foreign languages).

Propositon 3.2

(a) Similarity relations are reflexive and symmetric

(b) Informational inclusion relations are reflexive and transitive

(c) sim(A∪B)=sim(A)∩sim(B)

(d) in(A∪B)=in(A)∩in(B)

(e) If (o1,o2)∈in(A) and (o1,o3)∈sim(A), then (o2,o3)∈sim(A)

(f) If (o1,o2)∈in(A) and (o2,o1)∈in(A), then (o1,o2)∈ind(A)

(g) ind(A)⊆in(A)

(h) ind(A)⊆sim(A).

Conditions 3.1(b) and 3.2(c),(d) say that the relations determined by a collection of attributes are intersections of the corresponding relations determined by members of the collection. It means that the more properties we have at our disposal to characterize objects, the finer is the respective relationship between the objects. From the point of view of a formalization, the conditions show how important is the intersection operator in logics for knowledge representation. One of the advantages of the relational proof systems is that within the relational framework we can easily deal with ∩, although it is not axiomatizable by means of a modal formula.

Now we define an information logic IL. The language of IL is obtained from the modal language described in section 2 as follows. Set VARPROP is an infinite, denumerable set of propositional variables. Let a set AT of attributes be fixed. We admit the family of accessibility relation constants CONAC={ind(A), sim(A), in(A)}ₐ ₐₜ. The set of relational operations is OPAC={∩}. Hence the set EAC of relational expressions representing the accessibility relations consists of the relational terms obtained from the constants by means of ∩. Set OPPROP of propositional operations consists of the classical operations and the modal operations [R] and ⟨R⟩ for R∈EAC.

Models of the language are systems of the form M=(OB,AT, {ind(A),sim(A),in(A)}ₐ⊆ₐₜ,m) where OB is a nonempty set of objects, AT is a nonempty set of attributes, ind(A) are equivalence relations in OB, sim(A) are reflexive and symmetric relations in OB, in(A) are reflexive and transitive relations in OB, and moreover, the relations satisfy conditions 3.2(e),(f),(g),(h).

The standard Hilbert-style aximatization of logic IL is not known. Clearly, for formulas built with modal operators determined by relations ind(A) we should assume the axioms of logic S5, for sim(A) the axioms of logic B, and for in(A) the axioms of logic S4 (Prior (1967), Segerberg (1971), Goldblatt (1987)). Conditions 3.2(e),(g),(h) can be axiomatized by the following formulas, respectively:

A1I. ⟨in(A)⟩[sim(A)]F→[sim(A)]F

A2I. [in(A)]F→[ind(A)]F

A3I. [sim(A)]F→[ind(A)]F

The rules of inference should be modus ponens and F/[ind(A)]F, F/[sim(A)]F, F/[in(A)]F. However, condition 3.2(f) and intersection of the accessibility relations are not axiomatizable.

The relational proof system for logic IL consists of all the rules given in section 2, and the following:

(ref R) $\dfrac{\text{K, xR(A)y, H}}{\text{K, xIy, xR(A)y, H}}$ for R=ind, sim, in

(sym R) $\dfrac{\text{K, xR(A)y, H}}{\text{K, yR(A)x, H}}$ for R=ind, sim

(tran R)

$$\dfrac{\text{K, xR(A)y, H}}{\text{K, xR(A)z, xR(A)y, H}\qquad\text{K, zR(A)y, xR(A)y, H}}$$

where z is a variable and R=ind, in

(sim in)

$$\dfrac{\text{K, xsim(A)y, H}}{\text{K, zin(A)x, xsim(A)y, H}\qquad\text{K, zsim(A)y, xsim(A)y, H}}$$

where z is a variable

(ind in 1)

$$\dfrac{\text{K, xind(A)y, H}}{\text{K, xin(A)y, xind(A)y, H}\qquad\text{K, yin(A)x, xind(A)y, H}}$$

(ind in 2) $\dfrac{\text{K, xin(A)y, H}}{\text{K, xind(A)y, xin(A)y, H}}$

(ind sim) $\dfrac{\text{K, xsim(A)y, H}}{\text{K, xind(A)y, xsim(A)y, H}}$

The fundamental sequences in the relational proof system for IL are (f1),(f2),(f3). It is easy to see that due to the respective assumptions on properties of accessibility relations all the above rules are admissible. For example, admissibility of (ref ind) follows from reflexivity of indiscernibility relations, admissibility of (sim in), (ind in 1), (ind in 2), (ind sim) follows from the relationships 3.2(e),(f),(g),(h), respectively.

Example 3.1

We show that formula $F=[ind(A)]p\rightarrow[ind(A)\cap ind(B)]p$ is valid in logic IL. We have $t(F)=-(p;1/ind(A)^{-1})\cup p;1/(ind(A)\cap ind(B))^{-1}$.

$$xt(F)y$$
$$\Big|\,(\cup)$$
$$x-(p;1/ind(A)^{-1})y,\ xp;1/(ind(A)\cap ind(B))^{-1}y$$
$$\Big|\,(/)\ \text{new variable:}z\ (^{-1})$$
$$x-(p;1/ind(A)^{-1})y,\ x-(ind(A)\cap ind(B))z,\ zp;1y$$
$$\Big|\,(-\cap)$$
$$x-(p;1/ind(A)^{-1})y,\ x-ind(A)z,\ x-ind(B)z,\ zp;1y$$

$$\diagup\ (/)\ \text{new variable:=z}\ (^{-1})\qquad\diagdown$$

xind(A)z,...   z-(p;1)y,...

fundamental   fundamental

5. Relational formalization of reasoning about knowledge

In the present section we introduce the relational deduction rules for an epistemic logic EIL. The logic is obtained from logic IL by restricting the set of accessibility relations to indiscernibility relations only, and by assuming that the relations are determined by sets of knowledge agents, instead of the sets of attributes. If A is a set of agents, then the intuitive interpretation of the fact that objects o1 and o2 stand in relation ind(A) is that knowledge of the agents from A does not enable them to distinguish between o1 and o2.

For an object o let ind(A)o be the equivalence class of ind(A) generated by o: ind(A)o={o'∈OB: (o,o')∈ind(A)}.

Given a set X⊆OB of objects and a set A of agents, we define sets of positive and negative instances of X (Orlowska 1983):

POS(A)X=∪{ind(A)o: ind(A)o⊆X}

NEG(A)X=OB−∪{ind(A)o: X∩ind(A)o≠∅}.

Thus an object o is a member of POS(A)X iff it belongs to X together with all the objects o' which are indiscernible from o by agents A. Similarly, an object o is a member of NEG(A)X iff neither o nor any object o' which is indiscernible from o by agents A belong to set X. Indiscernibility relation ind(A) serves as a filter through which the agents A recognize the objects. The equivalence classes of ind(A) are the smallest, indivisible pieces that can pass through that filter.

We define knowledge of agents A about objects from a set X as follows (Orlowska 1989):

K(A)X=POS(A)X∪NEG(A)X.

That definition reflects the intuitive interpretation of knowledge as the ability to distinguish between an object which is claimed to be known and the other objects. In other words, knowing a thing means that we grasp it as a distinguished entity, distinct from the other entities in the given domain. The discussion of various concepts of knowledge can be found in Hintikka (1962), Halpern (1986), Lenzen (1978).

Knowledge of agents A about X is said to be complete iff K(A)X=OB, and incomplete otherwise.

Let a set AGT of knowledge agents be given. The formulas of logic EIL are built with propositional variables, the classical propositional operations, and the epistemic operations K(A) for A⊆AGT. We also admit operators K(A∪B) determined by indiscernibility relation ind(A∪B). Models of the language are systems M=(OB,AGT,{ind(A)}A ⊆ AGT,m), where OB is a nonempty set of objects, AGT is a nonempty set of knowledge agents, with every set A of agents there is associated an indiscernibility relation ind(A) which is an equivalence relation, and meaning function m assigns sets of objects to propositional variables. Semantics of the formulas built with the classical operations is defined as in section 3, and for formulas built with the epistemic operators we have:

M,o sat K(A)F iff for all o' if (o,o')∈ind(A), then M,o' sat F, or for all o'' if (o,o'')∈ind(A), then not M,o'' sat F.

It follows that operators K(A) are definable in terms of the modal operators of logic IL:

M,o sat K(A)F iff M,o sat [ind(A)]F∨[ind(A)]¬F.

The Hilbert-style axiomatization of logic EIL is not known. From the results presented in Valiev (1988) the axiomatization of the restricted language without ∩ is the following:

A0E. All formulas having the form of a tautology of the classical propositional calculus

A1E. (F∧K(A)F∧K(A)(F→G))→K(A)G

A2E. K(A)(K(A)F→F)

A3E. K(A)F↔K(A)¬F

The rules of inference are modus ponens and F/K(A)F.

Knowledge operators K(A) can be treated as the relational operators:

K(A)P=P/ind(A)⁻¹∪(-P)/ind(A)⁻¹.

The decomposition rules for those operators are as follows:

(K)
$$\frac{H, \ xK(A)Py, \ H'}{H, \ x\text{-}ind(A)z, \ zPy, \ x\text{-}ind(A)t, \ t\text{-}Py, \ H'}$$
where z and t are variables which do not appear in any formula above the line

(-K)
$$\frac{H, \ x\text{-}K(A)Py, \ H'}{H1 \quad H2 \quad H3 \quad H4} \quad \text{where}$$
H1=H, xind(A)z, x-(P/ind(A)⁻¹)y, H'
H2=H, z-Py, x-(P/ind(A)⁻¹)y, H'
H3=H, xRt, x-(-P/ind(A)⁻¹)y, H'
H4=H, tPy, x-(-P/ind(A)⁻¹)y, H'
z and t are arbitrary variables

The relational proof system for logic EIL consists of all the rules from section 1, rules (ref ind), (sym ind), (tran ind) from section 3, rules (K), (-K), and the following two rules which enable us to decompose indiscernibility relations of the form ind(A∪B) into the intersection of ind(A) and ind(B).

(ind ∪)
$$\frac{K, \ xind(A∪B)y, \ H}{K, \ xind(A)y, \ H \qquad K, \ xind(B)y, \ H}$$

(-ind ∪)
$$\frac{K, \ x-ind(A∪B)y, \ H}{K, \ x-ind(A)y, \ x-ind(B)y, \ H}$$

The fundametal sequences are (f1),(f2),(f3).

Example 4.1

We give a relational proof of formula F=K(A)(K(A)p→p). Its translation is t(F)=K(A)(-K(A)(p;1)∪p;1).

$$xt(F)y$$
$$\big| \ (K) \ \text{new variables}:=z,t$$

x-ind(A)z, z(-K(A)(p;1)∪p;1)y, x-ind(A)t, t-(-K(A)(p;1)∪p;1)y

$$\big| \ (∪)$$

x-ind(A)z, z-K(A)(p;1)y, zp;1y, x-ind(A)t, t-(-K(A)(p;1)∪p;1)y

/ (-∪) (--) \

x-ind(A)z, z-K(A)(p;1)y,   x-ind(A)z, z-K(A)(p;1)y,

zp;1y, x-ind(A)t   zp;1y, x-ind(A)t,

tK(A)(p;1)y   t-(p;1)y

$\big|$(-K) $\big|$(-K)

new variables:=z,z   new variables:=z,z

H1   H2   H3   H4   G1   G2   G3   G4

where H1=x-ind(A)z, zind(A)z, z-(p;1/ind(A)$^{-1}$)y, zp;1y,

x-ind(A)t, tK(A)(p;1)y=H3

H2=x-ind(A)z, z-(p;1)y, z-(p;1/ind(A)$^{-1}$)y, zp;1y, x-ind(A)t,

tK(A)(p;1)y

H4=x-ind(A)z, zp;1y, z-(p;1/ind(A)$^{-1}$)y, x-ind(A)t, tK(A)(p;1)y

G1=x-ind(A)z, zind(A)z, z-(-(p;1)/ind(A)$^{-1}$)y, zp;1y, x-ind(A)t,

t-(p;1)y=G3

G2=x-ind(A)z, z-(p;1)y, z-(-(p;1)/ind(A)$^{-1}$)y, zp;1y, x-ind(A)t,

t-(p;1)y

G4=x-ind(A)z, zp;1y, z-(-(p;1)/ind(A)$^{-1}$)y, x-ind(A)t, t-(p;1)y.

Sequences H2 and G2 are fundamental. Below are given fundamental decomposition trees for the remaining sequences.

  H1   H4

  $\big|$ (ref ind)   / (-/) new variable:=z (-1) \

H1, zIz   zind(A)z,...   z-(p;1)y,..

fundamental $\big|$(ref ind)   fundametal

                           zIz,...

                        fundamental

The fundametal decomposition tree for G1 is obtained in a similar way as the tree for H1 by applying rule (ref ind) to zind(A)z.

```
                          G4
            /(-/) new variable:=t (⁻¹)  \
        zind(A)t,...                      tp;1y,...
      /  (tran ind)    \                  fundamental
     / new variable:=x  \
zind(A)x,...       xind(A)t,...
   |(sym ind)       fundamental
xind(A)z,...
fundamental
```

Example 4.2

We give a relational proof of formula $F=K(A)p \lor K(B)p \to K(A \cup B)p$.
Its relational representation is $t(F)=-K(A)(p;1) \cap -K(B)(p;1) \cup K(A \cup B)(p;1)$.

```
                        xt(F)y
                          |(∪)
        x-K(A)(p;1)∩-K(B)(p;1)y, xK(A∪B)(p;1)y
                      /   (∩)    \
x-K(A)(p;1)y, xK(A∪B)(p;1)y      x-K(B)(p;1)y, xK(A∪B)(p;1)y
  |(K) new variables:=z,t          |(K) new variables:=z,t
x-K(A)(p;1)y,                     x-K(B)(p;1)y,
x-ind(A∪B)z, zp;1y,              x-ind(A∪B)z, zp;1y,
x-ind(A∪B)t, t-(p;1)y           x-ind(A∪B)t, t-(p;1)y
  | (-ind ∪) twice                 | (-ind ∪) twice
x-K(A)(p;1)y,                     x-K(A)(p;1)y,
x-ind(A)z, x-ind(B)z, zp;1y,     x-ind(A)z, x-ind(B)z, zp;1y,
x-ind(A)t, x-ind(B)t, t-(p;1)y   x-ind(A)t, x-ind(B)t, t-(p;1)y
    |(-K)                             |(-K)
  new variables:=z,t               new variables:=z,t
H1   H2   H3   H4                G1   G2   G3   G4
```

H1=xind(A)z, x-ind(A)z,...   G1=xind(B)z,...,x-ind(B)z,...

H2=z-(p;1)y,...,zp;1y,...   G2=z-(p;1)y,...,zp;1y,...

H3=xind(A)t,...,x-ind(A)t,...   G3=xind(B)t,...,x-ind(B)t,...

H4=tp;1y,...,t-(p;1)y   G4=tp;1y,...,t-(p;1)y

All those sequences are fundamental.

References

Fariñas del Cerro,L. and Orlowska,E. (1985) DAL-a logic for data analysis. Theoretical Computer Science 36, 251-264.

Goldblatt, R. (1987) Logics of time and computation. Center for the Study of Language and Information, Stanford.

Halpern,J. (ed) (1986) Theoretical aspects of reasoning about knowledge. Morgan Kaufmann, Los Altos, California.

Hintikka,J. (1962) Knowledge and belief. Cornell University Press.

Lenzen,W. (1978) Recent work in epistemic logic. Acta Philosophica Fennica 30, 1-219.

Orlowska,E. (1983) Semantics of vague concepts. In: Dorn,G. and Weingartner,P. (eds) Foundations of logic and linguistics. Problems and their solutions. Selected contributions to the 7th International Congress of Logic, Methodology and Philosophy of Science, Salzburg 1983. Plenum Press, London-New York, 465-482.

Orlowska,E. (1985) Logic of nondeterministic information. Studia Logica XLIV, 93-102.

Orlowska,E. (1988) Logical aspects of learning concepts. Journal of Approximate Reasoning 2, 349-364.

Orlowska,E. (1989) Logic for reasoning about knowledge. Zeitschrift fur Mathematische Logik und Grundlagen der Mathematik 35, 559-572.

Orlowska,E. (1991a) Relational interpretation of modal logics. In: Andreka,H., Monk,D. and Nemeti,I. (eds) Algebraic Logic. North Holland, Amsterdam, to appear.

Orlowska,E. (1991b) Dynamic logic with program specifications and its relational proof system. Journal of Applied Non-Classical Logics, to appear.

Pawlak,Z. (1981) Information systems-theoretical foundations. Information Systems 6, 205-218.

Prior,A. (1967) Past, present and future. Oxford University Press, Oxford.

Segerberg,K. (1971) An essay in classical modal logic. University of Uppsala, Uppsala.

Valiev,M.K. (1988) Interpretation of modal logics as epistemic logics (In Russian). Proceedings of the Conference Borzomi'88, Moscow, 76-77.

Vakarelov,D. (1987) Abstract characterization of some knowledge representation systems and the logic NIL of nondeterministic information. In: Jorrand,Ph. and Sgurev,V. (eds) Artificial Intelligence II, Methodology, Systems, Applications. North Holland, Amsterdam.

Vakarelov,D. (1989) Modal logics for knowledge representation. Springer Lecture Notes in Computer Science 363, 257-277.

Formal Grammars and Cognitive Architectures

Gheorghe PĂUN

Institute of Mathematics

Str. Academiei 14, București

70109 Romania

Abstract

After briefly discussing the thesis that the cognitive processes (in mind, but also in artificial "intelligences") can be naturally modelled by symbol manipulations, hence by grammars, we present some formal language theory devices and results with motivation from and significance for cognitive psychology and AI (such as various types of conditional grammars and serial and parallel grammar systems).

1 Introduction

The present paper must be considered as a formal language theory view to a question/research area which deserves a whole monograph and which is both theoretically appealing, practically important and philosophically significant: in what extent the formal grammars are adequate and useful models of cognitive processes taking place in the human mind or in AI frameworks. We shall mainly discuss here the motivation point and, implicitly, the usefulness of such a modelling for language theory (mathematical developments suggested in this context), rather than the "practical" relevance of this approach. Some details are given about grammar systems, a branch of formal language theory aimed to provide a way to better understand cooperation and communication in distributed cognitive systems.

2 The Universality of the Language Structure

We start from a general remark, advocated by many authors and extensively illustrated, about ten years ago, by the Romanian school of mathematical linguistics leaded by professor Solomon Marcus: the language structure is present everywhere, from natural to human and social processes/activities. A language means, in short, *syntax*, a way to define the correct-ness of *phrases* over a given finite *vocabulary*, hence a *grammar*. Why the "correct behaviour" of "all" processes around us means a language, hence syntax? The question is not fair, as it implicitly assumed that, indeed, the language structure is universal, but its answer solves in some extent also this preliminary question. On the one hand, the syntax is in some sense inherent, as both the temporal and the spatial relationships mean formal structure; the causality relations make more substantive the assertion for the temporal dimension. On the other hand, we, as men, are able to see around us what we are more prepared to see. Chomsky states it sharply: linguistics and mental processes are virtually identical. Moreover, the evolution of science, by considering smaller and smaller units of systems, hence more and more abstract items, is more and more interested by formal structure, and this is visible from molecular genetics to physics, biology, linguistics and so on. Also historical arguments can be invoked: the algorithmic thinking, so present in our life, is related to strings manipulation, hence to language structures (and linguistic "machines").

3   Linguistics as a Pilot Science

A direct consequence of the previous assertions is that the (mathematical) linguistics becomes more and more a laboratory for building and testing tools which are then used in approaching quite non-linguistical phenomena. Detailed discussions and illustrations (at the level of 1974 year) can be found in a paper by S. Marcus, with the same title as this section (in vol. *Current Trends in Linguistics*, ed. by Th. A. Sebeok, Mouton, The Hague, Paris, 1974). Further illustrations can be found, for example, in Gh. Păun, *The Grammars of Economic Processes*, The Technical Publ. House, Bucureşti, 1980 (in Romania). For a general treatment of human and social *action* one can also see the book of M. Nowakowska, *Languages of Actions. Languages of Motivations*, Mouton, The Hague, Paris, 1973.

Another corollary of discussions in the previous section is that *learning means to acquire a grammar*, the title of another S. Marcus work (in vol. *Mathematical and Semiotical Models of Social Development*, ed. S. Marcus, The Publishing House of the Romanian Academy of Sciences, Bucureşti, 1986). Learning a language means, as Chomsky proves, to actuate, by experience, certain innate mechanisms of our brain, able of infinite creativity. We are led to the famous distinction between competence and performance. But the same happens with other brain competences: by experience we know a finite set of "correct behaviours" of "systems" around us: to really learn means to overpass the finiteness, hence to have rules of correct-ness, hence grammars. From formal language theory point of view we already reached in this way a powerful branch of it, that of grammatical inference (we do not insist here in this direction). From psycho-cognitive point of view we are led, however, to a "technical" question: how so many grammars find room in our brain, how they are used, why not interfere, how we choose the right one when we need one, and so on. A possible (metaphorical) answer can be provided by ...

4   ...A Grammatical Model of the Brain

The model has been proposed by C. Calude, S. Marcus, Gh. Păun ("The Universal Grammar as a Hypothetical Brain") in *Rev. Roum. Lingv.*, 5 (1979), 479-489, and discussed by C. Calude, Gh. Păun ("On the Adequacy of a Grammatical Model of the Brain"), in the same journal, 4 (1982), 343-351. The basic idea is to use a universal grammar, which, by having modified certain parameters, behaves as a given particular grammar; thus we need to have active in our brain only this sort of meta-competence and some additional (also universal) devices, in order to be able to simulate a large family of individual competences. Picturally, a model as in Figure 1 is obtained, which, taking the brain as a black box, can simulate it (of course, without any claim about the morphology or the psychology of the real brain ...).

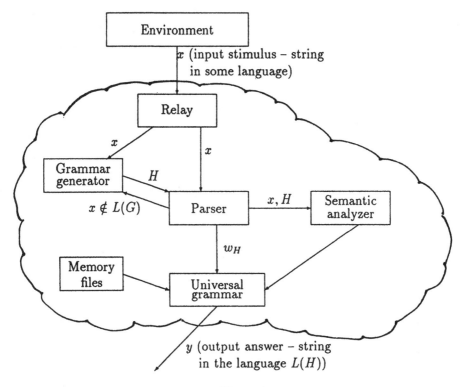

Figure 1:

Here are some explanations about this model. When an external stimulus (in the form of a string x in the language corresponding to some particular competence) comes into the brain, it activates, via an exciting *really*, the *grammar generator* and the *parser*. The grammar generator simply produces strings of the form $u_1 \longrightarrow v_1 / u_2 \longrightarrow v_2 / \cdots / u_n \longrightarrow v_n$, consisting of rewriting rules composing a grammar G. Denote by w_G such a string. The input string x and the code w_G come into the parser, which answers the question whether x belongs to the language generated by G or not. If the answer is negative, then a new grammar code is produced. The process continues until a grammar H is found such that x is in $L(H)$ (the right competence is identified). Now the *universal grammar* enters the stage. It receives w_H as input (starting string) and it works exactly as H (the activated competence). According to some information provided by the *semantic analyser* and some data in the *memory files*, the brain can answer, producing a string y in the language $L(H)$.

Technical details related to constructing the syntactic components of the model (the grammar codes generator, the parser, the universal grammar) are discussed in the above quoted papers. Here we only stress the fact that the model deals not with the concrete cognitive activity of the brain (learning, problem solving and so on), but with the brain behaviour in an environment in which "the language structure is universal".

5 Cognitive Processes and Chomsky Grammars

A working hypothesis related to the above discussed axiom about the universality of the language structure is that concerning the understanding of cognitive processes as rule based symbol manipulation processes. The idea is advocated in works of famous authors, as Newell and Simon, Minsky , Pylyshyn; recent developments can be found in a series of papers by J. Kelemen. Thus, a natural framework for the theoretical study of cognitive processes is the formal language/grammar theory, more precisely the Chomsky grammars. A strong *a posteriori* supporting argument for such an approach is its high explicative power, the intimate parallelism between certain cognitive psychology elements and Chomsky grammar elements: the rewriting rules model the long term memory, the sentential forms correspond to the short term memory, the terminals are surface cognitive units, where as the nonterminals are deep level cognitive units; the serial work of cognitive "processors" (rules, brain faculties, competences) corresponds to the sequential rewriting in grammars, the distinction between competence and performance is clear in the grammatical frame, and this is deeply involved in grammatical inference, the formal counterpart of learning. A further similar *a posteriori* argument concerns the thesis stating that the cognitive processes are "goal oriented". Well motivated in cognitive psychology, this thesis can be naturally modelled in formal grammar therms too, by the so-called...

6 ...Conditional Grammars

The notion is not new, it appears in many forms in the regulated rewriting area, a much investigated branch of formal language theory. The basic idea is to add to usual context-free rewriting rule certain *conditions*, such that a rule can be used for rewriting a given string only when the associated condition is controlled by information existing in the current state of the "problem" to be solved. Details about such a modelling of the goal orientation feature can be found in J. Kelemen, "Conditional grammars: motivations, definition, and some properties" (*Proc. Conf. Aut., Languages, Math. Systems*, ed. I. Peák, Budapest, 1984, 110-123). Variants of conditional grammars existed long time before. For instance, in 1968, I. Fris considered grammars in which each rule has a regular language associated and the rule can rewrite only sentential forms contained in this language. A variant is considered by J. Kral, in 1973: the associated language is the same for all rules. The generative power of these two types of conditional grammars is the same and it equals the power of context sensitive grammars, a sound result from cognitive theory point of view. An increased competence is obtained also for the so-called random context grammars introduced by van der Walt, in 1971: each rule has associated two sets of nonterminals and it can be applied to strings containing all symbols in one set (permitting symbols), but no symbol in the other set (forbidding symbols). In the above mentioned papers, J. Kelemen introduces an intermediate type of grammars, namely with rules having associated a string; a sentential form can be rewritten only if it contains the string. Generalizations to both permitting and forbidding sets of strings were considered by Gh. Păun, in 1985, and A. Meduna, in 1990. In many cases, again all the context sensitive languages are obtained. Mathematical details about all these classes of conditional grammars can be found in the monograph J. Dassow, Gh. Păun, *Regulated rewriting in Formal Language Theory*, Springer-Verlag, Berlin, Heidelberg, 1989.

Adding such conditions which restrict the rule applicability is useful from two points of view: the generative capacity of (nonrestricted) grammars is considerably increased and, moreover, the complexity of cognitive resources (the number of rules, for instance) needed for solving a given problem (for generating a language) can be essentially decreased. This conclusion is explicitly discussed for his conditional grammars by J. Kelemen, in "Measuring Cognitive Resources Use (A Grammatical Approach)" (*Computers and AI*, 8 (1989), 29-42), but it holds true for all the above mentioned types of conditional grammars. The technical result is of the following form: there are sequences of languages which request context-free grammars of arbitrarily increasing complexity, but can be generated by conditional grammars of bounded complexity (the complexity is evaluated by the number of nonterminal symbols, the number of rewriting rules, the total number of symbols appearing in rules and other similar known measures - see J. Gruska, "Descriptional Complexity of context-free Languages", *proc. Symp. Math. Found. Computer Sci.*, High Tatras, 1973, 71-83).

7 Distributivity and Cooperation

There are problems which cannot be solved by single problem solver, the so-called *social goals* which can be achieved only by a group of such problem solvers (possibly with a lower total cost). Thus system of problem solvers are necessary, which, anatomically speaking, means distributivity, the existence of several separated cognitive agents; in order to have a system, a body with global performances different from the sum of local performances, these agents must cooperative. Distributivity and cooperation are two quite important topics in the our days computer science, in AI and cognitive psychology. Computer nets, computers with parallel architecture, distributed data based and knowledge systems, computer conferencing and editing common texts, even the modularity of the mind (Chomsky, Fodor etc.) are practical aspects of this topic. An interesting remark concerning the relationship between formal language theory and its fields of application, true also in the case of conditional grammars, the grammatical model of the brain and other previously discussed points, is the fact that the theoretical tools were developed in many cases before the need for them was explicitly stated. The situation is the same for distributed cooperating systems: in 1978, R. Meersman and G. Rozenberg (*Proc. Symp. Math. Found. Computer Sci., Lect. Notes Computer Sci.*, 84, 1978, 364-373) considered "cooperating grammars systems", but with motivations related to the study of two level grammars used in formalizing (the syntax of) programming languages. The connections between such systems and AI notions (distributivity and cooperation) have been explicitly established ten years later.

8   Cooperating Distributed Grammar Systems

In short a cooperating grammar system, as defined by Meersman and Rozenberg, consists of n usual Chomsky grammars, working in turn, one in each moment (the other are waiting), on a common sentential form; each component, when enabled, rewrites the sentential form as long as possible; the order of entering the stage is controlled by a graph.

Let us compare this with the *blackboard model* described by P. H. Nii ("Blackboard Systems: the Blackboard Model of Problem Solving and the Evolution of Blackboard Architectures", *The AI Magazine*, Summer 1986, 38-53). Such a model consists of a set of knowledge sources, the current state of the problem solving process (the blackboard), in which the knowledge sources are making changes, and an external control of the order in which the knowledge sources are acting. The parallelism with cooperating grammar systems is clear and, adding the parallelism between basic items of cognitive processes and Chomsky grammars notions (Section 5), we are led to consider the cooperating grammar systems as natural formal models of blackboard systems. This was the starting point of an intensive investigation of grammar systems, beginning with the paper "On Cooperating/ Distributed Grammar Systems" (we shall abbreviate from now on by CDGS) of E. Csuhaj-Varjú and J. Dassow, published in 1990 (*J. Inform. Processing, EIK*, 26, 49–63). About a dozen of subsequent papers were written by E. Csuhaj-Varjú, J. Dassow, J. Kelemen, Gh. Păun, M. Gheorghe, S. Vicolov, A. Atanasiu, V. Mitrana about this subject. For an overview of approaches see Kelemen's article Syntactical Models of Distributed Cooperative Systems (*Journal of Experimental and Theoretical AI* 3 (1991), 1-10).

Some basic definitions about CDGS are given in the Annex A. Here we only informally describe some classes of such systems. The number of variants is impressive, most of them being "practically" motivated and some having only theoretical importance. The main problem is how to define the cooperational protocol. The components communicate only through the common sentential form (the content of the blackboard). When a component becomes active, how long it has to rewrite the current string, which component will be the next one? One can consider CDGS in which each component, when enabled, applies exactly k rules, or at most k, or at least k, for a given k, or any number of rules, or as many rules as possible. We can have conditions about the starting and the stopping moments (like in conditional grammars, hence a sort of goal oriented restriction), we can have a hypothesis/goal language, to which the sentential forms must belong, we can restrict the number of nonterminals present in the sentential forms (they denote "questions" to be solved in the next steps of the process), we can have or not an external control about the order of working and so on and so forth. The (theoretical) results are of the same type as for conditional grammars: a considerable increase in generative power (hence the *systems* — cognitive or grammatical — are really more powerful than their components) and the possibility to describe languages in an essential easier way (as grammatical complexity, hence ammount of cognitive resources).

A fundamental question, both practically and theoretically significant, is that concerning hierarchies. The nature loves hierarchies, it is said. What about CDGS: are $n+1$ components more powerful than n components? Intuitively, this must be the case. The question is not simple at all from formal point of view. For most types of CDGS it is still open. For systems with the derivation mode "as long as possible, rewrite", a strange result can be obtained, namely, that *three components are enough*: for each given CDGS, an equivalent CDGS can be constructed, consisting of at most three components. But the complexity of these components is comparable to the total complexity of the starting system. Thus, a double hierarchy can be defined if we take into account both the number of components and the maximum number of rewriting rules in these components (bounding one of these parameters, an infinite hierarchy is obtained according to the other one). An expected result, mathematically non-trivial, providing a proof for the trade-off between the number of problem solvers and their complexity. Many others similar results, with precise meaning in cognitive terms, can be found in the theory of CDGS.

9   Parallel Communicating Grammar Systems

In the previous type of grammar systems, the components work sequentially, only one in a given moment. This corresponds, it is claimed, to the way the brain components work (when solving problems). But in the present days computer science there are many cases when parallel systems are designed for solving complex tasks not (efficiently) solvable by a single "processor". Starting from another point: let as modify the blackboard model, by designing a *master* knowledge source, which is the only component allowed to modify the blackboard, and adding to each other component its own "notebook" on which it current string describing the state of the own sub-problem; the master decides when the global problem of the system is solved. A model of problem solving which we can call the *classroom* model is obtained in this way. Many actual applications of such a cognitive system can be listed: the team way of working on a research problem, the team way of producing any kind of document are problem solving processes of this type (the team is supposed to have a leader).

From the formal language point of view, this classroom model leads to parallel communicating grammar systems (PCGS, for short), introduced by Gh. Păun and L. Sântean ("Parallel Communicating Grammar Systems: the Regular Case", *Ann. Univ. Buc., Ser. Matem.-Inf.*, 38 (1989), 55-63). As a CDGS too, a PCGS consists of n usual Chomsky grammars, but all of them work on separate sentential forms (starting from their own axioms); special nonterminals are considered, which are not rewritten but replaced by the current strings generated by other components of the system (these *query* symbols identify the grammar which must satisfy them); the strings generated by the master grammar (one designed component) are considered as generated by the whole system.

the theory of PCGS is somewhat parallel to that of CDGS: variants, generative power, succinctness of describing languages (compared to context-free grammars), decision problems and so on. Also the results are similar: considerable increase of power, essential decrease of complexity. A specific question is that of synchronization: the components of a PCGS work simultaneously, each using one rule in each time unit; this assumes a universal clock there exists which marks the time. What about the case when this synchronization is not present? What type of more powerful synchronizing restrictions can be considered? The answer to the first question is the expected one: The synchronization is really useful, in the regular case (with respect to the Chomsky hierarchy) nothing more than regular languages can be obtained by using unsynchronized PCGS, where as in the context-free case the unsynchronized PCGS still generate non-context-free languages, but they are strictly weaker than the synchronized systems.

Also for PCGS there are about a dozen of papers (written but not all of them published yet) by Gh. Păun, L. Sântean, M. Gheorghe, J. Kari, J. Hromkovič. A synthesis of results about both CDGS and PCGS, including motivations and interpretations of results in terms of cognitive psychology and AI, can be found in the monograph *Grammar Systems. A Grammatical Approach to Cooperation and Distribution*, by E. Csuhaj-Varjú, J. Dassow, J. Kelemen, Gh. Păun (in preparation). Some basic definitions are provided in the Annex B.

10 Concluding Remarks

The formal language theory proves to be one of the most *adequate* frameworks for mathematically modelling and investigating questions of cognitive psychology and AI. Moreover, the grammatical models are not only suitable in this aim, but they are also *relevant*, provide significant results/conclusion, theoretically justifying common-sense remarks but also furnishing new insights, new statements. Sometimes, the formal instruments were developed before they were asked by such applications, but having the contact with this area of "practice", the formal language theory gained a serious amount of new ideas, motivations for new notions, new problems, new research topics. In what concerns the grammar system area, one can say that a real new branch of formal language theory is born in this way, both mathematically appealing and well motivated. (As a matter of fact, the very development of this theory corresponds to a sort of parallel distributed cooperating and communicating system of authors...)

Annex A:
Cooperating Distributed Grammar Systems –
– Basic Variants

A CDGS (of degree n, $n \geq 1$) is a system

$$\gamma = (G_1, G_2, \ldots, G_n)$$

where $G_i = (V_{N,i}, V_{T,i}, S_i, P_i)$, $1 \leq i \leq n$, are Chomsky grammars. Denote $V_i = V_{N,i} \cup V_{T,i}$. For $x, y \in V_i^*$, $x = x_1 u x_2$, $y = x_1 v x_2$, $x_1, x_2 \in V_i^*$, $u \longrightarrow v$ in P_i, we write $x \underset{G_i}{\Longrightarrow} y$. A k-steps derivation is denoted by $x \underset{G_i}{\overset{=k}{\Longrightarrow}} y$, $k \geq 1$. Denote too

$$x \underset{G_i}{\overset{\leq k}{\Longrightarrow}} y \quad \text{iff} \quad x \underset{G_i}{\overset{=k'}{\Longrightarrow}} y \text{ for some } k' \leq k$$

$$x \underset{G_i}{\overset{\geq k}{\Longrightarrow}} y \quad \text{iff} \quad x \underset{G_i}{\overset{=k'}{\Longrightarrow}} y \text{ for some } k' \geq k$$

$$x \underset{G_i}{\overset{*}{\Longrightarrow}} y \quad \text{iff} \quad x \underset{G_i}{\overset{=k'}{\Longrightarrow}} y \text{ for some } k' \geq 1$$

$$x \underset{G_i}{\overset{t}{\Longrightarrow}} y \quad \text{iff} \quad x \underset{G_i}{\overset{*}{\Longrightarrow}} y \text{ and there is not } z \in V_i^*, \text{ such that } y \underset{G_i}{\Longrightarrow} z$$

For $a \in \{= k, \geq k, \leq k, *, t\}$, denote

$$L_a(\gamma) \; = \; \{x \in V_{T,j}^* \; \mid \; S_{i_1} \underset{G_{i_1}}{\overset{a}{\Longrightarrow}} w_1 \underset{G_{i_2}}{\overset{a}{\Longrightarrow}} w_2 \ldots \underset{G_{i_r}}{\overset{a}{\Longrightarrow}} w_r,$$

$$r \geq 1, i_r = j, 1 \leq i_s \leq n, 1 \leq s \leq r\}$$

We have here only step limitations, no external control, start and stop conditions, or other type of additional feature.

Example: Consider $\gamma = (G_1, G_2)$, with

$$G_1 = \; (\; \{S_1, A, A', B, B', C\}, \{a, b, c\}, S_1,$$
$$\{S_1 \longrightarrow C, C \longrightarrow AB, A' \longrightarrow A, B' \longrightarrow B\})$$
$$G_2 = \; (\; \{S_2, A, A', B, B'\}, \{a, b, c\}, S_2,$$
$$\{A \longrightarrow aA'b, b \longrightarrow cB', A \longrightarrow ab, B \longrightarrow c\})$$

We obtain (the reader can verify):

$$L_{=2}(\gamma) \; = \; L_{\geq 2}(\gamma) = \{a^n b^n c^n \mid n \geq 1\}$$
$$L_{=k}(\gamma) \; = \; L_{\geq k}(\gamma) = \emptyset, \text{ for } k \geq 3$$
$$L_{=1}(\gamma) \; = \; L_{\geq 1}(\gamma) = L_*(\gamma) = L_t(\gamma) = \{a^n b^n c^m \mid n, m \geq 1\}$$

Note that in $= 2$ and ≥ 2 modes of derivation, the system γ produces a non-context-free language; remark the way in which the two grammars cooperate in order to obtain this increase of competence.

Annex B:
Parallel Communicating Grammar Systems –
– Basic Variants

A PCGS (of degree n, $n \geq 1$) is a system

$$\gamma = (G_1, G_2, \ldots, G_n)$$

where $G_i = (V_{N,i}, V_{T,i}, S_i, P_i)$, $1 \leq i \leq n$, are Chomsky grammars such that $V_{T,i} \cap V_{N,j} = \emptyset$, $1 \leq i, j \leq n$, and there is a set of special symbols, $K \subseteq \{Q_1, Q_2, \ldots, Q_n\}$, $K \subseteq \bigcup_{i=1}^{n} V_{N,i}$, called *query symbols* and used in derivations as follows. Denote by $|x|_K$ the number of occurences in x of symbols in K.

For $(x_1, x_2, \ldots, x_n)$, $(y_1, \ldots, y_n)$, $x_i, y_i \in V_i^*$, $1 \leq i \leq n$, we write $(x_1, x_2, \ldots, x_n) \Longrightarrow (y_1, \ldots, y_n)$ if one of the next two cases holds:

(i) $|x|_K = 0$, $1 \leq i \leq n$, and $x_i \underset{G_i}{\Longrightarrow} y_i$ for each i, excepting the case with $x_i \in V_{T,i}^*$, when $x_i = y_i$;

(ii) there is i, $1 \leq i \leq n$, such that $|x_i|_K > 0$; then for each such i we write $x_i = z_1 Q_{i_1} z_2 Q_{i_2} \ldots z_t Q_{i_t} z_{t+1}$, $t \geq 1$, for $z_j \in V_i^*$, $|z_j|_K = 0$, $1 \leq j \leq t+1$; if $|x_{i_j}|_K = 0$, $1 \leq j \leq t$, then $y_i = z_1 x_{i_1} z_2 x_{i_2} \ldots z_t x_{i_t} z_{t+1}$ $\left[\text{and } y_{i_j} = S_{i_j}, 1 \leq j \leq t\right]$; when, for some j, $1 \leq j \leq t$, $|x_{i_j}|_K > 0$, then $y_i = x_i$; for all i for which y_i is not specified above, we have $y_i = x_i$.

In words, an n-tuple $(x_1, x_2, \ldots, x_n)$ directly yields $(y_1, \ldots, y_n)$ if

(i) either no query symbol appears in $x_1, x_2, \ldots, x_n$ and then we have a componentwise derivation, $x_i \Longrightarrow y_i$ in G_i, $1 \leq i \leq n$, in the usual sense (one rule is used in each component), excepting the terminal components x_i which remain unchanged, or

(ii) in the case of query symbols appearing, a *communication* step is performed as these symbols impose: each occurrence of Q_j in x_i is replaced by x_j, providing x_j does not contain query symbols; after communicating, the grammar G_j resumes working from its axiom; the communication has priority over effective rewriting, no rewriting is possible when at least a query symbol is present.

The language generated by γ is

$$L(\gamma) = \left\{ x \in V_{T,1}^* \mid (S_1, S_2, \ldots, S_n) \overset{*}{\Longrightarrow} (x, u_2, \ldots, u_n), u_i \in V_i^*, 2 \leq i \leq n \right\}.$$

In words, we start from the n-tuple of axioms and proceed by repeated rewriting and communication steps until the component G_1 (this is the master grammar) produces a terminal string (no care is paid to other components).

In the previous definition, all components can introduce query symbols; we say that γ is *non-centralized*. If $K \cap V_{N,i} = \emptyset$, for $2 \leq i \leq n$, that is only G_1 can introduce query symbols, then we obtain a *centralized* PCGS. Moreover, the above definition refers to *returning* PCGS (after communicating, each component whose string has been sent to another component returns to axiom). A PCGS is *non-returning* if the words in brackets, [and $y_{i_j} = S_{i_j}$, $1 \leq j \leq t$], in the definition, are erased. That is, after communicating, the components continue to process the current strings. Four basic classes of PCGS are obtained in this way.

Example: Consider $\gamma = (G_1, G_2, G_3)$, with

$$G_1 = (\ \{S_1, S_2, S_3, Q_2, Q_3\}, \{a, b, c\}, S_1,$$
$$\{S_1 \longrightarrow aS_1, S_1 \longrightarrow aQ_2, S_2 \longrightarrow bQ_3, S_3 \longrightarrow c\})$$
$$G_2 = (\ \{S_2\}, \{b\}, S_2, \{S_2 \longrightarrow bS_2\})$$
$$G_3 = (\ \{S_3\}, \{c\}, S_3, \{S_3 \longrightarrow bS_3\})$$

Here is a terminal derivation in γ:

$$(S_1, S_2, S_3) \overset{*}{\Longrightarrow} (a^k S_1, b^k S_2, c^k S_3) \quad \Longrightarrow (a^{k+1} Q_2, b^{k+1} S_2, c^{k+1} S_3)$$
$$\Longrightarrow (a^{k+1} b^{k+1} S_2, u_2, c^{k+1} S_3) \quad \Longrightarrow (a^{k+1} b^{k+2} Q_3, u_2', c^{k+2} S_3)$$
$$\Longrightarrow (a^{k+1} b^{k+2} c^{k+2} S_3, u_2', u_3) \quad \Longrightarrow (a^{k+1} b^{k+2} c^{k+3}, u_2'', u_3'),$$

for $k \geq 0$, u_2, u_2', u_2'', u_3, u_3' depending on the fact whether γ is returning or non-returning. Therefore

$$L(\gamma) = \{a^n b^{n+1} c^{n+2} \mid n \geq 1\}$$

a non-context-free language, generated by the most restrictive type of PCGS: centralized (returning or non-returning, this does not matter), with regular components. Please note the synchronization of components, as well as the way they communicate.

Efficient Simulations of Nondeterministic Computations and their Speed–up by the Ring of Cooperating Machines

Juraj Wiedermann

VUSEI-AR, Dúbravská 3, 842 21 Bratislava

Czecho-Slovakia

Abstract: Using the formalism of Turing machines, efficient deterministic, both sequential and parallel, realizations of nondeterministic computations are described and analyzed. It is shown that when simulating nondeterministic computations by the ring of cooperating Turing machines, the speed–up linear in the number of machines can be achieved as compared to the case of a single machine. Using examples it is argued that the methods and results described in the paper can be applied to more realistic models of computations and thus are of practical significance e.g. in artificial intelligence for solving various combinatorial search problems.

1. Introduction

1.1 The significance of nondeterminism. Nowadays there are certainly no doubts concerning the immense theoretical significance of the concept of nondeterminism that has enabled to formalize certain fundamental notions and to define certain fundamental problems in the theoretical foundations of informatics.

From the practical point of view nondeterminism can be seen as a conceptual tool for the design and description of (nondeterministic) algorithms. In theory such a tool is usually represented by the model of nondeterministic or alternating Turing machine, while in practice it manifests itself in the disguise of logic, functional or equational programs, respectively, or as term rewriting systems.

The standard example illustrating the power of nondeterminism is given by so–called combinatorial search problems often encountered in various AI applications. In these applications it is required to find in a large set ('search space') an element satisfying a certain condition. While the nondeterministic algorithm for this problem simply 'guesses' the element sought and verifies the condition it has to satisfy, the deterministic algorithm usually has to search a significant portion of the search space in order to find the right element.

1.2 Sequential realization of nondeterminism. In general the efficient realization of nondeterministic computation systems on deterministic devices, no matter whether sequential or parallel ones, calls for designing of anefficient simulation of nondeterministic computations by deterministic ones. This of course is the traditional problem in complexity theory which has been studied in the formalism of automata theory since the emergence of the notion of nondeterminism in the early sixties. At that time the straightforward simulation algorithm (called 'backtrack algorithm') of exponential time complexity $O(c^{T(n)})$ has been devised, enabling to simulate a nondeterministic Turing machine of time complexity $T(n)$, essentially by examining all of its possible computational paths [1,3]. There are numerous variants of this algorithm used in practice known under the name of branch-and-bound search , α–β pruning, backtrack with search re-arrangement [2], etc.

From that time on no asymptotically faster algorithm has been found and the corresponding formalized problem of proving àt least the existence of asymptotically more efficient algorithms presents currently the central open problem not only in informatics but in mathematics as well.

For instance the problem of deciding whether there exist deterministic algorithms of polynomial time complexity for the class of nondeterministic computations of the same asymptotic complexity presents the well–known P–NP problem that also has its analogues in other complexity classes.

In spite of this,a certain success has been achieved in the design of efficient probabilistic algorithms for certain concrete NP-complete problems (see e.g. since 1981 the series of papers by Johnson [4]). Unfortunately the results do not carry over to a general case of nondeterministic computations.

As far as the general case is concerned, recently it has been observed that by tuning the proof of the known fact in theory that $NSPACE(S(n)) \subseteq DTIME(c^{S(n)})$ one can obtain an alternative simulation algorithm that is exponential w.r.t. space, rather than time, used by the nondeterministic computation [9]. In frequent cases when there is an asymptotic difference between time and space complexity of the simulated nondeterministic machine, this change leads to a dramatic improvement over previous simulation which was exponential w.r.t. time complexity of the machine simulated.

Nevertheless there are instances of problems where not even this algorithm presents any asymptotic improvement.

Therefore, failing to find efficient sequential algorithms for simulating nondeterministic computations the attention of researchers has turned to parallel algorithms.

1.3 Parallel realization of nondeterminism. The problem of an efficient (deterministic) parallel realization of nondeterministic algorithms has been most often investigated within the framework of artificial intelligence where researchers have tried to speed up the solutions of various concrete NP–complete problems (usually called *combinatorial search problems* in this context — like the traveling salesman problem, knapsack problem, Hamiltonian path problem, satisfiability problem, etc.) by implementing them on diverse multiprocessor systems.

Along these lines one particularly stimulating work was done in Paderborn [6,7,11] where the multiprocessor system with the simplest topology has been investigated — namely a ring–structured network of personal computers in the role of asynchronous processors. Via two serial ports each computer in the ring could communicate with its neighbours, i.e., could exchange information with its predecessor or successor in the ring. The distributed program allowed the parallel execution of sequential backtracking algorithm.

The general conclusion from this experiment and from all the previous ones was that the parallelism did indeed help in speeding up the corresponding sequential algorithms and that the rate of speed–up depended both on the number of processors and on the algorithm used to solve the problem at hand.

Another area where the problems of efficient parallel realizations of nondeterministic algorithms have been studied is the theory of logic programming [8].

1.4 The uniform approach to realization of nondeterministic computations. When one wishes to concentrate on the pure, quite well formalizable problem which the problem of efficient realization of nondeterministic computation undoubtedly is, it appears that the above described experiments or studies suffer by a non–uniform, non–systematic approach: different solutions described so far are only hardly comparable since concrete and often different problems have been considered, and the models of target machines on which the algorithms have been implemented varied in a wide range. Therefore it seems that it could be of interest to try to find a common, uniform framework for investigating these problems from a complexity point of view.

The purpose of this paper is to provide such an approach based on models of computations well established in complexity theory — on Turing machines.

1.5 Form and contents of the paper. The paper consists of two main parts, represented by Section 2 and 3, respectively.

In Section 2 two basic sequential algorithms for simulating nondeterministic computations as briefly mentioned in part 1.2 are presented: a simple backtracking algorithm based on the depth–first search strategy of exploration of a corresponding solution space, and a somewhat more elaborated algorithm, based on the breadth–first search exploration of the solution space. Both algorithms are analyzed both from time and space complexity point of view. In the close of this section the differences between these algorithms are discussed and their practical significance is illustrated.

The previous algorithms create a basis for forthcoming considerations in Section 3 where their efficient distributed implementation on a ring of Turing machines is described and analyzed. For that purpose the formal definition of a ring of Turing machines is given. It is shown that using such a ring, it is possible within the wide range to speed up the computation of corresponding simulating algorithms by a factor linear in number of processors, as compared to the case of a single processor. Again, in the close of this section some practical aspects of both previous distributed algorithms are discussed.

2. Efficient sequential simulation of nondeterministic computations

2.1 Depth–first simulation (sequential backtracking). First we shall present the standard text–book result [1,3] describing the simulation of an off–line nondeterministic Turing machine (NTM) (i.e., of a NTM that has a separate read–only input tape) on a deterministic off–line Turing machine (DTM) with the complexity exponential w.r.t. the time complexity of the simulated machine.

Theorem 1. *Let M be an off–line NTM of known time complexity $T(n)$ and of space complexity $S(n)$; let $T(n)$ be a fully time constructible function. Then there exists a constant $c > 1$ and an off–line DTM $\bar{M}$ such that $\bar{M}$ simulates M in time $O(T(n)c^{T(n)})$ and in space $O(T(n))$.*

Proof (Sketch): Certainly there exists a constant $d > 0$ such that in any situation M has at most d possibilities to prolong its computation. This means that any sequence of length $T(n)$ of M's instructions can be represented by a string of length $T(n)$ over the alphabet $\Sigma = \{0, 1, \ldots, d - 1\}$.

$\bar{M}$ simulates M on the input x of length n in the following way: $\bar{M}$ generates subsequently in the lexicographic order all the possible strings over Σ of length $T(n)$. There are $d^{T(n)}$ of such strings and for each of them $\bar{M}$ verifies by deterministically simulating the corresponding sequence of instructions whether it describes an accepting computation.

One such string can be generated and verified in time and space $O(T(n))$ which altogether leads to the estimate as in the claim of our theorem.

∎

Note that in the above simulation often the entire string of instructions need not be generated in order to verify its correctness. Rather, each time a new instruction in the string is generated, one can verify immediately whether it presents a correct prolongation of some computation, or not. In the positive case the next instruction can be generated, otherwise the next instruction is tried on the same position in the string, or we return to the previous position (a so–called *backtrack*) and we try the next instruction on that position.

It is easily seen that the idea of the above algorithm can be straightforwardly realized e.g. on RAMs. Then the real efficiency of backtracking in practice depends on the frequency of how often and how soon we recognize that a certain computational path generated so far leads to a nonaccepting state. This, however, depends on a concrete algorithm that is being simulated and therefore cannot be reflected in our complexity analysis.

2.2 Breadth–first simulation. The idea of breadth–first simulation is quite simple [13]. The strategy involved is to see the running computation of a NTM as a collection of individual DTMs running synchronously in parallel, each of them following a possible computational path of the original machine. With this conception it may easily happen that at the same time different machines follow the same computation path or follow the path encountered already before (i.e., the computation of the corresponding machine is cycling in that case). To avoid these redundant computations the whole collection of machines is inspected after each computational step and the machines performing, or repeating, identical computations are eliminated from further consideration. For fast detection of duplicates or cycling computations sorting is used. The details are given in the proof of thefollowing theorem (the proof presented here differs from the original one [13] in that it includes, moreover, the idea of cycle detection and elimination).

Theorem 2. *If M is an off–line k–tape NTM of time complexity $T(n)$ and of space complexity $S(n)$ then there is a constant $c > 0$ and an off–line DTM $\bar{M}$ such that $\bar{M}$ simulates accepting computations of M in $O(nT(n)M^2(n)S^k(n)c^{S(n)})$ time and $O(n.S^k(n).M(n)c^{S(n)})$ space, where $M(n) = max\{S(n), log\, n\}$.*

Proof (Sketch): The simulation of M by $\bar{M}$ can be conceptually partitioned into $T(n)$ cycles; during each cycle $\bar{M}$ simulates one step of M.

For that purpose after the end of the i–th cycle the following invariant is preserved on certain two (let us say on the first and on the second) of $\bar{M}$'s working tapes:

— the first tape T_1 contains the set T_1 of (distinct) instantaneous descriptions (IDs) that can be achieved by M in i steps without any cycling starting from the initial ID of M, for $i = 1, 2, \ldots, T(n)$.

— the second tape T_2 contains the set T_2 of all IDs that M has entered so far, i.e., during its first $i - 1$ steps.

Note that $T_1 \cap T_2 = \emptyset$.

For efficiency reasons the IDs both on T_1 and T_2, respectively, are kept sorted in lexicographic order.

The simulation starts with initial ID of M written on T_1, and with T_2 empty. Now, for $i = 1, 2, \ldots$ the following cycle consisting of four phases is performed:

Phase 1: (restoring the invariant on T_1)
The IDs from T_1 are merged with these of T_2 and the resulting sorted sequence is placed onto T_2.

Phase 2: (updating the IDs)
The IDs on T_1 are sequentially scanned and in the $(i + 1)$–st cycle and ID C replicates itself into j other IDs that follow from C if, in ID C, M has j choices for the next move. The new IDs corresponding to the $(i + 1)$–st move of M are sequentially written onto another working tape of $\bar{M}$ — let us say tape T_3.

Phase 3: (eliminating duplicates)
The IDs on T_3 are lexicographically sorted using e.g. the balanced two–tape merge sort; at this occasion identical ID's are eliminated. The resulting sorted sequence finds itself still on T_3.

Phase 4: (eliminating cycles and restoring the invariant on T_1)
Using the process similar to merging those IDs from T_3 that do not occur on T_2 are copied back to T_1; in this way cycles are eliminated from further computation.

Note that at the end of Phase 4 we have $T_1 \cap T_2 = \emptyset$.

The above process is repeated until either an accepting ID is generated in Phase 3 (which should occur after performing at most $T(n)$ cycles due to our assumption), or the computation runs for ever without accepting (eventual cycling on all paths can be detected — in that case T_1 becomes empty after some cycle).

As far as the simulation efficiency of an accepting computation is concerned, we first estimate its space complexity which is dominated by the length of three tapes T_1, T_2, and T_3, respectively.

If a k-tape off–line NTM is of space complexity $S(n)$, then the number of distinct IDs which M can enter when started on an input of length n is bounded by $q(t + 1)^{kS(n)} S^k(n).n$, where q denotes the number of states of M, the second factor bounds the number of possible work–tapes contents for a working alphabet of size t, the next factor bounds the number of possible working head positions, and the last one the number of possible input–head positions.

Since the IDs on T_1 and T_2 are all distinct, and the size of each instantaneous description is bounded by $M(n) = O(max\{S(n), log\, n\})$, the length of either of these tapes cannot exceed $O(nS^k(n)M(n)c^{S(n)})$, with $c = q(t + 1)^k$. However, since the number of possible moves of M from each ID is bounded, the above estimate holds also for the length of T_3 after Phase 2.

The time complexity of simulation during any cycle is dominated by the cost of Phase 3 where T_3 is being sorted. It is known that using the balanced two–way merge sort a tape can be sorted using an approximatively logarithmic number of passes over data.

Hence, our tape can be sorted in $O(nS^k(n)M^2(n).c^{S(n)})$ time and this estimate bounds at the same time the complexity of one cycle of our simulation.

Since altogether at most $T(n)$ cycles are to be performed when simulating any accepting computation, the time estimate as in the claim of the theorem follows. ∎

Although it might seem that the factor $n.T(n).S^k(n).M^2(n)$ plays a minor role in the presence of the exponential factor $c^{S(n)}$ in the expression for the efficiency of the simulation, it is not always so: consider e.g. the case of logarithmic, or even sublogarithmic space complexity, which our simulation admits!

Also note that in the proof of the previous theorem the elimination of duplicate IDs is essential since without it we could easily generate up to $O(c^{T(n)})$ IDs for some machines. Moreover, although cycle elimination could not have been taken into account in the efficiency estimation of the final algorithm (and indeed, to achieve our estimate it is not necessary at all — see the original paper [13]), in some practical situations it could be if great importance from the complexity point of view.

2.3 Comparing the two simulations. The simulations from Theorem 1 and Theorem 2 differ in two important aspects.

First, as noticed in part 1.2, the simulation from Theorem 1 is exponential w.r.t. time of the machine simulated, while the simulation from Theorem 2 is exponential w.r.t. space of the machine simulated. At the same time, however, the latter simulation requires exponential space w.r.t. space complexity of the machine simulated, whereas this need not be the case in the former simulation from Theorem 1.

Second, the simulation from Theorem 1 required $T(n)$ to be known beforehand, and to be fully time–constructible. This assumption was necessary in the proof of Theorem 1 in order to know and to compute the bound on the length of the sequence of instructions to be generated and verified. Without knowing this bound, it could happen that the simulation would run forever, exploring a certain computational path deeper and deeper without ever backtracking to some neighbouring path on which an accepting computation could occur.

The assumption of knowing $T(n)$ beforehand and that of its constructibility can be avoided e.g. by trying the simulation for $T(n) = 1, 2, 4, 8, \ldots$ etc. Then, when there is an accepting computation, it will be eventually discovered and the simulation will terminate and, clearly, the original running time will not by asymptotically affected due to the exponential growth of the upper bounds in the above mentioned sequence of trials.

On the other hand, the simulation from Theorem 2 does not require any special assumption on $T(n)$.

By the way, in the theory of logic programming the issue just discussed — i.e., that of knowing or not knowing $T(n)$ beforehand — plays an important role, since logic programs present typical examples of nondeterministic algorithms with no a priori known complexity.

Therefore, when evaluating a logic program with the help of depth–first strategy known as 'SLD resolution', that is generally used e.g. in Prolog, the correct termination cannot be guaranteed since the above strategy is analogous to that of the simulation from Theorem 1 and no upper bound on $T(n)$ is known. Of course, this is no more true for the strategy analogous to the simulation from Theorem 2, that has been often considered in the theory of logic programs, including redundant computation detection and elimination [5].

2.4 Applications. From the methodological point of view the simulation described in Theorem 2 has one perhaps surprising side effect — namely, it assures that one possible way in designing time–efficient deterministic algorithms for solving e.g. NP-complete problems is looking not for time–efficient, but rather for space–efficient nondeterministic algorithms for the problem at hand, followed by their subsequent efficient simulation according to Theorem 2. Following this theorem, to obtain the deterministic solution of subexponential time complexity, it is enough to design only the corresponding non-deterministic algorithm of sublinear space complexity.

This approach is illustrated in [13] where a nondeterministic Turing machine algorithm for the knapsack problem (which is known to be NP-complete) of space complexity $O(\sqrt{n})$ is designed to give rise via breadth–first simulation to a deterministic algorithm of time complexity $O(c^{\sqrt{n}})$. Note that the realization of the same algorithm via depth–first search could lead to an algorithm of time complexity $O(c^n)$, since the above nondeterministic algorithm performs up to $O(n)$ nondeterministic steps!

The last result seems to represent currently the best known deterministic worst-case algorithm on Turing machines for this problem.

A similar approach has also led to the design of a knapsack algorithm of both time and space complexity of order $O(2^{\sqrt{n}})$ on a RAM with a logarithmic cost criterion [12].

The above facts nicely demonstrate the practical significance of a seemingly 'totally unpractical' approach to the design of efficient algorithms with the help of Turing machines.

3. Simulating nondeterministic computations by a ring of cooperating machines

3.1. Parallelism vs. nondeterminism. The well–known theory, and as a matter of fact all the previous results indicate that the sequential simulation of nondeterministic computations is in general of exponential complexity which means that its practical realization requires an exorbitant amount of time, and sometimes also that of space. In the complexity theory there is an important open problem to what extent parallelism can help in solving these problems.

On the one hand, in theory from the so-called *parallel computation thesis* it follows that NP–complete, or even $PSPACE$–complete problems can be solved in polynomial parallel time using exponential parallelism, i.e., one which uses an exponential amount of processors. On the other hand, it is clear that from time, space, construction, and economical reasons, respectively, the exponential parallelism cannot be realized in practice.

Let us take, therefore, another viewpoint: what would happen if we could use only a bounded number, or with growing n only a slowly increasing number — let us say $p = p(n)$ — of processors. Then, of course, as compared with the case of a single processor, we would like to achieve a p–fold speed–up.

In what follows we shall prove that certain ways of simulating sequential nondeterministic algorithms by a bounded, or slowly increasing, number of processors indeed do posses this desirable property, i.e., the simulation will be p–times faster when p cooperating processors, working in parallel, will be used.

3.2. The ring machine. As our model of cooperating processors we shall introduce a so-called *ring of p Turing machines* (p–RTM), $p > 1$. The corresponding Turing machines will be all either deterministic, or nondeterministic, as we shall need (a p–RTM, or a p–RNTM).

For each TM in a p–RTM its two *neighbours* — the left one and the right one — are defined. Each TM in the ring is extended by so-called *communication tapes* — the left and the right input tape, respectively, and similarly the left and the right output tape, respectively. The right communication input tape of any machine M is at the same time the left output tape of M's right neighbour; the right communication output tape of M is at the same time the left input tape of M's right neighbour. The left communication tapes are organized in an analogous manner (see Fig. 1).

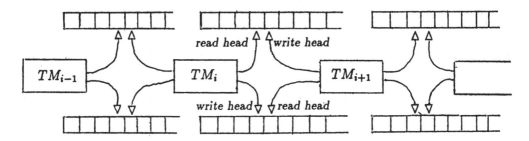

Fig. 1. A part of a ring of Turing machines.

The communication between neighbour machines is realized via communication tapes: what is written onto a communication output tape of some machine that can be read by the corresponding neighbour machine.

A distinguished machine from among the machines in the ring — the so-called *leader machine* — is equipped also by standard input and output tape.

At the beginning of a computation, the input is written on the input tape, and the output tape is empty.

The leader starts the computation and subsequently, via communication tapes, activates all the other machines in the ring.

All the machines in the ring are identical — i.e., they follow the same program. Any machine can test whether it is a leader. The machines work synchronously and the computation of the ring ends successfully when all its machines enter the accepting state (not necessarily at the same time).

3.3. Complexity measures. For any p–RTM, all complexity measures are considered as functions of both p and n, where p denotes the number of processors in the ring at hand, and n denotes the length of the input.

The *time complexity* of an RTM is given by the number of synchronous parallel steps that, for the given input, have been performed during computation, i.e., from the start of computation by the leader machine until the halting of the last machine in the ring.

The *space complexity* of a computation of any RTM is determined by the maximal space complexity from among all machines belonging to the ring.

The *transmission complexity* (or *transmission time*) of a computation of any RTM is defined as the total number of (parallel) computational steps during which any machines in the ring communicate, i.e., as the total number of (parallel) read/write operations (so-called *messages*) on all communication tapes during the computation.

Note that we are not interested in the total number of messages received or sent — what we are interested in is the total time of sending or receiving messages in the ring.

3.4 Discussion of the model. The model of a ring of Turing machines just described presents quite an idealized model of ring computer networks (e.g. that of a local network of personal computers) as used in real life.

In our model there is no difference made between the price of communication (transmission time) and the price of computation itself that occurs, so to speak, 'inside' the machines. Moreover, Turing machines themselves certainly do not represent the most realistic model of computations.

On the other, hand the model captures the basic features of computer networks — the absence of a common global memory and the communication with the help of message exchanges. Also — what is of utter importance for our purposes — this model enables a natural simulation of sequential Turing machine computations on a network of such machines and hence to study the relation among time, space and transmission complexity as a function of network size.

Later on we shall see that the results obtained for this particular model can be applied also under practical circumstances.

3.5. Nondeterministic parallel simulation. In the sequel we shall require the input and the working alphabet of the machine simulated to be of the same cardinality as the corresponding alphabets of simulating machines, since otherwise arbitrary fixed linear speed–up could be achieved simply by transiting to a machine with a larger working alphabet (making use of a so–called *speed–up theorem* [3]). In this case the essence of the problem at hand — i.e., the exploitation of parallelism offered by a fixed number of processors — would vanish. However, the above assumption is superfluous when p as a function of n is considered.

First we shall deal with the pure theoretical problem of simulating an NTM by an RNTM — just to investigate the power of parallel nondeterminism. We shall see that the simulation with linear speed–up is possible in that case.

Theorem 3. *Let M be an NTM of time complexity $T(n)$ and of space complexity $S(n)$. Then there is a p-RNTM R such that R simulates M in time $O(T(n)/p + S(n) + n + p)$, in space $S(n)$, and in transmission time $O(S(n) + p + n)$, for arbitrary p, $1 \leq p \leq T(n)$.*

Proof (Sketch): Let us divide conceptually the accepting computation of M into p time intervals of equal length. The idea of simulation is as follows. The i–th machine M_i of R will simulate the i–th time interval of M's computation starting from an instantaneous description (ID), that M_i has nondeterministically guessed and stored at the beginning of its computation.

At the end of simulation each machine verifies at its right neighbour whether its computation terminated in such an ID in which this neighbour has started its own computation.

The computation of each M_i takes time $T(n)/p$, the generation of starting ID in each M_i takes time $S(n)$ similarly as the final verification of correctness guessing. The time necessary for a signal to come from the leader to the last machine is p, and obviously the time complexity cannot be smaller than n — the length of the input.

Thus the total computation time estimate follows since the computations of individual machines overlap in time with a constant delay.

The machines in R do not communicate except at the very beginning of a computation when activating individual machines, and at the very end when correctness of local guesses is being verified. Obviously, this can be done in transmission time $O(S(n) + p + n)$.

∎

It is clear that the speed–up achieved in previous simulation grows with p increasing and its maximum is achieved when the total time complexity is asymptotically equal to the total transmission time — i.e., for $p = O(\sqrt{T(n)})$.

This illustrates the speed–up limits for ring networks. The result is to be compared with an analogous result — from parallel computation theses it follows that alternating Turing machines, which can be thought of as a kind of tree structured network of NTMs, can simulate an NTM of space complexity $S(n)$ in time $O(S^k(n))$, for a fixed k. In this case, however, the network consists of an exponential number of processors. Thus it seems that the larger maximal speed–up of an RNTM is prevented by its simple topology.

3.6. Parallel depth–first search. When it comes to the simulation of an NTM by an RDTM, algorithms from Theorem 1 or Theorem 2 come into consideration. These algorithms must be implemented in a distributed manner in such a way that the potential of ring parallelism is optimally used.

Further we shall show that both algorithms can be implemented on a ring in such a way that within certain bounds the speed–up proportional to the number of processors will be achieved.

Theorem 4. *Let M be an NTM of time complexity $T(n)$ and of space complexity $S(n)$; let $T(n)$ be a known fully time–constructible function. Then there is a constant $c > O$ and a p–RDTM R such that R simulates M in time $O(T(n)c^{T(n)}/p + p + n)$, in space $O(S(n) + T(n))$ and in transmission time $O(p + T(n))$, for any $p \geq 1$.*

Proof (Sketch): Let us consider the following distributed implementation of the algorithm from Theorem 1: imagine the set of all instruction sequences of M of length $T(n)$ lexicographically sorted, and partitioned into p subsets of equal cardinality. Now, let the i–th machine M_i of R verify the i–th subset of sequences, for $i = 1, 2, \ldots p$.

The individual machines can subsequently, with a constant delay, start their computations and each of them has to verify at most $c^{T(n)}/p$ instruction sequences of length $T(n)$. From these considerations the claims of Theorem 4 follow.

∎

Obviously the optimal speed–up is achieved for $p = O(\sqrt{T(n)c^T(n)})$. Note already now the particularly small transmission time as compared to the total computation time.

3.7. Parallel breadth–first search. The distributed realization of the simulation algorithm from Theorem 2 gives rise to the following theorem:

Theorem 5. *Let M be a k-tape NTM of time complexity $T(n)$ and of space complexity $S(n) \geq \log n$. Then there is a constant $c > O$ and a p–RDTM R such that R simulates accepting computations of M in time $O(nT(n)S^{k+2}(n)c^{S(n)}/p + nT(n)S^{k+1}c^{S(n)} + p)$, in space $O(S^{k+1}(n)c^{S(n)}/p)$, and in transmission time $O(T(n)S^{k+1}(n)c^{S(n)} + p)$, for any $p \geq 1$.*

Proof (Sketch): Consider the following distributed version of algorithm from the Theorem 2: The sorted tapes T_1 and T_2, respectively, of the original machine M, are partitioned into p pieces of equal length and the i-th piece of each tape, i.e., tapes $T_1^{(i)}$ and $T_2^{(i)}$, respectively — is kept in machine M_i of R, for $i = 1, 2, \ldots p$.

Each machine M_i performs one computational cycle over its data in the manner described in the proof of Theorem 2 and ends with some sorted sequences of IDs written on tape $T_1^{(i)}$ and $T_2^{(i)}$, respectively.

Now the contents of tapes $T_2^{(i)}$, for $i = 1, 2, \ldots p$, must be sorted in parallel in a distributed manner. This can be done by the following modification of the parallel odd–even exchange sorting algorithm.

First, the machines with odd index send the respective IDs (from tapes $T_2^{(i)}$, $i = 1, 3, 5, \ldots$) to their right neighbours. In even–indexed machines the original sequence, and that just obtained are merged together, duplicates are eliminated, and the first half of the resulting sorted sequence is sent back to the respective left neighbours. The same process repeats itself for the even–indexed machines.

It can be shown [ ] that after performing p of such sorting cycles in which alternatively sort odd- and even- numbered machines, respectively, the resulting sequence will be sorted and uniformly distributed in all machines M_i.

Similarly it is proceeded with tapes $T_1^{(i)}$, $i = 1, 2, \ldots p$.

From the proof of Theorem 2 it follows that the length of tapes $T_1^{(i)}$ and $T_2^{(i)}$ in each machine M_i is bounded by $O(nS^{k+1}(n)c^{S(n)}/p)$. Then sorting the tapes within each machine, and cycle and duplicate elimination, can be done in $O(nS^{k+2}(n)c^{S(n)}/p)$ parallel time.

Both the time and transmission complexity of one iteration of odd–even parallel sort is of the order $O(nS^{k+1}(n)c^{S(n)}/p)$ (since sending the corresponding sequences, and their merging can be done in linear time), giving $O(nS^{k+1}(n)c^{S(n)})$ time complexity altogether.

Since any simulation leading to accepting state will end after repeating the previous complex process at most $T(n)$ times the resulting complexity characteristics as stated in the claim of the theorem follow.

$\blacksquare$

From the complexity analysis of the previous algorithm it can be inferred that the speed–up is still linear in p, but this time the range within which the speed–up occurs is much smaller than before — viz. it is bounded by $p = O(S(n))$.

3.8. Practical considerations. As seen from the claims of the respective theorems, it seems that for practical purposes — say on a ring of PCs — the parallel simulation from Theorem 4 should be preferred, although the simulation from Theorem 5 is asymptotically faster. Namely, in the former simulation, even in the case of very slow transmission rate between the computers, the total transmission time is negligible as compared to the complexity of 'internal' computations, for a fixed number of processors.

In the case of simulation from Theorem 5 in practice the whole gain in speed by using several processors can be 'washed out' by the intensive data traffic between computers.

The vitality of the above ideas has been demonstrated for the case of the traveling salesman problem by Vornberger [11] who observed the speed–up proportional to the number of processes using a ring consisting of 1, 2, 4, 8 and 16 PC's, respectively. Monien, Speckenmeyer and Vornberger [6] have arrived to similar conclusions for the satisfiability problem, and Monien, Vornberger [7] for the Hamiltonian cycle problem.

Our results indicate that the experimental results observed by the before–mentioned authors can be expected to hold in general — i.e., for the parallel deterministic simulation of any nondeterministic computation using a ring of cooperating processors.

4. Conclusion

4.1. Summary. In the paper we have presented fundamental methods of efficient deterministic, both sequential and parallel, simulations of nondeterministic computations.

As far as sequential simulation is concerned, two algorithms have been presented: depth–first combinatorial search, and breadth–first combinatorial search. While the complexity of the former algorithm was exponential w.r.t. the time complexity of the nondeterministic algorithm simulated, the time complexity of the latter algorithm as well as its space complexity was exponential w.r.t. the space complexity of the nondeterministic algorithm simulated. These complexity characteristics predetermine the use of the latter algorithm, or that of similar ones, on computers with a large internal memory, like PCs. Moreover, it has been shown that both algorithms can be implemented in a distributed manner on a ring–type network of processors where a speed–up proportional to the number of processors is to be expected.

In their clean form, roughly as they have been presented, but realized on more realistic models of computations, the simulation algorithms can be used in practice e.g. when solving difficult combinatorial search problems. The distributed implementation allows for faster obtaining the solutions, or allows for solving larger problem instances. In the near future one can count on using similar algorithms also when realizing programming systems that make use of nondeterminism in place of specification tools (e.g, in systems of logic programming, executable formal specifications, etc.).

References

[1] Aho, A.V. — Hopcroft, J.E. — Ullman, J.D.: The Design and Analysis of Computer Algorithms. Addison–Wesley, Reading, Mass., 1974
[2] Bitner, J,R, — Reingold, E.M.: Backtracking program techniques. *Comm. ACM*, Vol. **18**, 1975, pp. 651–655
[3] Hopcroft, J.E. — Ullman, J,D,: Introduction to Automata Theory, Languages, and Computation. Addison–Wesley, Reading, Mass., 1979
[4] Johnson, D.S. The NP-completeness column: an ongoing guide, *Journal of Algorithms*, since December 1981
[5] Lloyd, J.W.: Foundations of Logic Programming. Springer–Verlag,1984
[6] Monien, B.— Speckenmeyer, E. — Vornberger, O.: Superlinear speed–up for parallel backtracking. *Technical report*, No. 30, Dept. of Comp. Sci., University of Paderborn, 1983
[7] Monien, B. — Vornberger, O.: The ring machine. *Computers and Artificial Intelligence*, Vol. **6**, No.3, 1987, pp. 195—208
[8] Shapiro, E.Y.: Alternation and the computational complexity of logic programs. *Journal of Logic Programming*, No. 1, 1984, pp. 19–33
[9] van Emde–Boas, P.: Machine models and simulations. In: van Leeuwen, J. (Ed.), *Handbook of Theoretical Computer Science*, North–Holland, 1990
[10] van Leeuwen, J.: Distributed computing. *Technical Report*, Dept. of Comp. Sci., Utrecht University, 1983
[11] Vornberger, O.: Implementing branch–and–bound in a ring of processors. *Technical Report* No. 29, Dept. of Comp. Sci., University of Paderborn, 1986
[12] Wiedermann, J.: Solving the knapsack problem in subexponential time. *Proc. 13–th SOFSEM*, Liptovský Ján, 1986, pp. 251–253, (in Slovak)
[13] Wiedermann, J.: Fast simulation of nondeterministic Turing machines with application to the knapsack problem. *Computers and Artificial Intelligence*, Vol. **8**, No. 6, 1989,pp. 591–596

Part II
Selected Contributions

A semantic characterization of disjunctive relations

Michael Freund

Department of Mathematics, University of Orleans
45067, Orléans, FRANCE.

e-mail: freund@fror101

We study here some semantic properties of a specific kind of preferential relations, those which satisfy the rule of *Disjunctive Rationality*. This rule means essentially that if one may draw a conclusion from a disjunction of premisses, then one should be able to draw this conclusion from at least one of these premisses taken alone.

The main result we obtain in this paper is the construction of a representation theorem for such relations. We show indeed that there exists a certain kind of models which induce precisely all preferential operations satisfying the rule of *Disjunctive Rationality*. These models themselves reveal to be particularly easy to handle. They are just subsets of the set of all worlds of the language supporting the given relation, equipped with an ordering which is *filtered*: if two worlds n and m satisfy a proposition x and are not minimal for that property, then there exists a world p less than m and less than n that satisfies x.

1- Background

We let $\mathfrak{L}$ be the set of well formed formulas over a set of propositional variables. The classical propositional connectives will be denoted by $\neg$, $\wedge$, $\vee$, $\rightarrow$ and $\leftrightarrow$. With the language $\mathfrak{L}$, semantics is provided by the set $\mathfrak{E}$ of all assignments of truth values to the propositional variables. Elements of $\mathfrak{E}$ will be refered to as **worlds** and the satisfaction relation between a world m and a formula x is defined as usual and written $m \vdash x$.

Thus, $m \vdash a \vee b$ iff $m \vdash a$ or $m \vdash b$, and $m \vdash \neg a$ iff not($m \vdash a$).

For every subset A of $\mathfrak{L}$, we write $m \vdash A$ iff m satisfies all the elements of A.

The classical consequence operation attached to $\mathfrak{L}$ and $\mathfrak{E}$ will be denoted by Cn. Thus, for any subset A of $\mathfrak{L}$, Cn(A) is the set of all formulas of $\mathfrak{L}$ which are logically entailed by A. For any formula x, we write Cn(x) for Cn({x}). Recall that Cn is a Tarski consequence operation: it satisfies

1) Reflexivity $\qquad$ $A \subseteq Cn(A)$
2) Idempotence $\qquad$ $Cn[Cn(A)] = Cn(A)$
3) Monotonicity $\qquad$ If $A \subseteq B$, then $Cn(A) \subseteq Cn(B)$.

Moreover, Cn is **compact**: For any subset A of $\mathfrak{L}$ and $x \in Cn(A)$, there exists a **finite** subset A_0 of A such that $x \in Cn(A_0)$.

A subset A of $\mathfrak{L}$ is called **consistent** if $Cn(A) \neq \mathfrak{L}$ or, equivalently if there exists a world m such that m satisfies A. The compactness of Cn implies that A is consistent if and only if all finite subsets of A are consistent.

Preferential relations

A **preferential relation** on $\mathfrak{L}$ is a relation $|\sim$ which satisfies the following rules:

1) Reflexivity
$\quad$ a $|\sim$ a
2) Left Logical Equivalence
$\quad$ If $Cn(a) = Cn(b)$ and a $|\sim$ c, then b $|\sim$ c
3) Right Weakening
$\quad$ If $b \in Cn(a)$ and c $|\sim$ a, then c $|\sim$ b
4) Cut
$\quad$ If a∧b $|\sim$ c and a $|\sim$ b, then a $|\sim$ c
5) Or
$\quad$ If a $|\sim$ c and b $|\sim$ c, then a∨b $|\sim$ c
6) Cautious Monotonicity
$\quad$ If a $|\sim$ b and a $|\sim$ c, then a∧b $|\sim$ c.

Given such a relation, we shall denote by $C_{\sim}(a)$ -or $C(a)$ when there is no ambiguity- the set of all conequences of a formula a, that is the set of all b such that a $|\sim$ b. The above rules imply then that $C(a)$ is closed w.r.t. Cn, that is $Cn[C(a)] = C(a)$.

We recall finally that an element a of $\mathfrak{L}$ is said to be $|\sim\sim$-consistent if **false** $\notin C(a)$. In the remainder of this paper, we will say that a **world** m is $|\sim\sim$-consistent iff all the formulas it satisfies are $|\sim\sim$-consistent.

Lemma 1 *Let a be an element of $\mathfrak{L}$ and m a world that satisfies $C(a)$. Then m is $|\sim\sim$-consistent.*

Proof: Suppose indeed that b is a formula such that $m \vdash b$. If b were not $|\sim\sim$-consistent, we would have b $|\sim$ **false**, hence b $|\sim$ a by **Right Weakening** and a∧b $|\sim$ **false** by **Cautious Monotonicity**. By **Right Weakening** again, this would imply that a∧b $|\sim$ b→**false**. But one has a∧¬b $|\sim$ b→**false**, so, using **Or** and **Left Logical Equivalence**, we would get a $|\sim$ b→**false**.
Now m is supposed to satisfy $C(a)$, so we would have $m \vdash$ b→**false**, and therefore $m \vdash$ **false** since we had $m \vdash$ b, which

is impossible.∎

In the terminology of [5], a world m is said to be **normal** for a if m satisfies C(a).

Lemma 2 *An element x is $|\sim$-consistent if and only if there exists a $|\sim$-consistent world m such that m satisfies x.*

Proof: The "only if " part of the lemma is obvious. Let us prove the converse:

Suppose that x is a $|\sim$-consistent formula. Then C(x) is consistent: otherwise there would exist a finite subset B = {b$_1$, b$_2$, ..., b$_k$} of C(x) such that **false** ε Cn(B). If β is the conjunction of all b$_i$'s, we would then have β ε C(x) and **false** ε Cn(β), so **false** ε C(x) contradicting, the fact that x is a $|\sim$-consistent formula.

C(x) is therefore a consistent set, so there exists a world m such that m ⊨ C(x), and we conclude by lemma 1.∎

Preferential models

The above inference rules have been defined and studied in [5]. There, the authors define a *Preferential Model* in the following way:

Let W = (S, ≺, l) be a triple where (S, ≺) is an ordered set and l a mapping from S in the set of worlds 𝒲 of 𝔏.

For any element a of 𝔏, denote by a* be the set of all elements s of S such that s "satisfies" a, i.e. such that l(s) ⊨ a. W is said to be a **preferential model** when the following condition of **smoothness** is fulfilled: for any a ∈ 𝔏 and s in a*, either s is minimal in a* or there exists an element t in S, t ≺ s, and minimal in a* .

Such a preferential model defines a preferential relation $|\sim_W$ by:

(def)   a $|\sim_W$ x . iff all minimal elements of a* satisfy x.

Let us write s ⊨$_≺$ a when s satisfies a and is minimal in a*. Then the preferential relation induced by the preferential model W can be written

a $|\sim_W$ x iff s ⊨$_≺$ a implies s ⊨ x.

An important and non-trivial theorem stated in [5] shows that conversely, for any preferential relation $|\sim$, there exists a preferential model W such that $|\sim$ = $|\sim_W$. Apart from the semantic interpretation it provides, this kind of representation theorem shows itself to be particularly useful because a lot of properties of preferential relations can be established and proven using this semantic characterization. Most of the results of the present work concerning preferential relations will be proven only by means of preferential models.

Disjunctive relations

A preferential relation $|\!\sim$ is said to be **disjunctive** if it satisfies the following property of **disjunctive rationality**:

(D.R.) **If** $a \lor b \mathrel{|\!\sim} c$, **then either** $a \mathrel{|\!\sim} c$, **or** $b \mathrel{|\!\sim} c$.

This property expresses the fact that if we can draw a conclusion from a disjunction of premisses, then we should be able to draw this conclusion from one at least of these premisses.

The definition and some elementary properties of disjunctive relations can be found in [5] and [6]. The purpose of this paper is to provide a semantic characterization of this family of relations.

2–Standard and filtered models

Let $W = (S, l, \prec)$ be a preferential model and $|\!\sim$ the preferential relation defined by W. We say that W is a **standard model** for $|\!\sim$ if

1) S is the subset of $\mathscr{E}$ consisting of all $|\!\sim$-consistent worlds and $l = i_{\mathscr{E}}$ is the canonical injection from S to $\mathscr{E}$.
2) For any formula a, an element s of S is minimal in a' if and only if s satisfies $C_\_(a)$.

The interest of standard models is double: on one hand, it is clearly more convenient to deal with sets of worlds rather than arbitrary sets of "states" as defined in [6]. On the other hand, lemma 1 together with condition 2) above shows that *a world m is normal for α if and only if m is minimal in α^*,* and this provides a direct link between the set of worlds and the order relation "$\prec$".

It would be interesting to characterize **all** the preferential relations that can be defined by means of a standard model. It can be easily proven that for such relations one should have $C_\_(a \lor b) \subseteq Cn[C_\_(a) \cup C_\_(b)]$ for any elements a and b of $\mathscr{L}$, but we do not know whether the converse is true and we do not know either what are the preferential relations that satisfy this condition. Nevertheless we shall see that in the case of **disjunctive relations**, this notion of standard model is suitable to establish a representation theorem, provided we equip the standard model with a special kind of order which we will describe now.

Definition *A preferential model $W = (S, l, \prec)$ is said to be **filtered** if whenever two states s and t satisfy a formula x without being minimal in x^*, there exists a state r, $r \prec s$ and $r \prec t$, such that $r \mathrel{\vdash} x$.*

This condition is *weaker* than the condition of **rankedness** defined and studied by Kraus-Lehmann in [5], which can be defined by the property: " *if p and q are two states which are not comparable, then for any state u, u ≺ p implies u ≺ q* ".

Let us show indeed that a ranked model is filtered: Suppose that s and t satisfy x and are not minimal in x*. If t and s are not comparable, let r be a state minimal in x* such that r ≺ s. Such a state exists because s is not minimal in x*. Since the model is supposed to be ranked, we have then r ≺ t as desired. If now t and s are comparable, we have for instance t ≺ s and we can chose r minimal in x* such that r ≺ t.

Ranked models are used to characterize a special family of preferential relations, those which satisfy the condition of **rational monotony**, first defined by D.Makinson. Details can be found in [5] and especially in [6], where the authors prove that rational relations are precisely those which can be defined by ranked models.

As shown by D.Makinson, the condition of rational monotony is strickly stronger than the condition of disjunctive rationality. Makinson's example can be used in the following way to show that the condition, for a model, to be filtered, is strictly weaker than the condition of being ranked:

Let $\mathfrak{L}$ be the propositional calculus on the three variables p_0, p_1, p_2 and $W = (T, l, \prec)$ the preferential model with set of states $T = \{s_0, s_1, s_2\}$. The order is just $s_1 \prec s_2$, and the label function l maps s_i onto the world which satisfies only p_i.

It is then trivially true that W is filtered, since, given any formula x, there do **not** exist two elements of T which are not minimal in x*. But the model is not ranked because s_0 and s_2 are not comparable, $s_1 \prec s_2$, but $s_1 \nprec s_0$.

3—The order induced by a preferential relation

If we are given a **monotonic relation**, that is a preferential relation $|\sim$ such that for any a and b in $\mathfrak{L}$, $C_-(a) \subseteq C_-(a \wedge b)$, the relation " $a \leq b$ **iff** $b |\sim a$ " is an order on $\mathfrak{L}$, the meaning of which is clear: a formula a can be said to be less exceptional than b if a is a consequence of b. But the relation thus defined is no longer transitive if $|\sim$ is not supposed to be monotonic. In the general preferential case, the authors of [5] considered an order which amounts to
" $a \leq b$ **iff there exists** $b' \in Cn(b)$ **such that** $b' |\sim a$ ". In other words, a is said to be less exceptional than b if there exists a logical consequence b' of b such that $b' |\sim a$. This order was thoroughly used by Kraus-Lehmann-Magidor in their proof of a representation theorem for preferential relations, but revealed insufficient to provide a representation theorem for stronger relations like, for instance, the rational ones.

In a recent paper [4], P.Gärdenfors and D.Makinson considered the family of rational relations $|\sim$ such that the set of consistent elements is equal to the set of |~-consistent elements. They showed that these relations can all be induced by **a standard ranked** model, and that, associated with these relations, an "expectation" ordering can be defined on $\mathfrak{L}$. The paper of Gärdenfors and Makinson is at the origin of the present work.

We let now |~ be an arbitrary preferential relation and $(T, 1, \prec)$ a fixed K.L.M model for |~.

We define a relation among the set of all |~-consistent formulas of $\mathfrak{L}$ in the following way:

(Def) " $a < b$ iff $a \lor b$ |~ ¬b "

In other words, a formula a is "smaller" than a formula b if b is not consistent with C(a∨b). (This relation was used in [5] by Lehmann and Magidor, and proven to be an order when |~ is supposed to be **rational**.)

To clarify a bit the meaning of this relation, we first make the following

Remark : $a < b$ iff ¬b ∈ C(a) and a ∈ C(a∨b).

Indeed, suppose that $a < b$. Let t be a state of T such that $t \models_{\prec} a \lor b$. Then $t \models ¬b$, so t does not satisfy b and it must therefore satisfy a. This shows first that a∨b |~ a and secondly, using the hypothesis a∨b |~ ¬b and Cautious Monotonicity, that a |~ ¬b.
Conversely, suppose that a ∈ C(a∨b) and ¬b ∈ C(a). Then any state t minimal in (a∨b)˙ satisfies a. It is therefore minimal in a˙ and hence satisfies ¬b, and this proves that a∨b |~ ¬b.∎

Note that in particular we have $a < b$ as soon as a |~ ¬b and b |~ a . This enables us to illustrate the meaning of the relation "<" in the following way:
Suppose a car has an automatic speed limiting system which corrects instantly its speed so it does not exceed 80 miles an hour. Let b stand for "The speed exceeds 80 miles an hour" and a stand for "The limiting system is working". Then it is sensible to expect a from b, and from a it is sensible too to derive ¬b. So we see that in this case we do have "a < b", which can be interprated as "*a is the rule and b the exception*".

Let us check now that the relation defined above is a strict partial order:

Lemma 3 *The relation "<" is irreflexive and transitive.*

Proof: This relation is clearly irreflexive since all the considered elements are |~-consistent.
Let us prove transitivity:
Suppose $a < b$ and $b < c$. We want to show that $a < c$, that is ¬c ∈ C(a∨c).
Let s be a state in T such that $s \models_{\prec} a \lor c$. Suppose that

s ⊢ c. Then s ⊢ b∨c. Since ¬c ∈ C(b∨c), we do not have s ⊢< b∨c. Therefore there exists a state t ≺ s such that t ⊢< b∨c. It follows that t ⊢< b, hence t ⊢ a∨b. But we have ¬b ∈ C(a∨b), so we do not have t⊢< a∨b, and there exists a state r in S such that r ≺ t and r ⊢< a∨b. Now, we have r ≺ t and r does not satisfy b, so this implies that r ⊢ a, and therefore that r ⊢ a∨c. But this is impossible because we have r ≺ s and s ⊢< a∨c. ∎

If α is a $|\sim$-consistent formula and β a formula such that $\alpha \mid\sim \beta$, β is $|\sim$-consistent. Therefore, the set of all elements β such that $\alpha \mid\sim \beta$ and $\alpha\vee\neg\beta \mid\sim \alpha$ is the set of $|\sim$-**consistent** elements β that satisfy these two conditions. We shall denote by α^+ this set. Hence

Notation: *For any $|\sim$-consistent formula α, α^+ is the set of all the elements β such that $\alpha < \neg\beta$.*

The set α^+ turns out to be very similar by its properties to the set $C(\alpha)$. It was first introduced in [1] under the name of "trace", and revealed quite useful, there as well as in [2], to prove some heavy theorems concerning infinite inference operations. In the present work, though, we will only need the following simple characterization of α^+, due to D.Lehmann :

Lemma 4: *Let α be a $|\sim$-consistent formula and β an element of $\mathcal{L}$. Then $\beta \varepsilon \alpha^+$ iff $\alpha' \mid\sim \beta$ for every formula $\alpha' \varepsilon Cn(\alpha)$.*

Proof: If $\beta \varepsilon \alpha^+$, we have $\alpha \mid\sim \beta$, so by **Right Weakening** $\alpha'|\sim \beta$ for any formula $\alpha' \varepsilon Cn(\alpha)$.
  Suppose conversely that $\alpha' \mid\sim \beta$ for any α' in $Cn(\alpha)$ and let us prove that $\beta \varepsilon \alpha^+$. Taking $\alpha' = \alpha$ shows first that $\alpha \mid\sim \beta$. It remainds to check that $\alpha\vee\neg\beta \mid\sim \alpha$.
  Let s be a state in T minimal in $(\alpha\vee\neg\beta)^+$. We want to prove that s satisfies α. Note that $\alpha\vee\neg\beta \varepsilon Cn(\alpha)$, so taking $\alpha' = \alpha\vee\neg\beta$ shows that $\alpha\vee\neg\beta \mid\sim \beta$. Therefore s must satisfy β. But we had s ⊢ $\alpha\vee\neg\beta$, so s, which does not satisfy $\neg\beta$, must satisfy α as desired. ∎

An immediate consequence of Lemma 4 is the following

Corollary: *Let α and β be $|\sim$-consistent formulas and x and y be elements of $\mathcal{L}$. Then*

1) $x\wedge y \varepsilon \alpha^+$ iff $x \varepsilon \alpha^+$ and $y \varepsilon \alpha^+$.
2) $\alpha^+\cup\beta^+ \subseteq (\alpha\vee\beta)^+$
3) $\alpha \mid\sim x$ iff $\alpha\rightarrow x \varepsilon \alpha^+$

Proof: Straightforward, using lemma 4 above. **Note that property 3 can be expressed by the simple equality** $C(\alpha) = Cn[\alpha\cup\alpha^+]$. It shows that in a certain way the sets α^+ behave very much like the sets $C(\alpha)$ "pointed" from the elements which are equivalent to α.

Let us underline that, for the results we have so far obtained, the **only hypothesis made on** $|\sim$ **is that it is a preferential relation**. With the same hypothesis, we will now carry on our order relation "<" on the subset S of $\mathcal{E}$ consisting of all $|\sim$-consistent worlds.

For any elements m,n in S, define:

" m ≤ n **iff there exists u such that** n ⊢ u **and** m ⊢ u⁺ "

We write " m < n " iff m ≤ n and n $\not\le$ m.

Lemma 5 *The relation " < " is irreflexive and transitive.*

Proof: Irreflexivity is immediate. Let us prove transitivity:

Suppose m < n and n < p. Then

(i)   There exists u such that n ⊢ u and m ⊢ u⁺
(ii)   There exists no v such that m ⊢ v with n ⊢ v⁺ .
(iii) There exists u′ such that p ⊢ u′ and n ⊢ u′⁺.
(iv)   There exists no v′ such that n ⊢ v′ and p ⊢ v′⁺.

We want to prove that m < p.

a) First, we claim that m ⊢ u′⁺. For otherwise, there would exist an element w such that ¬w > u′ and m ⊢ ¬w.
Let v = ¬w. Observe that v⁺ ⊆ u′⁺, for if a ∈ v⁺, we have ¬a > v > u′. Since n ⊢ u′⁺, we have n ⊢ v⁺. But m ⊢ v, and this contradicts (ii).

Hence we have proven that there exists u′ such that p ⊢ u′ and m ⊢ u′⁺, and this shows that m ≤ p.

b) Suppose now that there exists an element w such that m ⊢ w and p ⊢ w⁺ By (ii), n does not satisfy w⁺ . So there exists z ∈ w⁺ such that n ⊢ ¬z. By (iv), p does not satisfy (¬z)⁺, so there exists an element t ∈ (¬z)⁺ such that p ⊢ ¬t.
But ¬t > ¬z > w, so t ∈ w⁺, so p ⊢ t, a contradiction.

We have therefore proven that there exists no element w such that m ⊢ w and p ⊢ w⁺. This, together with a) shows that m < p. ∎

Proposition 1 *If m and n are elements of S such that m ⊢ a and m < n, then n does not satisfy C(a).*

Proof: We noted, in the proof of the corollary of Lemma 4 that if a is $|\sim$-consistent, C(a) = Cn(a,a⁺). Suppose that a is a formula such that m ⊢ a. Then a is $|\sim$-consistent. If n satisfied C(a), n would satisfy a⁺, so we would have n ≤ m, a contradiction. ∎

The triple $(S,<,i_{\mathcal{E}})$ is not in general **stoppered**. We shall see in the next paragraph that it is, though, **if** $|\sim$ **satisfies disjunctive rationality**. The key lemma we shall use to prove it is nevertheless valid for any preferential relation:

Lemma 6: If $a|\sim b$ and $a|\sim v$, then $(a\wedge\neg b)\vee v \ |\sim \neg a\vee b$.

Proof: Let s be a state in T such that $s \vDash_\prec (a \wedge \neg b) \vee v$, and suppose that $s\vDash a$. We want to show that $s \vDash b$. If s is minimal in $a^\cdot$, then $s \vDash b$. If this is not the case, there exists a state t in T such that $t \prec s$ and $t \vDash_\prec a$. Since $a |\sim v$, $t \vDash v$, so $t \vDash (a \wedge \neg b) \vee v$, which is impossible by the choice of s. ∎

A consequence of this proposition, is the following technical result:

Proposition 2: *Let a be an element of $\mathcal{L}$ and n a world of S which satisfies a but does not satisfy $C(a)$. Then for any element v such that $n \vDash v^+$, $\neg v$ is consistent with $C(a)$.*

Proof: Suppose indeed that $\neg v$ is not consistent with $C(a)$. Then we would have $v \in C(a)$, that is $a |\sim v$. Let b be a formula such that $a |\sim b$. Then, by Lemma 6, $(a\wedge\neg b)\vee v |\sim \neg a\vee b$, that is $\neg a\vee b \ \varepsilon \ v^+$. But n is supposed to satisfy this latter set, hence $n \vDash \neg a\vee b$ and therefore $n \vDash b$ since we had $n \vDash a$.

This shows that if $\neg v$ is not consistent with $C(a)$, n satisfies all formulas b such that $a |\sim b$. The world n satisfies therefore $C(a)$, contradicting our hypothesis. ∎

4-Representing disjunctive relations

We now turn back to deductive relations. The aim of this section is to prove the following:

<u>Theorem</u> : A preferential relation is disjunctive if and only if it can be defined by a standard filtered model.

Let us prove first the easy part of the theorem and show that if $|\sim$ is given by a filtered (not necessarily standard) model $(T,l,\prec)$, then $|\sim$ satisfies disjunctive rationality:

We have to prove that if $x \in C(a \vee b)$ and $x \notin C(a)$, then $x \ \varepsilon \ C(b)$. Since $x \notin C(a)$, there exists a state s in T such that $s \vDash_\prec a$ but s does not satisfy x. The state s satisfies therefore $a\vee b$ but is not minimal in $(a \vee b)^\cdot$ since, otherwise, it would satisfy x.

To prove that $x \in C(b)$, we let r be a state such that $r \vDash_\prec b$. We want to show that $r \vDash x$. We have $r \vDash a\vee b$ and we claim that in fact $r \vDash_\prec a\vee b$. Otherwise, by the fact the model is **filtered**, there would exist a state $t \prec r$ and $t \prec s$ such that $t \vDash a\vee b$. But this is impossible, for if t satisfies a, we cannot have $t \prec s$, and if t satisfies b we cannot have $t \prec r$. This shows that $r \vDash_\prec a\vee b$, and therefore $r \vDash x$, as desired. ∎

To prove the "only if" part of the theorem,, we suppose we are given a **disjunctive preferential** relation $|\sim$.

We let $W = (S, <, i_{\mathscr{E}})$ be the triple where S stands for the set of all $|\sim$-consistent worlds, "$<$" is the order relation on S defined in the preceeding paragraph and $i_{\mathscr{E}}$ is the canonical injection from S to $\mathscr{E}$. We shall prove that **W is a standard filtered model for** $|\sim$.

We first notice that in the case of disjunctive relations, the union of a finite number of sets α^+ can be simply expressed using disjunction. More precisely, we have the

Lemma 7: *For any* $|\sim$-*consistent elements a and b,* $(a \lor b)^+ = a^+ \cup b^+$.

Proof: We know already by the Corollary of Lemma 4 that $a^+ \cup b^+$ is a subset of $(a \lor b)^+$. Let us prove the converse inclusion:
If $z \in (a \lor b)^+$, we have $(a \lor b) \lor \neg z \mid\sim z$, which can be written $(a \lor \neg z) \lor (b \lor \neg z) \mid\sim z$. Now, $|\sim$ is supposed to satisfy Disjunctive Rationality, so we have $a \lor \neg z \mid\sim z$ or $b \lor \neg z \mid\sim z$ and therefore $z \in a^+$ or $z \in b^+$ as desired. ∎

Corollary *For every* $|\sim$-*consistent elements a and b,* $(a \lor b)^+$ *is equal either to* a^+ *or to* b^+.

Proof: Suppose indeed that $(a \lor b)^+$ is not equal to a^+. Then, by the above lemma, there exists a formula $x \in (a \lor b)^+$ such that $x \notin a^+$.
Let z be any element of $(a \lor b)^+$. We have then $z \land x \in a^+$ by the corollary of Lemma 4. But $z \land x \in (a \lor b)^+$ by this same corollary, so, using Lemma 7 above, we see that $z \land x \in b^+$, which shows that $x \in b^+$.
We have therefore proven that if the set $(a \lor b)^+$ is not equal to a^+, then it must be included in b^+. This, together with the above lemma, concludes the proof. ∎

The following result generalizes proposition 2 of section 3. It will enable us to prove that W is stoppered:

Proposition 3: *Let n be an element of S which satisfies a formula a, but does not satisfy C(a). Then the set*
$X_n(a) = C(a) \cup \{ \neg v; n \vdash v^+ \}$ *is consistent.*

Proof: Suppose indeed that $X_n(a)$ is not consistent. Then there exists a finite number of elements v_i, $1 \le i \le n$, such that n satisfies v_i^+ and $v = v_1 \lor v_2 \lor ... \lor v_n \in C(a)$.
By lemma 7, we see then that $n \vdash v^+$. But proposition 2 shows then that $\neg v$ must be consistent with C(a) and this contradicts the fact that $v \in C(a)$. ∎

Corollary: *The model W is stoppered.*

Proof: Let indeed n be an element of S which satisfies a formula α. We consider two cases:

a) n satisfies $C(\alpha)$.
Then by Lemma1 and proposition 1, n is minimal in α^+ and there

is nothing to prove.

b) n does not satisfy $C(\alpha)$.
Then by proposition 3, there exists a world m such that m satisfies $X_n(\alpha)$. Note that it follows then that m satisfies $C(\alpha)$, so $m \in S$, by lemma 1.
We claim that m is minimal in a^* and that m < n.
The first assertion is a direct consequence of proposition 1. To prove the second one, we have to show that
1) There exists u such that $n \vdash u$ and $m \vdash u^+$ and
2) There exists no v such that $m \vdash v$ and $n \vdash v^+$.

For 1) take $u = a$ and note that since $m \vdash C(a)$, $m \vdash a^+$ by the property 3) stated in the corollary of Lemma 4.
For 2), note that if $n \vdash v^+$, then $\neg v \in X_n(\alpha)$, so $m \vdash \neg v$. ■

To achieve the proof of theorem 1, we have to check that **W** is filtered and that the relation it defines is precisely the relation $|\sim$.

Proposition 4 *The relation $|\sim_W$ defined by W is equal to $|\sim$.*

Proof: We have to prove that for any element a, a $|\sim$ b if and only if a $|\sim_W$ b.
We observe first that if $m \vdash_< a$, $m \in S$, then $m \vdash C(a)$: otherwise, the preceeding construction shows that m would not be minimal. Therefore a $|\sim$ b implies a $|\sim_W$ b.
Conversely, suppose we have a $|\sim_W$ b. If a world m satisfies $C(a)$, we know that $m \in S$. Furthermore, m is minimal in a^* by proposition 1, so m satisfies b. Hence, we see that $b \varepsilon Cn(C(a))$, and since this latter set is equal to $C(a)$, we have a $|\sim$ b as desired. ■

We have proven so far that the model **W** is stoppered and defines $|\sim$. Note that an immediate consequence of proposition 1 and 4 and of the definition S is the fact that this model is **standard**. We prove now that this model is **filtered**:

Proposition 5 *Suppose that m and n satisfy a $|\sim$--consistent formula a but do not satisfy $C(a)$. Then the set $X_n(a) \cup X_m(a)$ is consistent.*

Proof: If this set were not consistent, there would exist a finite number of elements $v_1, \ldots, v_k, w_1, \ldots, w_l$, with $n \vdash v_i^+$ and $m \vdash w_j^+$ such that $x = v_1 \lor \ldots \lor v_k \lor w_1 \lor \ldots \lor w_l \in C(a)$.
By the corollary of lemma 7, we see that x^+ is equal to one of the sets v_i^+ or w_j^+. We have then $n \vdash x^+$ or $m \vdash x^+$. In either case, proposition 2 shows that $\neg x$ should then be consistent with $C(a)$, contradicting the fact that $x \in C(a)$. ■

Corollary *The model W is filtered.*

Proof: Let indeed m and n be worlds of S which satisfy a but are not minimal in a⁺. Then neither of these worlds satisfies C(a), as follows from proposition 1.

Let p be a world which satisfies $X_n(a) \cup X_m(a)$. Such a world exists by the above proposition, and $p \in S$. An argument similar to the one used in the proof of the corollary of proposition 3 shows then that $p < n$ and $p < m$ as desired.∎

We have thus found a standard filtered model which defines |~, and completed therefore the proof of our representation theorem for disjunctive relations. Moreover, as we prove in a forthcoming paper, this same model can be used to represent **rational** relations : indeed it turns out that the order we defined among the elements of S is **ranked** as soon as the given relation |~ is rational.∎∎

References

[1] **M.Freund.** (To appear) *Supracompact inference operations.*

[2] **M.Freund and D.Lehmann.** (Manuscript) *Nonmonotonic inference operations.*

[3] **M.Freund,D.Lehmann and D.Makinson.** *Canonical extensions to the infinite case of finitary nonmonotonic inference relations.* (Arbeitespapiere der GMD n°443: Proceedings of the Workshop on Nonmonotonic Reasoning,133-138,1990)

[4] **P.Gärdenfors and D.Makinson.** (To appear) *Nonmonotonic Inference based on Expectations.*

[5] **S.Kraus,D.Lehmann and M.Magidor** *Nonmonotonic reasoning, preferential models and cumulative logics.* (Artificial Intelligence 44, 167-207, 1990)

[6] **D.Lehmann and M.Magidor.** *Rational logics and their models: a study in cumulative logic.* (Technical Report TR 88-16, Leibnitz Center for Computer Science,Dept. of Computer Science, Hebrew University,Jerusalem, November 1988)

Execution of defeasible temporal clauses for building preferred models

Anthony Hunter

Department of Computing, Imperial College,
180 Queen's Gate, London SW7 2BZ, UK

Email: abh@doc.ic.ac.uk

Abstract

Important forms of reasoning such as planning and scheduling can be both temporal and non-monotonic in nature. For a logic-based solution, such reasoning requires (1) the specification of components of a plan, or schedule, in a language that is sufficiently expressive and natural for notions of time and defeasibility, and (2) the facility to demonstrate the satifiability of the specification. For (2) we effectively need to build a model of the specification. Here we present a non-monotonic temporal logic NTE that is a development of executable temporal logics (Gabbay 1989) and the non-monotonic atemporal logic KAL (Hunter 1990). The development of NTE is intended as an initial attempt at addressing the requirements of (1) and (2) above.

Introduction

Temporal reasoning is closely tied to non-monotonic reasoning, yet attempts to harness non-monotonic formalisms for facilitating appropriate temporal reasoning have been limited. Important forms of reasoning that can be both temporal and non-monotonic include planning, scheduling and predicting. In this report we attempt to develop a non-monotonic version of an approach to temporal logic reasoning.

For the temporal reasoning we utilize an alternative view on temporal logics, that of executable temporal logics (Gabbay 1989), and develop a non-monotonic variant called NTE, standing for Non-monotonic Temporal Execution. The traditional view on temporal logic is of declarative statements about the world, or about possible worlds, over time. These relate the truth of propositions in the past, in the present, and in the future, and so have been considered for issues of specification of real-time systems. An alternative view on temporal logic is to consider the logic in terms of a declarative past, and an imperative present and future, based on the intuition that a statement about the future can be an imperative, initiating steps of action to ensure it becoming true. A temporal logic specification can thus be used by translating it into an executable form:

[a1]   ANTECEDENT ABOUT THE PAST   ⇒   CONSEQUENT ABOUT THE PRESENT AND FUTURE

For example, if we take time as points on the line of natural numbers, then atomic propositions are true or false at points on that line. Execution is undertaken at each point in succession. Executable temporal logic statements can be executed at a point t on that line by reviewing which atomic propositions were true in the past, i.e points less than t, and then taking appropriate action to satisfy the specification. If the appropriate action is to undertake some action in the present then that action must be undertaken now, to ensure satisfaction of the specification. If the appropriate action is to undertake some action in the future then the instruction must be passed on to the points in the future. Executable temporal logics that have been developed include USF (Gabbay 1989), MetateM (Barringer 1990), and MML (Barringer 1991). All specifications of USF and MetateM are reducible to a finite conjunction of [a1]'s.

If we write specifications of the form [a2] and if the antecedent refers to the past, and the consequent refers to the present and future, then we can execute such specifications so as to construct a model of the specification.

[a2] $\wedge_i A_i \rightarrow \vee_j B_j$

Suppose then that we have a specification S in the form of a finite conjunction of clauses [a2] such that each A_i and B_j is either a positive literal or negated literal. The executing agent tries to execute S in such a way as to build a model of S. It must make S true dynamically at each point in time. So at any time point, it will consider each clause [a2]. If $\wedge_i A_i$ is true then it must make the disjunction $\vee_j B_j$ true. It can do this by making any one of the disjuncts B_j true. The choosing of which B_j to make true is a subtle (though typically decidable) problem, and the agent will take into account several factors, such as its commitment to make the other clauses true, possible future deadlocks, and the environment at the time. There may be more than one valid choice at each time point, and the actual choice may vary over time.

Such a view point seems to parallel aspects of how an intelligent agent may operate. At a point in time, an agent has data on the past, and acts on the basis of this data. Effectively, NTE will be non-monotonic version of a simple executable temporal logic. Execution of NTE will be developed, in part, along the lines of USF and MetateM.

For the non-monotonic reasoning we adopt an approach to defeasible reasoning developed in the atemporal logic KAL (Hunter 1990). In defeasible reasoning, clauses represent general rules that can be defeated by more specialized defeasible rules. The work reported here is based on clausal logic, but (1) replaces the classical connective of material implication with that of defeasible implication, and (2) includes explicit orderings on defeasible clauses so as to facilitate selection of clauses according to context. We make inferences from an ordered set of clauses by only using the highest ordered clause that has the antecedent satisfied. We motivate the explicit ordering by the requirement within artificial intelligence for a "natural" representation of defeasible knowledge such as [b1] and [b2].

[b1]   "A match lights if struck"
[b2]   "A match doesn't light if the match is wet"

However, if the following assertions, [b3] and [b4], are represented in the knowledge-base with [b1] and [b2], we do not want to infer a contradiction.

[b3]   "A match is struck"
[b4]   "A match is wet"

To address this we need a mechanism to select the most appropriate clause based on a specified explicit ordering. In KAL and in NTE, we make inferences from an ordered set of clauses by only using the highest ordered clause that has the antecedent satisfied. Such a formalism would avoid the unnatural representation of listing all exceptions to each clause as extra conditions in the antecedent.

Outline of KAL for non-monotonic reasoning

Before presenting a syntax and semantics for NTE, we provide an outline of KAL. Knowledge in a KAL knowledge-base is in the form of a set of defeasible rules, eg [b5] and [b6], with '←' denoting implication, together with a set of assertions:

[b5] $\neg A \leftarrow$
[b6] $A \leftarrow B$

To indicate how the ordering works, we will consider the underlying semantics. We can consider a knowledge-base, based on classical propositional logic, corresponding to a partial model that can be immediately identified, and from this we wish to build the rest of the model. For example, a partial model of a knowledge-base consisting of the clause [b6], together with the assertion B, could be represented as a part row of a truth table [t1]:

[t1]	line	B	$A \leftarrow B$
	1	T	T

Using [t1], together with the "computation" rule modus ponens, we could construct a full model. This we represent as another truth table row [t2]:

[t2]	line	A	B	$A \leftarrow B$
	1	T	T	T

However, in KAL we may wish to have a knowledge-base that includes potentially contradictory clauses. To illustrate the problem, we can take a knowledge-base consisting of the [b5] and [b6] together with the assertion B and form a partial row of a truth-table [t3]:

[t3]   line   B   ¬A ← A ← B
       1   T   T   T

As before, if we now use [t3] together with modus ponens, we could attempt to construct a full model. Though it is obvious that we have a problem, in that we wish to put both A and ¬A true in the model. However, if we put some ordering on the clauses such that if the antecedents of both clauses are satisfied, then we can select the consequent of the highest clause, and hence reject the lower ordered clause. Effectively, clause [b6] has been used to defeat clause [b5]. The full row of the truth table corresponding to this is [t4].

[t4]   line   A   ¬A   B   ¬A ← A ← B
       1   T   ⊥   T   ⊥   T

However, this "solution" has introduced another problem: we have changed the truth-value of one of the clauses in the corresponding partial model. Effectively we have made one of the clauses in the knowledge-base false. To elaborate on this new problem, if we form a full truth table, we can see that the two clauses [b5] and [b6] in our knowledge-base are only ever both true in the model when both A and B false. This means that whenever we have an ordering of clauses, with complimentary literals as consequents of two of the clauses, we can not have both clauses true in a model if both their antecedents are true.

We can address this new problem by defining a new propositional connective that combines [b5] and [b6], and that furthermore represents the ordering on the clauses, and so we can represent the clauses in the knowledge-base and in the model, as the combination of the clauses. The basic syntactic idea is to represent the rules as an ordered n-tuple. An ordered n-tuple can be defined as an ordered pair where the first component is an ordered $(n - 1)$ tuple, and the second component is a KAL clause. In an ordered n-tuple, the n th item (clause) has the highest priority. So for example an ordered pair can be represented as <a, b>, an ordered triple as <<a, b>, c>, and an ordered quadruple as <<<a, b>, c>, d>. In each example, a is the first item in the ordered tuple, and b is the second item in the ordered tuple. The highest ordered item in each example is b, c and d, respectively. We can use an ordered n-tuple to represent the ordering of n defeasible rules by making each item in the n-tuple represent a defeasible rule. So for example taking the defeasible rules [b5] and [b6] from the below, the following ordered pair can represent the required ordering:

[b7]   <¬A ←, A ← B>

As discussed below, the interepretation of [b7] we require of this is that ¬A ← is the first item in the ordering, and hence if B is not satisfied, then ¬A ← is the defeasible consequent. So as above if we have a knowledge-base consisting of [b7] and the assertion B, we have the partial model [t6]:

[t6]   line   B   <¬A ←   A ← B>
        1   T   T

So using [t6] together with a sound inference rule, we can constructs model [t7] from a tuple. To do this, we take the highest ordered clause and include the consequent of the clause in the model if the antecdent is satisfied, and otherwise repeat the process on the next highest clause:

[t7]   line   A   ¬A   B   <¬A ←   A ← B>
        1   T   ⊥   T   T

In KAL, the ordering, or priority, to clauses is such that in an ordered n-tuple, the n th clause has the highest priority, the (n-1) th item has the second highest priority, and so on. So in [7] A ← B has the highest priority and ¬A ← has the lowest priority. We can describe the ordering of the n-tuple as a function, denoted as *, such that for any ordered n-tuple < W, X ← Y > can be rewritten, recursively, as *(W, X, Y) where W is an (n - 1) tuple, X is a literal and Y is a conjunction of i literals with i ⩾ 0. Representing the ordering of tuples as a function allows the representation of the semantics as follows:

W	X	Y	*(W, X, Y)
T	T	T	T
T	⊥	⊥	T
T	T	⊥	T
T	⊥	T	⊥
⊥	T	T	T
⊥	⊥	⊥	⊥
⊥	T	⊥	⊥
⊥	⊥	T	⊥

The truth-table of the operator reflects the intuition that for an ordered n-tuple if the antecedent of the n th item, i.e. Y, is assigned T, then the n tuple, i.e. *(W, X, Y), is only assigned T if the consequent X is assigned T, irrespective of the assignment of the remainder of the tuple, ie. W. However, if Y is assigned ⊥, then the n-tuple is only assigned T if W is assigned T, irrespective of the assignment of X.

We can relate an example of an ordered tuple [b8] to the truth-table, where we denote W as ¬match_lights, X as match_lights and Y as match_struck:

[b8]   <¬match_lights ←, match_lights ← match_struck>

Hence if match_struck is assigned T in the database together with the tuple, then match_lights is assigned T. However if match_struck is assigned ⊥, then ¬match_lights is assigned T.

Furthermore, we can view this approach is a form of model preference logic. We prefer models that satisfy the higher-ordered clauses in each tuple over models that satisfy the lower-ordered clauses but not the higher-ordered clauses.

Overview of NTE for non-monotonic temporal reasoning

As in KAL, clauses in NTE are represented as elements in ordered n-tuples. However, in NTE, the clauses in the same tuple are not restricted to the same literal as consequent. Therefore, if an inference can be made from a tuple the ordering effectively represents the preference for the inference to be a particular literal. For example, we may have two clauses [c4] and [c5], and may require a representation such that if the antecedent to [c5] is satisfied, then consequent to [c5] is inferred and the consequent of [c4] is not inferred irrespective of whether, or not, the antecedent of [c4] is satisfied:

[c4] $A \leftarrow B$

[c5] $C \leftarrow D$

So as in KAL, we could represent such a preference over clauses by an n-ordered tuple [c6]. We require a semantics such that if D is true in the model, C is true, irrespective of the truth-values of A or B.

[c6] $< A \leftarrow B, C \leftarrow D >$

As presented this generalization can lead to inconsistencies. However, we are also interested in introducing a temporal operator which we shall use to restrict the range of clauses allowed, and as a result circumvent these inconsistencies. If A is a literal, positive or negative, the formula OA is to be interpreted as at the next point in time the literal A will be true. The literals A and OA will be termed a present literal and a future literal respectively. Nesting of the operator will be permitted and will be abbreviated by a power index. For example, OOA will abbreviate to O^2A. We will also allow a negative power index. So for example O^{-1} will be interpreted as at the previous point in time, literal A was true. We term O^{-i}, where $i \in \mathbb{N}$, as a past literal. The clauses will be restricted to the antecedent being a conjunction of past literals, and the consequent to be either a present literal or a future literal. An examples of an NTE clause is [c7] below:

[c7] $OA \leftarrow B$

Obviously the addition of temporal operators will necessitate a semantics based on a temporal structure. Moreover it is intended that the rule-base will correspond to a temporal logic specification and that a model is constructed to meet the specification. Model construction will be a development on the approach of executable temporal logic (Gabbay 1989).

In NTE we can consider the specification for an agent as a knowledge-base that comprises a database of facts about the past, and a rule-base of ordered n-tuples that provide instructions on the next course of action. Such an approach would require a further restriction on the nature of the clauses: The agent should have the capability to carry out the instructions so that the literals in the instructions can be made true in the model and hence the knowledge-base made true in the model. The rule-base comprises one, or more, ordered n-tuples, and the database comprises of a set of literals, that are true in the model, for each point in time. It is an implicit assumption that there is an 'always' operator in front of all clauses. As an example an agent may have just the tuple [c8] in the rule-base and the facts in the database as represented in [c9]:

[c8] $< OA \leftarrow O^{-2}B, OC \leftarrow O^{-1}D >$

[c9]   At t0   B is in the database
        At t1   D is in the database

If the agent is at t = 2, it needs to know which course of action, if any, to take at t2. From the knowledge-base, the antecedent of the clause 'OC ← D' is satisfied, hence the consequent OC becomes the instruction for the agent. Hence the agent must make OC true at t2, and thereby entering C into the database at t3. From an example such as this, we can see that given a specification such as in [c8], we are effectively model building: We are executing the specification in a manner that captures the execution intuition presented in the approach of executable temporal logics, and captures the model preference presented in the approach of KAL. In this paper, a version of NTE will be developed, with syntax and semantics presented together with an outline execution mechanism.

Syntax for NTE

The language is based on a denumerable set of non-logical letters Π, the logical operators $\wedge$, $\leftarrow$ and $\neg$, and the temporal operator O^n, which is interpreted as the next state operator, where n is a non-negative integer and corresponds to the power. Note that α is an abbreviation for $O^0\alpha$, and $\mathbb{N}$ is the set of natural numbers. Literals are defined as follows:

[d1]   If $\alpha \in \Pi$, then α is a (positive) present literal.
[d2]   If $\alpha \in \Pi$ and $n \in \mathbb{N}$, then $O^n\alpha$ is a (positive) future literal.
[d3]   If $\alpha \in \Pi$ and $n \in \mathbb{N}$, then $O^{-n}\alpha$ is a (positive) past literal.
[d4]   If α is a positive literal, then $\neg\alpha$ is a negative literal.

Clauses in NTE are defined as follows:

[e1] $\alpha \leftarrow$ is a clause if α is a future temporal literal

[e2] $\alpha \leftarrow \beta_1 \wedge \ldots \wedge \beta_r$ is a clause if α is a future or present literal and $\beta_1, \ldots, \beta_r$ are past literals

From the clauses the ordered n-tuples can be formed:

[f1] φ is a unary tuple, where a is a clause

[f2] $<\varphi_1, \varphi_2>$ is a binary tuple, where φ_1, φ_2 are clauses

[f3] $<<\varphi_1, \varphi_2>, \varphi_3>$ is a ternary tuple, where $\varphi_1, \varphi_2, \varphi_3$ are clauses

$\vdots$ $\vdots$

[fn] $<<\varphi_1, \ldots> \ldots, \varphi_n>$ is an n-tuple, where $n \geqslant 1$ and $\varphi_1, \ldots, \varphi_n$ are clauses and n is finite

Semantics for NTE

We define the semantic intepretation of propositional NTE in the monadic theory of $(\mathbb{N}, <)$. An assignment h is a function associating with each atom α, $\alpha \in \Pi$, a subset $h(\alpha)$ of $\mathbb{N}$.

[m1] $M(t) \models \alpha$ and α is atomic   iff $t \in h(\alpha)$

[m2] $M(t) \models \neg\alpha$ iff not $M(t) \models \alpha$

[m3] $M(t) \models \alpha \wedge \beta$ iff $M(t) \models \alpha$ and $M(t) \models \beta$

[m4] $M(t) \models \alpha \leftarrow \beta$ iff if $M(t) \models \beta$ then $M(t) \models \alpha$

[m5] $M(t) \models <\delta, \alpha \leftarrow \beta>$ iff $[M(t) \models \beta$ and $M(t) \models \alpha]$
                                              or $[M(t) \models \neg\beta$ and $M(t) \models \delta]$

       where δ denotes an n-1 tuple if $<\delta, \alpha \leftarrow \beta>$ denotes an n tuple

[m6] $M(t) \models O^n\alpha$ iff $M(t+n) \models \alpha$

From the defintions [m1] - [m6], it becomes apparant that a set of NTE tuples can be executed at t = n on the basis of the past, i.e. the points less than n, so as to construct the model inductively at n. We define a model M to be a triple $(\mathbb{N}, <, h)$.

Outline of Execution Mechanism for an NTE Rulebase

The logic NTE provides us with a language to express specifications of plans and schedules. We now need to consider in further detail how we can construct a model from a NTE specification. We envisage that the specification will be composed in terms of components of the plan. Given the following

   Ω a rulebase, (i.e. the specification), which is a set of ordered n-tuples

   Ψ_{n-1} a model constructed by time = n - 1

   Γ_n a set of commitments brought forward to time = n

the rulebase, and the model constructed so far, can be used as a specification for constructing the next piece of the model at n. Effectively Ψ_{n-1} is a database on the past. At t = n, Ψ_{n-1} contains (n - 1) sets of assertions where for each i, $1 \leqslant i \leqslant (n - 1)$, the set i is the set of assertions true in the model at t = i. The following is an outline of an interpreter in logic programming style, developed from (Barringer 1990):

$\text{Execute}(n, \Omega, \Psi_{n-1}, \Gamma_n) :- \text{Generate\_commitments}(\Omega, \Psi_{n-1}, F_n),$
$\qquad\qquad\qquad\qquad\quad \text{Find\_current\_commitments}(\Gamma_n, F_n, G_n, \Gamma_{n+1}),$
$\qquad\qquad\qquad\qquad\quad \text{Environmental\_input}(n) = E_n,$
$\qquad\qquad\qquad\qquad\quad \text{Build\_model}(E_n, G_n, \Psi_{n-1}, \Psi_n),$
$\qquad\qquad\qquad\qquad\quad \text{Execute}(n, \Omega, \Psi_n, \Gamma_{n+1}).$

Below is a brief discription of the predicates used:

$\text{Generate\_commitments}(\Omega, \Psi_{n-1}, F_n)$ takes the rules and the model constructed so far, which is effectively the history and returns a set of commitments, (i.e. the consequents of rules that have antecedents satisfied), that must be satisfied in the model in order for the specification to be met.

$\text{Find\_current\_commitments}(\Gamma_n, F_n, G_n, \Gamma_{n+1})$ takes the outstanding commitments from the past, Γ_n, together with the newly generated commitments Γ_n, and returns a set of propositions that must be made true by the system in the model at n, together with a set of propositions that form the revised set of outstanding commitments.

$\text{Environmental\_input}(n)$ is a function that takes the current time point and returns a set of propositions made true by the environment at n

$\text{Build\_model}(E_n, G_n, \Psi_{n-1}, \Psi_n)$ takes the sets of propositions E_n and G_n made true by the environment and system respectively, together with the model made up to n-1, and returns the model up to n

If a rulebase Ω is realizable in an environment, the execution mechanism should eventually build model of Ω in the environment. By realizable, we mean that the specification does not contradict itself, or contradict the environment.

As a simple example we show for the following specification,

$$< O\alpha \leftarrow, \neg O\alpha \leftarrow \beta >$$
$$< \neg O\beta \leftarrow, O\beta \leftarrow \delta >$$
$$< \neg O\delta \leftarrow, O\delta \leftarrow \neg\alpha \wedge \neg\beta \wedge \neg\delta >$$

the following part of an execution:

:

time t	¬α, ¬β, ¬δ	hold in the model
time t +1	α, ¬β, δ	hold in the model
time t +2	α, β, ¬δ	hold in the model
time t +3	¬α, ¬β, ¬δ	hold in the model

:

:

In the next section we provide a definition of a semantic tableau for NTE, and provide a mechanism for extracting a finitely-preferred model from the tableau, where a finitely-preferred model is the first part of a preferred model from 0 to some n ∈ ℕ. An execution mechanism can be defined to produce a finitely-preferred model M of the specification if, and only if, the finitely-preferred model M can be extracted from the tableau of the specification.

Semantic Tableau for NTE

The semantic tableau Smullyan (1968) provides a mechanical test for consistency and, hence inconsistency for classical logic. In order to prove that a set of propositional formulae, Δ, implies a propositional formula, G, the set union of Δ ∪ {¬G} is assumed to be satisfiable. From this satisfiability assumption, smaller subformulae are systematically derived. Using a binary tree structure, the set of formulae to be tested for consistency is put at the root of the tree. The tree is grown by apply tableau rules to branches of the tree until all the leaves are of the form of a proposition letter or it's compliment. A branch of the tree is closed if it contains a propositional letter and its compliment. A tree is closed if all the branches are closed. If a tree is closed, the original satisfiability assumption is inconsistent, and hence Δ implies G. Furthermore, a branch is complete if no more tableau rules apply, and tree is complete if all branches are complete. If a tree is complete, but not closed, then it is open. If a tree is open, the original satisfiability assumption is consistent. The approach of semantic tableau has been extended to modal and intuitionistic logics (Fitting 1983) and to temporal logics (Wolper 1985).

The details of the following version of the semantic tableau are not important since we could present a formalization of the semantic tableau based upon the usual definition for propositional temporal logic. However, since both the approaches of executable temporal logics, and KAL, incorporate a notion of preferred models, what is important in this exposition is the idea of pruning to extract preferred models.

Three types of rule are required: (1) Rules 1 to 4 correspond to some of the classical semantic tableau rules; (2) Rule 5 is to manipulate the ordered n-tuples; and (3) Rule 6 is to decompose the temporal formulae. In the following representation of the rules, the formulae above the line indicate a branch-point, and the formulae below the line, indicate the sub-formulae in the branches immediately below the branch-point The symbols ',', '|', and '||' are used to indicate the nature of the branching.

[s1] Two formulae seperated by ',' indicate the two formulae are in the same branch.

[s2] Two formulae seperated by '|' indicate the two formulae are in two different branches.

[s3] Three formulae seperated by '||' indicate the three formula are in three different branches.

In order to outline the decomposition of the temporal operators and to identify the open branches we introduce the following notions:

<u>Sub-tableau:</u> Given a set of assumptions (at a sub-root), a sub-tableau is generated by applying decomposition rules 1 to 5 exhaustively.

<u>World:</u> A world is the set of formulae along one branch of a sub-tableau. Furthermore, each branch has a corresponding world.

<u>Level:</u> The sub-tableau generated from the root of the tableau is in level 1. The application of rule 6 to a branch of a sub-tableau in level n results in temporal formulae and tuples being exported as assumptions to a new sub-root in level n+1, where rules 1 to 5 can be applied exhaustively again. For all n, if n is a level, then n $\in \mathbb{N}$. The operation + is the usual arithmetical addition over $\mathbb{N}$, and < is the usual ordering on $\mathbb{N}$.

<u>Literal:</u> If φ is a literal in a world w, and φ is of the form $O^i\alpha$, then $\varphi \in class(\alpha, w)$ holds, which is the equivalence class of literals of propositional letter α in world in w. The relation old_literals(w) is the set of literals imported into world w by rule 6, and the relation new_literals(w) is the set of literals generated within world w by rules 1 to 5.

<u>Contra-pair:</u> We describe a pair of literals $(O^i\alpha, \neg O^j\alpha)$ as a contra-pair in a world w if there is an equivalence class of literals in w s.t. $O^i\alpha \in class(\alpha, w)$ and $\neg O^j\alpha \in class(\alpha, w)$. For an equivalence class of literals in w, if there is more than one convergent pair, a <u>minimal contra-pair</u> is the pair with the smallest difference in index powers, i.e. the pair $(O^p\alpha, \neg O^q\alpha)$ s.t. $|p - q|$ is the smallest in the class.

<u>Convergent contra-pair:</u> A world w' in level n+1 has a convergent contra-pair if the previous world w in level n has a minimal contra-pair $(O^i\alpha, \neg O^j\alpha)$, and world w' has a minimal contra-pair $(O^p\alpha, \neg O^q\alpha)$, and $|i - j| > |p - q|$.

Given these definitions we define the decomposition rules as follows:

Rule 1	Rule 2	Rule 3	Rule4	Rule 5
$\neg\neg X$	$X \wedge Y$	$\neg(X \wedge Y)$	$X \leftarrow Y$	$<W, X \leftarrow Y>$
X	X, Y	$\neg X \mid \neg Y$	$X \mid \neg Y$	$W \parallel X \parallel \neg Y$

Rule 6: If in a world w at a level n, there are no more possible applications of decomposition rules 1 to 5, and w is not a loop-point, then all the prefixed wff $O^i\alpha$ are exported to a new world w' at level n+1 as $O^{i-1}\alpha$, together with all the tuples in w.

We label worlds in the tableau as closed, open or loop-point, and then label branches as open or closed:

<u>Loop-point</u>: If a world w is s.t. for each equivalence class of literals class(α, w) there are no contra-pairs in class(α, w) or there are no convergent contra-pairs in class(α, w), then w is a loop-point.

<u>Closed world</u>: If the world contains a literal, and its negation, then the branch is closed.

<u>Open world</u>: If the world contains no literal and its negation, and is not a loop-point, then the world is open. Obviously a loop-point is an open world.

We terminate the decomposition of the formulae in the branches when we reach a closed world, or a loop-point, or no further decomposition rules apply. However, decomposition can consistently be extended beyond a loop-point if a longer finitely-preferred model is required. A world at which we terminate the decomposition process is termed a leaf. A branch of the tableau is therefore the sequence of worlds from the root to a leaf. We label branches as open or closed starting from the leaf. We label the branch starting at the leaf as open if the leaf is open, and closed if the leaf is closed. We label the branches above branch-points inductively according to the following table.

Rule at branch point	Left branch	Middle branch	Right branch	Branch-point
1 or 2 or 6	n/a	open	n/a	open
1 or 2 or 6	n/a	closed	n/a	closed
3 or 4	open	n/a	open	open
3 or 4	open	n/a	closed	open
3 or 4	closed	n/a	open	open
3 or 4	closed	n/a	closed	closed
5	closed	closed	closed	closed
5	open	closed	closed	closed
5	closed	open	closed	open
5	open	open	closed	open
5	closed	closed	open	closed
5	open	closed	open	open
5	closed	open	open	closed
5	open	open	open	open

After having labelled the tree according to the above four labels, we identify the preferred branch(es) by pruning the tree. The remaining branch(es) are preferred branches. Tree pruning requires starting at the root and moving down the tree removing branches from rule 4 and rule 5 branch points according to the following table. The branch-points generated by the rules 1, 2, 3 or 6 are not prunned:

Rule at branch point	Left branch	Middle branch	Right branch	Delete subtree eminating
4	closed	n/a	open	left
4	open	n/a	open	left
4	open	n/a	closed	right
4	closed	n/a	closed	left & right
5	open	closed	closed	middle & right
5	closed	closed	closed	left, middle & right
5	open	open	closed	left & right
5	closed	open	closed	left & right
5	open	closed	open	middle & right
5	closed	closed	open	left & middle
5	open	open	open	middle & right
5	closed	open	open	left & middle

Having identified preferred branches of the tree, we outline the generation of a finitely-preferred model I from a preferred branch. I is the smallest initial section of a model s.t. for each positive literal $O^i\alpha$ in a branch in each level n, $(n+i)' \in h(\alpha')$ holds, where $(n+i)'$ and α' are identical to $n+i$ and α, respectively, apart from possibly name.

The following is an example of an NTE tableau, given the following specification: $\beta <- O^{-1}\alpha$, α, where $\beta <- O^{-1}\alpha$ is a unary tuple, B denotes branch-point and L denotes level.

$$[L1, Root]\, \beta <- O^{-1}\alpha,\ \alpha$$

$[L1, B1]\ \beta, \alpha$ $\qquad$ $[L1, B2]\ \alpha, \neg O^{-1}\alpha$

$$[L2, Sub\text{-}root]\, \beta <- O^{-1}\alpha,\ O^{-1}\alpha, \neg O^{-2}\alpha$$

$$[L2, B2.1]\, \beta, O^{-1}\alpha, \neg O^{-2}\alpha \qquad [L2, B2.2]\, \neg O^{-1}\alpha, O^{-1}\alpha, \neg O^{-2}\alpha$$

From this tableau, the branches ending at [L2, B2.2] is closed, and the branch ending at [L21 B1] is prunned. Hence the branch ending at [L2, B2.1] is the only preferred branch. Below is another example of a tableau, with the NTE specification: $\neg O^2\alpha <-$, $\neg\alpha$, where $\neg O^2\alpha <-$ is a unary tuple.

[L1, Root]	$\neg O^2\alpha <-,\ \alpha$
[L1, B1]	$\neg O^2\alpha, \alpha$
[L2, Sub-root]	$\neg O^2\alpha <-,\ O^{-1}\alpha$
[L2, B1]	$\neg O^2\alpha, O^{-1}\alpha$
[L3, Sub-root]	$\neg O^2\alpha <-,\ O^{-2}\alpha$
[L3, B1]	$\neg O^2\alpha, O^{-2}\alpha$

Since the above example has only one branch, which is closed, there is no preferred branch.

The NTE variant of the semantic tableau has the following property that is important for constructing a finitely-preferred model from a consistent NTE specification Δ:

THEOREM: The NTE semantic tableau generated from Δ is open
             iff ∃ M s.t. M is a finitely-preferred model of Δ

Outline of proof: [<=] If there is a finitely-preferred model then there is a model.
Hence it is straightforward to show that the tree is open. [=>] The semantic tableau
rules are decompositional. Furthermore, it is obvious that the rules 1 to 5
terminate after a finite number of steps. It can also be shown that rule 6 results in a
loop-point after a finite number of steps. By examination of cases, we can
undertake tree pruning, and produce at least one preferred branch for any open
semantic tableau for NTE, and hence a finitely-preferred model.

We can also show that for any NTE specification, if the semantic tableau generated
is closed, then there is no model of the specification, and hence no preferred
model.

Discussion

The development of NTE has shown that a language can be developed for
specifications that is sufficiently expressive to represent notions of time and
defeasibility, and furthermore has the facility to be used to build a preferred model
of the specification. Executable temporal logics provide a natural alternative to
programming languages, such as PROLOG, for temporal reasoning. Versions of
executable temporal logic include MetateM, which is a first-order US temporal
logic (Barringer 1990), MML, which is a meta-language for executable temporal
logic (Barringer 1991), and RDL, which is a language for expressing requirements
and designs for time-dependent systems (Gabbay 1991)

Even though both the temporal logic and the defeasible logics used to derive NTE
are relatively simple, the combination does indicate how we may proceed in using
more expressive temporal logics such as first-order US (Kamp 1968), and more
expressive defeasible logics such as first-order ordered logic (Laenens 1990).
Furthermore, the prototype implementations for executing US logic could
potentially be extended to NTE according to the outline mechanism presented.
Finally the NTE semantic tableau provides us with a mechanism for analysing
specifications, which with further developments, including defining a semantic
tableau based on the usual temporal logic formalization, allows for identifying
infinite preferred models.

A variety of approaches to developing non-monotonic systems for temporal
reasoning have been proposed (including Kowalski 1986, Shoham 1987, Bacchus
1989, Baker 1989). In comparing NTE with these other approaches, the primary
differences include: (1) As a development of the executable temporal logic
paradigm, we can view an NTE specification as a program that provides a clear
link between a declarative past and an imperative future; and (2) As a
development of a non-monotonic logic that uses explicit notion of preference of
rules, we can explicitly express preference for certain 'futures' on the basis of the

history of the execution. For a formal comparison of NTE with other approaches to non-monotonic temporal reasoning systems, we will use a first-order version.

Other ways that a defeasible logic approach may be appropriate in an executable temporal logic framework is in one or more of the following contexts: (1) incomplete model on the past; (2) default values for some, or all, the missing values in the partial model on the past; (3) changing commitments before, or during, discharge; and (4) dealing with failure of an agent to meet a specification. These constitute some current research questions.

Acknowledgements

This work was partly supported by ESPRIT under Basic Research Action 3096 (SPEC) and by the UK Science and Engineering Research Council under the Rule-Based Systems project. Special thanks are due to Dov Gabbay.

References

Bacchus F, Tenenberg J, & Koomen J (1989) A non-reified temporal logic, Proceedings of the First International Conference on the Principles of Knowledge Representation and Reasoning, Morgan Kaufmann

Baker A (1989) A simple solution to the Yale shooting problem, Proceedings of the First International Conference on the Principles of Knowledge Representation and Reasoning, Morgan Kaufmann

Barringer H, Fisher M, Gabbay D, Gough G & Owens R (1990) MetateM: A framework for programming in temporal logic, in REX Workshop on Stepwise Refinement of Distributed Systems, LNCS 430, 94 - 129, Springer

Barringer H, Fisher M, Gabbay D, & Hunter A (1991) Meta-Reasoning in Executable Temporal Logic, Proceedings of the Second International Conference on the Principles of Knowledge Representation and Reasoning, Morgan Kaufmann

Fitting M (1983) Proof Methods for Modal and Intuitionistic Logics, Reidel

Gabbay D (1989) The declarative past and imperative future: Executable temporal logic for interactive systems, in Banieqbal B, Barringer H, & Pnueli A, Proceedings of Colloquium on Temporal Logic in Specification, LNCS 398, 409 - 448, Springer

Gabbay D, Hodkinson I, and Hunter A (1991) Using the temporal logic RDL for design specifications, in Yonezawa A, Concurrency: Theory, Language and Architecture, LNCS 491, Springer

Hunter A, (1990) KAL: A linear-ordered logic for non-monotonic reasoning, Draft paper, Imperial College, London

Kamp J (1968) Tense Logic and the Theory of Linear Order, PhD thesis, UCLA

Kowalski R & Sergot M, (1986) A logic-based calculus of events, New Generation Computing, 4, 67 - 95

Laenens E & D Vermier (1990) A fixpoint semantics of ordered logics, Journal of Logic and Computation, 1, 159 - 185

Shoham Y (1987) Reasoning About Change, MIT Press

Wolper P (1985) The tableau method for temporal logic: an overview, Logique et Analyse, 110-111, 119-136

ON THE PHENOMENON OF FLATTENING "FLEXIBLE PREDICTION" CONCEPT HIERARCHY

Mieczyslaw A. Klopotek

Institute of Computer Science
Polish Academy of Sciences
00-901 Warsaw, P.O.Box 22, Poland

Abstract

The Incremental Concept Formation method, as described in [19] is claimed therein to be "able to formulate diagnostically useful categories even without class information", "given real world data on heart disease".We suggest in this paper that the method does not derive categories from the data but from primary and derived attribute selection by showing that equal treatment of all attributes leads to a flat (one level) concept hierarchy.

1 Introduction

Cluster analysis (or numerical taxonomy [1, 2]) evolved from strong belief into the paradigm "similar things should behave similarly" [1]. The clustering depends heavily on attribute selection and attribute pondering [2], usually encapsulated into some similarity/ dissimilarity measure. If we do not want to obtain an arbitrary clustering (domain structure), some kind of "objective" or "natural" similarity measure should be found. One possibility is to take the proportion between the number of attributes values of which are shared by two objects to the number of all attributes defined for those objects. Under such a seemingly plausible similarity definition Watanabe [3] has shown that if we take into account all available primary (observed) and derived attributes, then any object is equally similar to any other ("Ugly Duckling Effect"). To avoid interpretation problems, Michalski and Stepp [4]—[7] proposed an attribute pondering called "conceptual clustering" (simplicity of logical cluster description in terms of primary attributes ponders cluster utility). Though criticized e.g. by Dale [8], this approach evolved further into Incremental Concept Formation in [9]—[19] and other works aiming at maximizing mutual prediction capability between class membership and attribute value (so-called "flexible prediction" [18]).

In [19, p. 58], , it is claimed that "given real world data on heart disease, an incremental concept formation system CLASSIT (a mutation of Fisher's COB-WEB [18]) was able to formulate diagnostically useful categories even without class information".

In this paper we question this claim by showing that the Incremental Concept Formation (ICF) line of clustering systems around COBWEB, CLASSIT etc. is also subject to the "Ugly Duckling Effect". That means once observables are selected, the derived domain (cluster) structure is immune to data presented to it given we consider all possible derived attributes. Flavor of data manipulation is still retained.

The rest of the paper is organized as follows: Section 2 reminds the COB-WEB clustering algorithm and presents an example showing necessity of using derived attributes to obtain useful categories. Section 3 contains the proof of the "Ugly Duckling Theorems" for the COBWEB algorithm using complete set of derived attributes. Section 4 discusses some consequences of those theorems.

2 Incremental Concept Formation by COBWEB

The basic COBWEB algorithm comprises learning and classification steps [12, 19]. An important role is played by the score evaluation function, which can be summarized [20] for both COBWEB and CLASSIT at each node as follows:

$$eval = \tag{1}$$

$$M^{-1} \left(\sum_{C_i} p(C_i) \left(\sum_{A_j \in D} \sum_{V_{jk}} p(A_j = V_{jk}/C_i)^2 + \sum_{A_j \in \mathcal{R}} \int f_{A_j/C_i}(x)^2 dx \right) \right.$$
$$\left. - \sum_{A_j \in D} \sum_{V_{jk}} p(A_j = V_{jk})^2 - \sum_{A_j \in \mathcal{R}} \int f_{A_j}(x)^2 dx \right)$$

with:

D —set of discrete attributes

$\mathcal{R}$ —set of continuous attributes

$p(C_i)$ —probability that an object belongs to Sub-category C_i

$p(A_j = V_{jk})$ —probability that the discrete parameter A_j takes the value V_{jk} .

$p(A_j = V_{jk}/C_i)$ —the respective conditional probability within the sub-category C_i

$f_{A_j}(x)$ —density function of the continuous parameter A

$f_{A_j/C_i}(x)$ —respective conditional density within the Sub-category C_i

M —number of sub-categories.

Obviously, we can discretize continuous functions into sufficiently fine intervals and attribute/value comparisons $A_j = V_{jk}$ may be treated as predicates on objects of the type $P_{A_j=V_{jk}}(X)$ (X— the object under consideration). So we can rewrite the above formula (1) as:

$$eval = \tag{2}$$

$$M^{-1}\left(\sum_{C_i} p(C_i) \sum_{P_{A_j=V_{jk}}} p(P_{A_j=V_{jk}}(X) = true/C_i)^2\right.$$

$$\left. - \sum_{P_{A_j=V_{jk}}} p(P_{A_j=V_{jk}}(X) = true)^2\right)$$

Let us analyze the performance of such an Incremental Concept Formation (ICF) algorithm for a certain application domain. Let us define predicates over a set of chemical organic compounds:

$P_1(X) - X$ possessing a hydrophobic group

$P_2(X) - X$ not possessing a hydrophobic group

$P_3(X) - X$ possessing a hydrophile group

$P_4(X) - X$ not possessing a hydrophile group

Obviously, compounds matching $P_1 \wedge P_3$ are surfactants, and non-surfactants are those matching $P_2 \vee P_4$. Now let us consider the behavior of COBWEB operating with these predicates. Let it be presented with a set of 10 surfactants and 10 hydrophobic non-surfactants. We may assume for sure that two classes are created:C_1 - surfactants and C_2 - (hydrophobic) non-surfactants. Let us now present it with a hydrophile non-surfactant. If the new instance were classified into the class C_1 (surfactants), we get the value of evaluation function:

$$\begin{aligned} eval_1 &= 1/2 * (11/21(10^2/11^2 + 1^2/11^2 + 11^2/11^2 + 0^2/11^2) \\ &\quad +10/21(10^2/10^2 + 0^2/10^2 + 0^2/10^2 + 10^2/10^2) \\ &\quad -(20/21)^2 - (1/21)^2 - (11/21)^2 - (10/21)^2) = 0.251 \end{aligned}$$

while classifying into C_2 (hydrophobic non-surfactants) would give:

$$\begin{aligned} eval_2 &= 0.5 * (10/21(10^2/10^2 + 0^2/10^2 + 10^2/10^2 + 0^2/10^2) \\ &\quad +1/21(10^2/11^2 + 1^2/11^2 + 1^2/11^2 + 10^2/11^2) \\ &\quad -(20/21)^2 - (1/21)^2 - (11/21)^2 - (10/21)^2) = 0.208 \end{aligned}$$

and creation of an entirely new class C_3 for the new instance results in:

$$\begin{aligned} eval_3 &= 0.3333 * (10/21(10^2/10^2 + 0^2/10^2 + 10^2/10^2 + 0^2/10^2) \\ &\quad +10/21(10^2/10^2 + 0^2/10^2 + 0^2/10^2 + 10^2/10^2) \\ &\quad +1/21(0^2/1^2 + 1^2/1^2 + 1^2/1^2 + 0^2/1^2) \\ &\quad -(20/21)^2 - (1/21)^2 - (11/21)^2 - (10/21)^2) = 0.197 \end{aligned}$$

Hence, COBWEB would select C_1 (surfactants) as the proper class for the new instance of a (hydrophile) non-surfactant which is intuitively non-sense. Eventually, after further instances, a class of hydrophile non-surfactants would appear at the top level of the hierarchy, but it will never merge with the other non-surfactants. It would behave in the same way even if the proportions between C_1 and C_2 were different (e.g. 10 : 5) The trouble lies in the fact, that the COBWEB method does search for conjunctive concepts, and not for disjunctive ones. However, in nature, the non-surfactants may also be partially described in conjunctive form, if we take such predicates as water hardness ion binding property or washing capacity etc. But we could discover non-surfactants simply by redefining the predicates and taking $P_a = P_1 \wedge P_3$, $P_b = P_2 \vee P_4$ instead of P_1, P_2, P_3, P_4.

It is obvious then, that discovery of a concept depends on presence or absence of logical expression describing it among the derived attributes. Hence we could naively ask the following question: Given a data set and the set of all conceivable derivations from primary (i.e. observable) predicates. Does the COBWEB discover the relevant structure of the domain based on the data ? The subsequent section gives the answer NO.

3 Ugly Duckling Effect for Incremental Concept Formation

Let us assume that the given data set is described by a set of primary predicates $P_1, \ldots, P_z$ applicable to each object (z finite). Let us assume further that we know constraints ruling the predicates (e.g. if P_1 and P_2 describe colors "is black" and "is white" resp. it may be reasonable to assume $P_1 \rightarrow \tilde{\ } P_2$ and $P_2 \rightarrow \tilde{\ } P_1$ etc.). Let C denote the set of all constraints. Let us define the set $\mathcal{U}$ of all atomic formulas not violating the constraints from C:
$\mathcal{U} = \{A = Q_1 \wedge Q_2 \ldots \wedge Q_z | Q_i = \text{either } P_i \text{ or } \tilde{\ } P_i , \forall_{C \in C} A \rightarrow C\}$
Let us define the set $\mathcal{P}$ of all conceivable predicates derived from atomic formulas as follows:

1. $\emptyset \in \mathcal{P}$ ($\emptyset$ - the always false predicate)

2. $\forall_{A \in \mathcal{U}} A \in \mathcal{P}$ (all atomic formulae belong to $\mathcal{P}$

3. $\forall_{P_1, P_2 \in \mathcal{P}} P_1 \vee P_2 \in \mathcal{P}$ (all disjuncts belong also therein)

4. No other predicates belong to P.

It is obvious that under these conditions every element of $\mathcal{P}$ matches the constraints from C.

Theorem 1 *If the set of primary predicates is fine enough to distinguish any two visited instances of the universe (fine grid assumption) the evaluation function (2) at any node for any element takes the form:*

$$eval = M^{-1}N^{-1}2^{G-2}(M-1)$$

with

N — *cardinality of all the examples at the given level*

G —*total number of ground (= atomic) predicates*

$n_i = card(C_i)$ —*cardinality of sub category C_i under the fine-grid assumption. It is the same as the number of ground predicates covered by the class C_i.*

Proof: Given in Appendix A.

Let us consider now the more complicated case of indistinguishable instances:

Theorem 2 *If the set of primary predicates is not fine enough to distinguish any two instances of the universe (rough grid assumption) then the evaluation function at any node for any element takes the form:*

$$eval = \overline{M}^{-1}(2^{-2} * 2^{card(\mathcal{P})})N^{-1}\left(\left(\sum_{C_i} n_i^{-1} \sum_{p_x=1}^{s_M} p_x^2 n_{i_{p_x}} \right) - N^{-1} \sum_{p_x=1}^{s_M} p_x^2 N_{p_x} \right)$$

s_M —*maximum number of elements in a grid point*

n_{i_s} —*number of grids in class C with s elements*

N_s —*overall number of grids with s elements*

N —*total number of elements*

G —*total number of ground predicates*

$n_i = card(C_i)$ —*cardinality of class C_i. Under the fine-grid assumption it is the same as the number of ground predicates covered by the class C_i.*

Proof: Given in Appendix B.

4 Discussion

The fine grid case Ugly Duckling theorem of the previous section has severe consequences:

(1) expression of evaluation function does depend on the number of existing classes, but NOT on distributions of particular predicates or joint distributions of some of them,

(2) hence attribute values have no influence on "derived" domain structure.

In fact,

COROLLARY 1 *The fine grid case evaluation function will create a flat hierarchy pushing each instance into a separate class being a leaf attached directly to the root node.*

PROOF: Let us consider the situation in which we have N objects distributed into M classes. The $N + 1^{st}$ object is coming in. If it were qualified into one of classes already present, we would get:

$$eval_{case1} = M^{-1}(N + 1)^{-1}2^{G-2}(M - 1)$$

If it were pushed into a separate class, we would obtain:

$$eval_{case2} = (M + 1)^{-1}(N + 1)^{-1}2^{G-2}(M + 1 - 1)$$

We claim that $eval_{case1} < eval_{case2}$ and hence, at every decision point, a new instance will be pushed into into a separate class, and hence by induction each instance will be assigned a separate class. In fact:

$$eval_{case1} < eval_{case2}$$

because: $M^{-1}(N + 1)^{-1}2^{G-2}(M - 1) < (M + 1)^{-1}(N + 1)^{-1}2^{G-2}(M + 1 - 1)$
because: $M^{-1}(M - 1) < (M + 1)^{-1}(M + 1 - 1)$
because: $(M - 1) * (M + 1) < M * M$
because: $M^2 - 1 < M^2$
because: $-1 < 0$
Q.e.d. $\square$

COROLLARY 2 *Any effort of pondering (by dropping or replicating) of primary (observable) attributes will only ponder the evaluation function, but not change relative weights of alternative class assignments*

PROOF: If we replicate (or drop as long as the instances remain distinguishable) the primary predicates, only the number G of all theoretically possible elementary (atomic) predicates will be possibly changed. But the factor 2^{G-2} is a constant for a given predicate space and hence has no impact on class assignment procedure. Hence class assignment remains unchanged. Q.e.d. $\square$

In case of rough grid assumption the following can be shown: if the coverage of grids is "balanced" then we will have the effect of Corollary 1 (each grid set pushed into separate category) otherwise small cardinality grids will be pushed together into joint categories. We can easily arrive at the apparently amazing conclusion that losing information (by dropping some observables and hence making some objects indistinguishable) leads to provision of structural information on the domain (the concept hierarchy stops being flat). Furthermore, we can state that the effects reported in [19] concerning information losses when using "non-relevant" variables may in fact stem from using variables revealing different aspects of the domain under consideration.

All the phenomenons mentioned above outline clearly limitations and potentials behind ICF. The domain structure derived by ICF has nothing in common with any "natural" classification of the data. It detects rather "syndromes" behind given sets of attributes.

5 Conclusion

Selection and pondering of DERIVED attributes (and not necessarily primary ones) has an immense impact on the derived conceptual structure obtained via ICF COBWEB (so called "flexible prediction") algorithm. Hence one relies heavily on the intuition of the expert giving the list of DERIVED attributes which may be related to the concept structure of interest. In fact, nothing like "without guidance of class information" claimed in [19] exists. Rather the ICF extracts explicitly the class information hidden implicitly in derived attributes used - it is suitable for detection of "syndromes".

Appendix A

Proof of Theorem 1 *If the set of primary predicates is fine enough to distinguish any two visited instances of the universe (fine grid assumption) the evaluation function (2) at any node for any element takes the form:*

$$eval = M^{-1}N^{-1}2^{G-2}(M-1) \tag{3}$$

with

N — *cardinality of all the examples at the given level*

G —*total number of ground (= atomic) predicates*

$n_i = card(C_i)$ —*cardinality of sub category C_i under the fine-grid assumption. It is the same as the number of ground predicates covered by the class C_i.*

PROOF: Let us consider the following fragment of the eval-function (2)

$$\left(\sum_{P_j \in \mathcal{P}} p(P_j = true/C_i)^2 \right) = \tag{4}$$

$$= \left(\sum_{P_j \in \mathcal{P}} \frac{card(P_j = true/C_i)^2}{card(C_i)^2} \right) = card(C_i)^{-2} \left(\sum_{P_j \in \mathcal{P}} card(P_j = true/C_i)^2 \right)$$

As the granulation by the set of primary predicates is fine enough, the $card(P_j = true/C_i)$ for any $P_j \in \mathcal{U}$ takes either the value 1 (if there exists an instance in C_i, for which P_j holds) or 0 (otherwise). Any other predicate P_j is a disjunct of some atomic predicates $P_{a1_j} \vee P_{a2_j} \vee \ldots \vee P_{asj_j}$ and $card(P_j = true/C_i) = card(P_{a1_j} = true/C_i) + card(P_{a2_j} = true/C_i) + \cdots + card(P_{asj_j} = true/C_i)$. So let us consider the class of predicates P_j such that $card(P_j = true/C_i) = k$, $1 \leq k \leq n_i = card(C_i)$. Let us count how many of them exist: they are disjuncts of k atomic predicates out of the n_i atomic predicates which are true in class C_i and a number of predicates out of the $G - n_i$ atomic predicates false in C_i. (Let us say that a predicate P is true in class C iff there exists an instance $X \in C$ such that $P(X) = true$ for that instance. Otherwise P is false in C). So their number is:

$$\binom{n_i}{k}\binom{G-n_i}{0} + \binom{n_i}{k}\binom{G-n_i}{1} + \cdots + \binom{n_i}{k}\binom{G-n_i}{G-n_i} =$$

$$= \binom{n_i}{k} 2^{G-n_i} \tag{5}$$

Hence the above expression (4) may be transformed to:

$$= n_i^{-2} \left(\sum_{k=1}^{n_i} \binom{n_i}{k} 2^{G-n_i} k^2 \right) = n_i^{-2} 2^{G-n_i} \left(\sum_{k=1}^{n_i} \binom{n_i}{k} k^2 \right) = \tag{6}$$

By definition $\binom{n}{k} = \binom{n-1}{k-1} n/k$ hence:

$$= n_i^{-1} 2^{G-n_i} \left(\sum_{k=1}^{n_i} \binom{n_i-1}{k-1} k \right) =$$

$$= n_i^{-1} 2^{G-n_i} \left(\sum_{k=1}^{n_i} \binom{n_i-1}{k-1} (k-1) + \sum_{k=1}^{n_i} \binom{n_i-1}{k-1} \right) \tag{7}$$

As:

$$\sum_{k=1}^{n_i} \binom{n_i-1}{k-1} = \sum_{k=0}^{n_i-1} \binom{n_i-1}{k} = 2^{n_i-1} \tag{8}$$

holds, we obtain from (7):

$$n_i^{-1} 2^{G-n_i} \left(\sum_{k=1}^{n_i-1} \binom{n_i-1}{k} k + 2^{n_i-1} \right) \tag{9}$$

$$= n_i^{-1} 2^{G-n_i} \left((n_i - 1) \sum_{k=1}^{n_i-1} \binom{n_i - 2}{k - 1} + 2^{n_i-1} \right) =$$

$$n_i^{-1} 2^{G-n_i} \left((n_i - 1) 2^{n_i-2} + 2^{n_i-1} \right) =$$

$$= n_i^{-1} \left((n_i - 1) 2^{G-2} + 2 * 2^{G-1} \right) = n_i^{-1} (n_i + 1) 2^{G-2} =$$

Since $p(C_i) = n_i/N$ we obtain substituting (9) into the eval-function: (2):

$$M^{-1} \left(\left(\sum_{C_i} (n_i/N) * n_i^{-1} (n_i + 1) 2^{G-2} \right) - N^{-1} (N + 1) 2^{G-2} \right) =$$

$$= M^{-1} \left(N^{-1} \sum_{C_i} \left((n_i + 1) 2^{G-2} \right) - N^{-1} (N + 1) 2^{G-2} \right) =$$

$$M^{-1} N^{-1} 2^{G-2} \left(\sum_{C_i} (n_i + 1) - (N + 1) \right)$$

$$= M^{-1} N^{-1} 2^{G-2} \left((N + M) - (N + 1) \right) = M^{-1} N^{-1} 2^{G-2} (M - 1) \qquad (10)$$

which was to be proven. □

Appendix B

Proof of Theorem 2 *If the set of primary predicates is not fine enough to distinguish any two instances of the universe (rough grid assumption) then the evaluation function at any node for any element takes the form:*

$$eval = \overline{M}^{-1} (2^{-2} * 2^P) N^{-1} \left(\left(\sum_{C_i} n_i^{-1} \sum_{p_x=1}^{s_M} p_x^2 n_{i_{p_x}} \right) - N^{-1} \sum_{p_x=1}^{s_M} p_x^2 N_{p_x} \right)$$

s_M —*maximum number of elements in a grid point*

n_{i_s} —*number of grids in class C with s elements*

N_s —*overall number of grids with s elements*

N —*total number of elements*

G —*total number of ground predicates*

$n_i = card(C_i)$ —*cardinality of class C_i. Under the fine-grid assumption it is the same as the number of ground predicates covered by the class C_i.*

PROOF: Let us consider the following fragment of the eval-function:

$$\left(\sum_{P_j \in \mathcal{P}} p(P_j = true/C_i)^2 \right) = \left(\sum_{P_j \in \mathcal{P}} \frac{card(P_j = true/C_i)^2}{card(C_i)^2} \right) =$$

$$= card(C_i)^{-2} \left(\sum_{P_j \in \mathcal{P}} card(P_j = true/C_i)^2 \right)$$

As the granulation by the set of primary predicates is not fine enough to distinguish any two samples, the $card(P_j = true/C_i)$ for any $P_j \in U$ takes either the value 0 (if there exists no instance in C_i, for which P_j holds) or 1, or 2 or $\ldots$ or $s_M < N$ (if the class C_i contains 1 or more instances resp. matching the predicate P_j). Any other predicate P_j is a disjunct of some atomic predicates $P_{a_{1_j}} \vee P_{a_{2_j}} \vee \ldots \vee P_{a_{s_{j_j}}}$ and $card(P_j = true/C_i) = card(P_{a_{2_j}} = true/C_i) + card(P_{a_{2_j}} = true/C_i) + \cdots + card(P_{a_{s_{j_j}}} = true/C_i)$. So let us consider the class of predicates P_j such that k_1 atomic predicates are matched by exactly 1, k_2— by 2 instances , $\cdots, k_{s_M}$— by s_M predicates, k_0—by no instance, $0 \leq k_i \leq N$, and let $l_1, \ldots, l_h$ be those indices out of $1, \ldots, s_m$ for which $k_{l_g} \geq 1, 1 \leq g \leq h$. Let us count how many of them exist: they are disjuncts of k_{l_1} atomic predicates out of the n_{l_1} atomic predicates which are true for l_1 instances each in class $C_i, \ldots$, of k_{l_h} atomic predicates out of the n_{l_h} atomic predicates which are true for a l_h instances each in class C_i and a number of predicates out of the $G - n_{l_1} - n_{l_2} - \cdots - n_{l_h}$ atomic predicates false for every instance in C_i. So their number is:

$$\prod_{p=0}^{h} \binom{n_{l_p}}{k_{l_p}} \sum_{k=0}^{G-n_{l_1}-\cdots-n_{l_h}} \binom{G - n_{l_1} - \cdots - n_{l_h}}{k} =$$

$$2^{G-\sum_{p=0}^{h} n_{l_p}} \prod_{p=0}^{h} \binom{n_{l_p}}{k_{l_p}} = 2^{G-n_{l_1}-n_{l_2}-\cdots-n_{l_h}} \prod_{p=0}^{h} \binom{n_{l_p}}{k_{l_p}}$$

Hence :

$$card(C_i)^{-2} \left(\sum_{P_j \in \mathcal{P}} card(P_j = true/C_j)^2 \right) =$$

$$= n_i^{-2} \sum_{k_1,k_2,\ldots,k_{s_M} 0 \leq k_p \leq n_{i_p}} \left(\sum_{p=1}^{s_M} k_p * p \right)^2 \prod_{p=0}^{s} \binom{n_{i_p}}{k_p} =$$

$$=$$
$$n_i^{-2} \sum_{k_1,k_2,\ldots,k_{s_M} 0 \leq k_p \leq n_{i_p}} \left(\sum_{p_x=1}^{s_M} k_{p_x}^2 * p_x^2 + \sum_{0 \leq p_1 < p_2 \leq s_M} 2k_{p_1}k_{p_2}p_1 p_2 \right) \prod_{p=0}^{s} \binom{n_{i_p}}{k_p}$$

$$= n_i^{-2} \sum_{k_1,k_2,\ldots,k_{s_M} 0 \leq k_p \leq n_{i_p}} \left(\sum_{p_x=1}^{s_M} k_{p_x}^2 * p_x^2 \prod_{p=0}^{s} \binom{n_{i_p}}{k_p} \right.$$
$$\left. + \sum_{0 \leq p_1 < p_2 \leq s_M} 2k_{p_1}k_{p_2}p_1 p_2 \prod_{p=0}^{s} \binom{n_{i_p}}{k_p} \right) =$$

$$= n_i^{-2} \sum_{k_1,k_2,\ldots,k_{s_M} 0 \leq k_p \leq n_{i_p}} \left(\sum_{p_x=1}^{s_M} k_{p_x}^2 * p_x^2 \prod_{p=0}^{s} \binom{n_{i_p}}{k_p} \right)$$

$$+n_i^{-2} \sum_{k_1, k_2, \ldots, k_{*M}, 0 \leq k_p \leq n_{i_p}} \left(\sum_{0 \leq p_1 < p_2 \leq *M} 2k_{p_1} k_{p_2} p_1 p_2 \prod_{p=0}' \binom{n_{i_p}}{k_p} \right) =$$

$$= n_i^{-2} \left(\sum_{p_x=1}^{*M} \sum_{k_{p_x}=1}^{n_{p_x}} k_{p_x}^2 * p_x^2 \binom{n_{i_p}}{k_p} \prod_{p=0 p \neq p_x}' 2^{n_{i_p}} \right)$$

$$+n_i^{-2} \left(\sum_{0 \leq p_1 < p_2 \leq *M} \sum_{k_{p_1}=0}^{n_{p_1}} \sum_{k_{p_2}=0}^{n_{p_2}} 2k_{p_1} k_{p_2} p_1 p_2 \binom{n_{i_{p_1}}}{k_{p_1}} \binom{n_{i_{p_2}}}{k_{p_2}} \prod_{p=0 \ p \neq p_1 \ p \neq p_2}' 2^{n_{i_p}} \right)$$

$$= n_i^{-2} \left(\sum_{p_x=1}^{*M} p_x^2 \prod_{p=0 p \neq p_x}' 2^{n_{i_p}} \sum_{k_{p_x}=0}^{n_{p_x}} k_{p_x}^2 \binom{n_{i_p}}{k_p} \right)$$

$$+n_i^{-2} \left(\sum_{0 \leq p_1 < p_2 \leq *M} \left(\prod_{p=0 \ p \neq p_1 \ p \neq p_2}' 2^{n_{i_p}} \right) 2n_{p_1} n_{p_2} p_1 p_2 \sum_{k_{p_1}=0}^{n_{p_1}} \sum_{k_{p_2}=0}^{n_{p_2}} \binom{n_{i_{p_1}}^{-1}}{k_{p_1}} \binom{n_{i_{p_2}}^{-1}}{k_{p_2}} \right) =$$

$$= n_i^{-2} \left(\sum_{p_x=1}^{*M} p_x^2 \left(\prod_{p=0 p \neq p_x}' 2^{n_{i_p}} \right) n_{p_x} (n_{p_x} + 1) 2^{n_{p_x}-2} \right)$$

$$+n_i^{-2} \left(\sum_{0 \leq p_1 < p_2 \leq *M} \left(\prod_{p=0 \ p \neq p_1 \ p \neq p_2}' 2^{n_{i_p}} \right) 2n_{p_1} n_{p_2} p_1 p_2 2^{n_{p_1}-1} 2^{n_{p_2}-1} \right) =$$

$$= n_i^{-2} \left(\sum_{p_x=1}^{*M} p_x^2 \left(2^{-2} * \prod_{p=0}' 2^{n_{i_p}} \right) n_{p_x} (n_{p_x} + 1) \right)$$

$$+n_i^{-2} \left(\sum_{0 \leq p_1 < p_2 \leq *M} \left(2^{-2} \prod_{p=0}' 2^{n_{i_p}} \right) 2n_{p_1} n_{p_2} p_1 p_2 \right) =$$

$$= n_i^{-2} \left(2^{-2} * \prod_{p=0}' 2^{n_{i_p}} \right) \left(\sum_{p_x=1}^{*M} p_x^2 n_{p_x} (n_{p_x} + 1) \right)$$

$$+n_i^{-2} \left(2^{-2} \prod_{p=0}' 2^{n_{i_p}} \right) \left(\sum_{0 \leq p_1 < p_2 \leq *M} 2n_{p_1} n_{p_2} p_1 p_2 \right) =$$

$$n_i^{-2} \left(2^{-2} * \prod_{p=0}' 2^{n_{i_p}} \right) \left(\left(\sum_{p_x=1}^{*M} p_x^2 n_{p_x} \right) + \left(\sum_{p_x=1}^{*M} p_x^2 n_{p_x} \right)^2 + \left(\sum_{0 \leq p_1 < p_2 \leq *M} 2n_{p_1} n_{p_2} p_1 p_2 \right) \right) =$$

$$= n_i^{-2} \left(2^{-2} * \prod_{p=0}' 2^{n_{i_p}} \right) \left(\left(\sum_{p_x=1}^{*M} p_x^2 n_{p_x} \right) + \left(\sum_{p_x=1}^{*M} p_x n_{p_x} \right)^2 \right) =$$

$$= n_i^{-2} \left(2^{-2} * 2^{card(\mathcal{P})} \right) \left(\left(\sum_{p_x=1}^{*M} p_x^2 n_{p_x} \right) + n_i^2 \right)$$

Hence the eval-function takes the form:

$$M^{-1} \left(\sum_{C_i} (n_i/N) * n_i^{-2} \left(2^{-2} * 2^{card(\mathcal{P})} \right) \left(\left(\sum_{p_x=1}^{*M} p_x^2 n_{i_{p_x}} \right) + n_i^2 \right) \right) -$$

$$-N^{-2} \left(2^{-2} * 2^{card(\mathcal{P})} \right) \left(\left(\sum_{p_x=1}^{*M} p_x^2 N_{p_x} \right) + N^2 \right) =$$

$$= \overline{M}^{-1} *$$
$$2^{card(\mathcal{P}-2)} N^{-1} \left(\left(\sum_{C_i} n_i^{-1} \left(\left(\sum_{p_x=1}^{*M} p_x^2 n_{i_{p_x}} \right) + n_i^2 \right) \right) - N^{-1} \left(\left(\sum_{.p_x=1}^{*M} p_x^2 N_{p_x} \right) + N^2 \right) \right) =$$

$$= \overline{M}^{-1} *$$
$$2^{card(\mathcal{P}-2)} N^{-1} \left(\left(\left(\sum_{C_i} n_i^{-1} \sum_{p_x=1}^{*M} p_x^2 n_{i_{p_x}} \right) + \sum_{C_i} n_i \right) - \left(\left(N^{-1} \sum_{p_x=1}^{*M} p_x^2 N_{p_x} \right) + N \right) \right) =$$

$$= \overline{M}^{-1} *$$
$$2^{card(\mathcal{P}-2)} N^{-1} \left(\left(\left(\sum_{C_i} n_i^{-1} \sum_{p_x=1}^{*M} p_x^2 n_{i_{p_x}} \right) + N - N^{-1} \sum_{p_x=1}^{*M} p_{ix}^2 N_{p_x} \right) - N \right) =$$

$$= \overline{M}^{-1} * 2^{card(\mathcal{P}-2)} N^{-1} \left(\sum_{C_i} n_i^{-1} \sum_{p_x=1}^{*M} p_x^2 n_{i_{p_x}} \right) - N^{-1} \sum_{p_x=1}^{*M} p_x^2 N_{p_x}$$

□

References

[1] Hartigan J.A., *Clustering Algorithms*, John WilleySons, New York 1975.

[2] Anderberg M.R., *Cluster Analysis for Applications*, Academic Press, New York, 1973

[3] Watanabe S., *Pattern Recognition*, Human and Machine, (1987)

[4] Michalski R.S., Stepp R.E., *Automated construction of classification, conceptual clustering versus numerical taxonomy*, IEEE Transactions on PAMI-5, 1983, pp. 396-410

[5] Michalski R.S., Stepp R.E., *III. Learning from observations: conceptual clustering*, in Machine Learning: An Artificial Intelligence Approach, Michalski R.S., Carbonell J.G., Mitchell T.M. (Eds), Morgan Kaufmann Publishers Inc., Los Altos, 1983

[6] Stepp R.E., Michalski R.S., *Conceptual clustering:Conceptual cluster- ing of structured objects*, Artificial Intelligence 28,1, 1986, pp. 43-70

[7] Stepp R.E., Michalski R.S., *Conceptual clustering: Inventing goal-oriented classifications of structured objects*, in Machine Learning II: An Artificial Intelligence Approach, Michalski R.S., Carbonell J.G.,Mitchell T.M.(Eds),Morgan Kaufmann Publ.Inc., Los Altos, 1986

[8] Dale B., *On the comparison of conceptual clusterimng and numerical taxonomy*, IEEE Transactions on PAMI-7, March 1985

[9] Gluck M., Corter J., *Information, uncertainty and the utility of categories*, Proc.7Ann.Conf.of the Cognitive Sci.,Ivrine CA (1985), 283-7

[10] Kolodner J.L., *Retrieval and Organizational strategies in conceptual memory:A computer model*, Lawrence Erlbaum Associates Publ,London 1984

[11] Fisher D., Langley P., *Approaches to conceptual clustering*, Proc. 9 IJCAI, Los Angeles, 1985, pp. 691-697

[12] Fisher D., *Knowledge acquisition via incremental conceptual clustering*, Machine Learning 2,2,1987, pp. 139-172

[13] Lebowitz M., *Experiments with incremental concept formation: UNIMEM*, Machine Learning 2,2, 1987, pp. 103-138

[14] Fisher D.H., *Conceptual clustering. learning from examples and inference*, Proc. 4 Intenationbal Workshop on Machine Learning, Irvine, Morgan Kaufaman, 1987, pp. 38-49

[15] Hadzikadic M., Yun D.Y.Y., *Concept formation by goal-driven context-dependent classification*, Proc.3 International Symposium on Methodologies for Intelligent Systems, Toronto, Italy, 1988, pp. 322-332

[16] Kodratoff Y., Tecusi G., *Learning based on conceptual distance*, IEEE trans. on PAMI-10,6 (1988), pp. 897-909.

[17] Hadzikadic M., Yun D.Y.Y., *Concept formation by incremental conceptual clustering*, Proc. IJCAI'89 Vol. 2, pp. 831-836

[18] Fisher D.H., *Noice-tolerant conceptual clustering*, Proc. IJCAI'89 Vol. 2, pp. 825-830

[19] Gennari J.H., Langley P., Fisher D., *Models of incremental concept formation*, Artificial Intelligence 40 (1989) 11-61

[20] S.T. Wierzchon, A. Pacan, M.A. Klopotek, *An Object-oriented Representation Framework For Hierarchical Evidential Reasoning*, in: Ph. Jorrand, V. Sgurev: "Artificial Intelligence IV: Methods, Systems And Applications", Proc. Conf.AIMSA'90, Albena, Bulgaria, 19-22 Sept.1990 r, North Holland, Amsterdam, New York, Oxford, Tokyo, 1990, pp. 239-248

Possibilistic Logic as a Logical Framework for Min-max Discrete Optimisation Problems and Prioritized Constraints

Jérôme LANG

Institut de Recherche en Informatique de Toulouse
Université Paul Sabatier – 31062 Toulouse Cedex – France
Tél.: (+33) 61.55.66.11 ext. 73.67 – E-mail: lang@irit.fr

Abstract

Possibilistic logic is basically a logic of uncertainty, but a significant fragment of it can also be seen as a logic for the representation of constraints with priorities. The gradation of inconsistency enables the definition of the "best" model(s) of a "partially inconsistent" set of possibilistic formulas. Many formal results have been proved for this fragment of possibilistic logic, including its axiomatisation. Besides, there are some well-adapted automated deduction procedures. Min-max discrete optimisation problems, and more generally problems with prioritized constraints, can be translated in this logical framework, and then solved by its automated deduction procedures.

1. Introduction

Possibilistic logic is basically a logic of uncertainty. Issued from Zadeh's possibility theory [22], it was initially developed by Dubois and Prade from 1987 [3]. It handles possibility and necessity measures on a logical (propositional or first-order) language, and has axiomatical and semantical bases; the main feature of possibilistic semantics is its tolerance to "partial" inconsistencies, which confers to it sort of a paraconsistent behaviour. Necessity degrees are usually seen as certainty degrees, i.e. measuring the certainty for a formula to be true, but they may also be seen as *priority degrees*, measuring to what extent the constraint represented by a given formula is necessary to be realized. In this paper we shall mostly focus on this interpretation of necessity measures; we shall give some formal results for the fragment of possibilistic logic involving only necessity measures, including its axiomatisation. Automated deduction procedures (already developed in other papers) are briefly recalled. Lastly we shall give some applications of possibilistic logic to the logical modeling and solving of min-max discrete optimization problems, and more generally to finding the best solution(s) of a class of problems with soft and prioritized constraints.

2. Possibilistic logic: introduction and formal aspects

2.1. Possibilistic logic: what for ?

Let $\mathcal{L}$ be a propositional or first-order logical language, consisting on well-formed formulae in the classical sense. Let Ω be the set of interpretations for this language. A *possibility distribution* on Ω is merely a function π from Ω to $[0,1]$; π is said to be *normalized* if and only if $\exists \omega \in \Omega$ such that $\pi(\omega) = 1$; the quantity $SN(\pi) = 1 - \text{Sup} \{\pi(\omega) \mid \omega \in \Omega\}$ is called *sub-normalization degree* of π, and is equal to 0 iff π is normalized.

A possibility distribution π on Ω induces two functions on the set $\mathcal{L}'$ of closed formulae of $\mathcal{L}$ to $[0,1]$, called *possibility and necessity measures* and denoted respectively by Π and N, defined by:

$$\Pi : \mathcal{L}' \to [0,1] \quad (\forall \varphi \in \mathcal{L}') \, \Pi(\varphi) = \text{Sup} \{\pi(\omega) \mid \omega \vDash \varphi\}$$

$$N : \mathcal{L}' \to [0,1] \quad (\forall \varphi \in \mathcal{L}') \, N(\varphi) = \text{Inf} \{1 - \pi(\omega) \mid \omega \vDash \neg\varphi\} = 1 - \Pi(\neg\varphi)$$

From an uncertainty-modeling point of view, $\Pi(\varphi)$ represents what its name suggests: the *possibility* of φ, measuring the extent to which φ is *compatible* with the knowledge of reference describing what we know about the real world; and $N(\varphi)$ represents the *certainty* of φ, measuring to what extent φ is *entailed* by the knowledge of reference. See for example [7].

We point out that possibilistic logic is not *fuzzy logic*, since fuzzy logic assigns *degrees of truth* taking their values in $[0,1]$ to *vague* (non-classical) formulae, measuring their *conformity* with the knowledge of reference, which is generally supposed to be complete. A statement like "it is 0.7-true that John is tall", which may be translated in fuzzy logic, by v (Tall (John)) = 0.7, expresses that the conformity of John's height with the interpretation of the vague predicate "tall" is 0.7, i.e. that (knowing John's height) it is rather true that John is tall. Whereas the statement "it is 0.7-certain that the contract will be signed" may be translated in possibilistic logic, by N (Signed (contract)) = 0.7, expressing that the *non-vague* event "the contract will be signed" is *rather certain*, and not *rather true*. See [5]. From a formal point of view, possibilistic logic is closer to *probabilistic logic* (see for instance [18]) than fuzzy logic: indeed probabilistic logic also assigns uncertainty degrees (probabilities) on a set of closed well-formed classical formulae. In the propositional case (which is the only one treated by Nilsson), the definition of a probability density on Ω (respectively, probability measure on $\mathcal{L}$) is very similar to the definition of a possibility distribution (respectively, possibility measure). Besides probabilistic and possibilistic logic may be unified in more general frameworks.

From a constraint-modeling point of view, Π (φ) represents the degree to which φ is allowed to be satisfied, and N (φ) the degree to which φ is necessary to be satisfied. This is sort of the numerical counterpart of a *deontic* logic; besides we emphasize that the analogy between the possibilistic terminology ("possibility" and "necessity") and the modal terminology is not fortuite; indeed some links between possibilistic logic and multi-modal logics have been established in [10]. Having in mind this interpretation of possibility and necessity measures, the relation N (φ) = 1 - Π ($\neg\varphi$) expresses that a formula is all the more necessary that its negation is forbidden. In the following we shall only deal with necessity degrees. We may also (in a very similar way) interpret the necessity degree of φ in terms of utility (of φ) or cost (of $\neg\varphi$): if N (φ) = α then the violation of the constraint expressed by φ, i.e. the satisfaction of $\neg\varphi$, costs α. We point out that the use of the terminology "cost" could be ambiguous since these costs are clearly non-additive. We give some properties of necessity measures with their interpretation:

(1) if φ is a tautology then N (φ) = 1, i.e. tautologies must necessary be satisfied;

(2) however, if φ is a contradiction, we generally do not have N (φ) = 0; this means that contradictions may sometimes be somewhat required to be satisfied! This surprising property is due to the fact that we did not require a possibility distribution to be normalised. If the normalisation condition is required then we always have N ($\perp$) = 0 where $\perp$ denotes the contradiction (this assumption is generally made). This choice is motivated by the need to model partial inconsistencies, as already mentioned in the introduction.

(3) $\forall\varphi \; \forall\psi$, N ($\varphi\wedge\psi$) = min [N ($\varphi$), N ($\psi$)], expressing that the simultaneous satisfaction of two formulas has the same importance that the satisfaction of the most important formula. However necessity measures are not compositional for disjunction, and we have only

(4) N ($\varphi\vee\psi$) $\geq$ max [N (φ), N (ψ)], hence the cost of the simultaneous violation of φ and ψ (indeed, N ($\varphi\vee\psi$) represents the cost of $\neg$ ($\varphi\vee\psi$), i.e. of $\neg\varphi\wedge\neg\psi$) cannot generally be computed from the cost of the violation of φ and the cost of the violation of ψ. Two extremal cases are:

(i) if φ and ψ are logically equivalent ($\varphi \equiv \psi$) then N ($\varphi\vee\psi$) = N (φ) = N (ψ)

(ii) if φ and ψ are opposite ($\varphi \equiv \neg\psi$) then N ($\varphi\vee\psi$) = 1 and however we may even have N (φ) = N ($\neg\varphi$) = 0, expressing that neither φ or $\neg\varphi$ is somewhat inviolable, or equivalently, that no constraint acts neither on φ or on $\neg\varphi$: we are in a situation of *complete indifference* about the satisfaction or the non-satisfaction of φ.

The use on *min* and *max* operators entails that the precise values of the necessity (or possibility) degrees is not so important, the essential being the ordering on the formulae induced by them: thus, necessity degrees may be seen as *priority degrees*, where N (φ) > N (ψ) expresses that the satisfaction of φ is more important than the satisfaction of ψ. This notion of ordering upon formulae is closed to Gärdenfors and Makinson's epistemic entrenchment [12]; the bridge has been established in [9]; the links with Shoham's preferential models is currently being studied. Lastly, we mention that possibilistic logic is compatible with the general model of Ginsberg [13] for bilattice-based logics.

2.2. Possibilistic logic: language, semantics, axiomatics

Let us define a *necessity-valued formula* as a pair $(\varphi \; \alpha)$, where φ is a classical propositional or first-order closed formula of $\mathcal{L}$, and α a valuation of $[0,1]$. A *necessity-valued knowledge base* is then defined as a finite set (in the conjunctive meaning) of necessity-valued formulae; these are the basic elements of the language of *necessity-valued logic*, which is a fragment of (general) possibilistic logic, which involves also possibility-valued formulae (general possibilistic logic is dealt with in [4], [5], [17]).

The necessity-valued formula $(\varphi \; \alpha)$ expresses that $N(\varphi) \geq \alpha$, i.e. that the satisfaction of φ is at least α-necessary. In particular, $(\varphi \; 1)$ expresses that φ must absolutely be satisfied and $(\varphi \; 0)$, which is a useless formula since it expresses only that $N(\varphi) \geq 0$ (which is always verified), expresses that we know nothing about φ (and, by default, satisfying φ will be considered as no necessary at all).

Let us consider now the semantical aspects of necessity-valued logic. Necessity-valued formulae will be interpreted by possibility distributions. Let π be a possibility distribution on Ω (π is not necessarily normalized), and $(\varphi \; \alpha)$ a necessity-valued formula. Then we define the notion of satisfaction by :

$$\pi \models (\varphi \; \alpha) \quad \text{iff} \quad N(\varphi) \geq \alpha$$

where N is the necessity measure induced by π.

If $\mathcal{F} = \{(\varphi_1 \; \alpha_1), ..., (\varphi_n \; \alpha_n)\}$ is a set of necessity-valued formulae then

$$\pi \models \mathcal{F} \quad \text{iff} \quad \forall i \in \{1, ..., n\}, \pi \models (\varphi_i \; \alpha_i)$$

Then, the notion of logical consequence is defined in a very natural way: $\mathcal{F}$ being a set of necessity-valued formulae and $(\varphi \; \alpha)$ a necessity-valued formula,

$$\mathcal{F} \models (\varphi \; \alpha) \quad \text{iff} \quad \forall \pi, \pi \models \mathcal{F} \text{ entails } \pi \models (\varphi \; \alpha)$$

i.e. the set of possibility distributions satisfying $\mathcal{F}$ is included in the set of possibility distributions satisfying $(\varphi \; \alpha)$.

Thus, the models of a set of necessity-valued formulae $\mathcal{F}$ are possibility distributions on the set Ω of all interpretations for $\mathcal{L}$. Measuring the consistency of $\mathcal{F}$ consists then in evaluating to what degree there is at least an interpretation completely possible for $\mathcal{F}$, i.e. to what degree the set of possibility distributions satisfying $\mathcal{F}$ contains normalized possibility distributions; the quantity $\text{Cons}(\mathcal{F}) = \text{Sup}_{\pi \models \mathcal{F}} \text{Sup}_{\omega \in \Omega} \pi(\omega)$ will be called *consistency degree* of $\mathcal{F}$, and its complement to 1,

$$\text{Incons}(\mathcal{F}) = 1 - \text{Sup}_{\pi \models \mathcal{F}} \text{Sup}_{\omega \in \Omega} \pi(\omega)$$
$$= \text{Inf}_{\pi \models \mathcal{F}} (1 - \text{Sup}_{\omega \in \Omega} \pi(\omega))$$
$$= \text{Inf}_{\pi \models \mathcal{F}} SN(\pi)$$

is called the *inconsistency degree* of $\mathcal{F}$. Necessity-valued logic extends classical logic in the following sense: let $F = \{\varphi_i, i = 1, ..., n\}$ be a set of classical formulae and let us associate to $\mathcal{F}$

the set of completely necessary necessity-valued formulae $\mathcal{F} = \{(\varphi_i \ 1), i = 1, ..., n\}$; then, it can be proved (easily) that if F is consistent then Incons $(\mathcal{F}) = 0$ and if F is inconsistent then Incons$(\mathcal{F}) = 1$. See []. Thus, necessity-valued logic enables the gradation of inconsistency: if Incons $(\mathcal{F}) = 0$ then $\mathcal{F}$ will be said *completely consistent*, if Incons $(\mathcal{F}) = 1$ then $\mathcal{F}$ will be said *completely inconsistent*, and if $0 <$ Incons $(\mathcal{F}) < 1$ then $\mathcal{F}$ will be said *partially inconsistent*. It comes down easily that

<u>Proposition 1</u> : Incons $(\mathcal{F}) = $ Inf $\{ N (\perp) \mid \pi \vDash \mathcal{F} \}$

where N is the necessity distribution induced by π.

<u>Proof</u> : $N (\perp) = $ Inf $\{1-\pi (\omega) \mid \omega \nvDash \perp\} = $ Inf $_{\omega \in \Omega} (1-\pi (\omega)) = SN (\pi)$, hence the result. $\square$

This equality achieves to justify the terminology "inconsistency degree" since Incons $(\mathcal{F})$ is the smallest necessity degree of the contradiction $\perp$ for all possibility distributions satisfying $\mathcal{F}$. Now we establish this important results:

<u>Proposition 2</u> : let $\mathcal{F} = \{(\varphi_1 \ \alpha_1), ..., (\varphi_n \ \alpha_n)\}$ be a set of necessity-valued formulae and let us define the possibility distribution $\pi^*_{\mathcal{F}}$ by

$\pi^*_{\mathcal{F}} (\omega) = $ Inf $\{ 1 - \alpha_i \mid \omega \vDash \neg\varphi_i, i = 1, ..., n\}$

then for any possibility distribution π on Ω, π satisfies $\mathcal{F}$ if and only if $\pi \leq \pi^*_{\mathcal{F}}$.

<u>Proof</u> : π satisfies $\mathcal{F}$ iff $(\forall i = 1,..., n) \ \pi \vDash (\varphi_i \ \alpha_i)$

         iff $(\forall i = 1,..., n) \ N (\varphi_i) \geq \alpha_i$ (N being the necessity measure induced by π)

         iff $(\forall i = 1,..., n) \ $ Inf $\{1-\pi (\omega) \mid \omega \vDash \neg\varphi_i\} \geq \alpha_i$

         iff $(\forall i = 1,..., n) \ (\forall \omega \vDash \neg\varphi_i) \ \pi (\omega) \leq 1 - \alpha_i$

         iff $\forall \omega \in \Omega, \pi (\omega) \leq $ Inf $\{ 1 - \alpha_i \mid \omega \vDash \neg\varphi_i, i = 1, ..., n\}$

         iff $\forall \omega \in \Omega, \pi (\omega) \leq \pi^*_{\mathcal{F}} (\omega)$. $\square$

<u>Proposition 3</u> : $\mathcal{F} \vDash (\varphi \ \alpha)$ iff $\pi^*_{\mathcal{F}} \vDash (\varphi \ \alpha)$

<u>Proof</u> : Due to the definition of the necessity measure induced by a possibility distribution, we have $\pi \leq \pi' \Leftrightarrow N \geq N'$ (where N and N' are respectively induced by π and π'). Then $\mathcal{F} \vDash (\varphi \ \alpha)$ *iff* $\forall \pi, \pi \vDash \mathcal{F}$ entails $N (\varphi) \geq \alpha$ *iff* $\forall \pi, \pi \leq \pi^*_{\mathcal{F}}$ entails $N (\varphi) \geq \alpha$ *iff* $\forall \pi, N \geq N^*_{\mathcal{F}}$ entails $N (\varphi) \geq \alpha$ *iff* $N^*_{\mathcal{F}} (\varphi) \geq \alpha$ *iff* $\pi^*_{\mathcal{F}}$ satisfies $\vDash (\varphi \ \alpha)$. $\square$

<u>Corollary 4</u> : Incons $(\mathcal{F}) = 1 - $ Sup $_{\omega \in \Omega} \pi^*_{\mathcal{F}} (\omega) = SN (\pi^*_{\mathcal{F}})$.

The proof is straightforward.

Then, computing the inconsistency degree of $\mathcal{F}$ reduces to compute the degree of subnormalisation of the possibility distribution $\pi^*_{\mathcal{F}}$. The quantity $\pi^*_{\mathcal{F}} (\omega)$ represents the compatibility degree of ω with $\mathcal{F}$. The following proposition leads to an important definition:

<u>Proposition 5</u> : the least upper bound in the computation of Incons $(\mathcal{F})$ is reached, i.e. there exist (at least) an interpretation ω^* such that $\pi^*(\omega^*) = $ Sup $_{\omega \in \Omega} \pi^*_{\mathcal{F}} (\omega)$.

Proof : (in the propositional case, this result is trivial, since Ω is finite). Since there are a finite number of necessity-valued formulae (and hence a finite number of valuations α_i), the definition of $\pi*_{\mathcal{F}}$ implies that $\pi*_{\mathcal{F}}(\omega)$ takes only a finite number of values when ω ranges along the (infinite in the first-order case) set of interpretations Ω. Hence the result. $\square$

Then, the interpretations $\omega*$ maximizing $\pi*$ will be called the *best models* of $\mathcal{F}$. They are the most compatible with $\mathcal{F}$ among the set of all interpretations Ω.

A lot of results can be proved about deduction in necessity-valued logic. They can be found in [17]. The most important one extends the deduction theorem to necessity-valued logic:

<u>Theorem 6</u> (deduction theorem):
$\mathcal{F} \cup \{(\varphi\ 1)\} \models (\psi\ \alpha)$ iff $\mathcal{F} \models (\varphi{\to}\psi\ \alpha)$

<u>Proof</u> :

($\Rightarrow$) $\mathcal{F} \cup \{(\varphi\ 1)\} \models (\psi\ \alpha)$

$\Rightarrow$ $N*_{\mathcal{F} \cup \{(\varphi\ 1)\}}(\psi) \geq \alpha$ (proposition 3)

$\Rightarrow$ $\text{Inf } \{1{-}\pi*_{\mathcal{F} \cup \{(\varphi\ 1)\}}(\omega),\ \omega \models \neg\psi\} \geq \alpha$

$\Rightarrow \forall\omega \models \varphi{\wedge}\neg\psi,\ \pi*_{\mathcal{F}}(\omega) \leq 1{-}\alpha$, since $\pi*_{\mathcal{F} \cup \{(\varphi\ 1)\}}(\omega) = \pi*_{\mathcal{F}}(\omega)$ if $\omega \models \varphi$

$\Rightarrow$ $N*_{\mathcal{F}}(\varphi{\to}\psi) \geq \alpha$

$\Rightarrow \mathcal{F} \models (\varphi{\to}\psi\ \alpha)$ (again proposition 3).

($\Leftarrow$) $\mathcal{F} \models (\varphi{\to}\psi\ \alpha)$

$\Rightarrow$ $\forall\pi \models \mathcal{F},\ N\ (\varphi{\to}\psi) \geq \alpha$

$\Rightarrow$ $\forall\pi \models \mathcal{F},\ N\ (\varphi) = 1$ implies $N\ (\psi) \geq \alpha$ since $N\ (\psi) \geq \min\ [N(\varphi), N(\varphi{\to}\psi)]$

$\Rightarrow$ $\forall\pi \models \mathcal{F} \cup \{(\varphi\ 1)\},\ N\ (\psi) \geq \alpha$

$\Rightarrow$ $\mathcal{F} \cup \{(\varphi\ 1)\} \models (\psi\ \alpha). \square$

<u>Corollary 7</u> (refutation theorem):
$\mathcal{F} \models (\varphi\ \alpha)$ iff $\mathcal{F} \cup \{(\neg\varphi\ 1)\} \models (\perp\ \alpha)$

<u>Proof</u> : let us apply theorem 6, replacing φ by $\neg\varphi$ and ψ by $\perp$:
$\mathcal{F}, (\neg\varphi\ 1) \models (\perp\ \alpha)$ iff $\mathcal{F} \models (\neg\varphi{\to}\perp\ \alpha)$, i.e. $\mathcal{F} \cup \{(\neg\varphi\ 1)\} \models (\perp\ \alpha)$ iff $\mathcal{F} \models (\varphi\ \alpha). \square$

Thus, if we want to know whether $(\varphi\ \alpha)$ is a logical consequence of $\mathcal{F}$ or not, it is sufficient to compute the inconsistency degree of $\mathcal{F} \cup \{(\neg\varphi\ 1)\}$, which is equal to the largest α such that $\mathcal{F} \models (\varphi\ \alpha)$.

The following result establishes that necessity-valued logic is axiomatisable.

<u>Theorem 8</u> : the following formal system is correct and complete with respect to necessity-valued logic, i.e. for any set $\mathcal{F}$ of necessity-valued formulae and for any necessity-valued formula $(\varphi\ \alpha)$ then $\mathcal{F} \vdash (\varphi\ \alpha)$ by this formal system iff $\mathcal{F} \models (\varphi\ \alpha)$.

Axioms schemata : those of classical logic, valued by 1 :

    (A1) $(\varphi \to (\psi \to \varphi)\ 1)$

    (A2) $((\varphi \to (\psi \to \xi)) \to ((\varphi \to \psi) \to (\varphi \to \xi))\ 1)$

    (A3) $((\neg\varphi \to \neg\psi) \to ((\neg\varphi \to \psi) \to \varphi)\ 1)$

    (A4) $((\forall x\ (\varphi \to \psi)) \to (\varphi \to (\forall x\ \psi)\ 1)$

               if x does not appear in φ and is not bound in ψ

    (A5) $((\forall x\ \varphi) \to \varphi_{x|t}\ 1)$ if x is free for t in φ

Inference rules :

    (GMP) $(\varphi\ \alpha)\ ,(\varphi \to \psi\ \beta) \vdash (\psi\ \min(\alpha, \beta))$

    (G) $(\varphi\ \alpha) \vdash ((\forall x\ \varphi)\ \alpha)$ if x is not bound in φ

    (S) $(\varphi\ \alpha) \vdash (\varphi\ \beta)$ if $\beta \le \alpha$

Proof : the complete proof is omitted for the sake of brievity; it can be found in [17]. We just give a sketch of the proof. It uses the following intermediary result:

$\mathcal{F} \vDash (\psi\ \alpha)$ iff $\mathcal{F}_\alpha^* \vDash \psi$, where $\mathcal{F}_\alpha^* = \{\varphi_i \mid (\varphi_i\ \alpha_i) \in \mathcal{F}, \alpha_i \ge \alpha\}$.

Since the formal system formed by the non-valued part of the axioms schemata and of the inference rules (except (S) whose non-valued part is trivial), there exist a proof of ψ from $\mathcal{F}_\alpha^*$ by this classical formal system. Then, considering again the valuations, the proof obtained by the previous one is a proof of $(\psi\ \gamma)$ from $\mathcal{F}_\alpha$ by the given formal system, with $\gamma \ge \alpha$. Lastly, using (S) we obtain a proof of $(\psi\ \alpha)$. The inference rule (GMP) is the *graded modus ponens*; it was proposed by Froidevaux and Grossetête for graded default theories in [11].

3.   Possibilistic logic: automated deduction

As we have pointed out in the previous section, everything reduces to a computation of the inconsistency degree, which constitutes the key problem of this section. After having briefly introduced clausal forms in necessity-valued logic, we shall present two automated deduction methods: first, *resolution* (for first-order necessity-valued logic) and then, *semantic evaluation* (working only for propositional necessity-valued logic). These automated deduction procedures are not the subject of this paper and have been more completely described in other papers: [5,17] for clausal form, [3,4,8,17] for resolution, [16,17] for semantic evaluation. The last sub-section deals with *generalized clauses* which are useful for representing some kinds of discrete optimisation problems.

3.1. Clausal form and resolution

A *necessity-valued* clause is a necessity-valued formula $(c\ \alpha)$ where c is a first-order clause. A *necessity-valued clausal form* is a finite set of necessity-valued clauses. If $(\varphi\ \alpha)$ is a necessity-valued formula and $\{c_1, ..., c_n\}$ a clausal form of φ then a clausal form of $(\varphi\ \alpha)$ is

$\{(c_1\ \alpha), ..., (c_n\ \alpha)\}$; if $\mathcal{F}$ is a set of necessity-valeud formulas then the set of necessity-valued clauses $\tilde{C}$ obtained by replacing each necessity-valued formula by one of its clausal forms, is a clausal form of $\mathcal{F}$, and is proved to have the same inconsistency degree as $\mathcal{F}$ (see [17]). The resolution rule for necessity-valued possibilistic logic is: $(c_1\ \alpha_1), (c_2\ \alpha_2) \vdash (c'\ \min(\alpha_1, \alpha_2))$, where c' is any resolvent of clauses c_1 and c_2. Possibilistic resolution for necessity-valued clauses is proved in [4] to be sound and complete for refutation, i.e. if Incons $(\tilde{C}) = \alpha$ then there is a deduction of $(\perp\ \alpha)$, called an α-refutation, from $\tilde{C}$, and this refutation is optimal, i.e. there is no β-refutation from $\tilde{C}$ where $\beta > \alpha$.

3.2. Semantic evaluation

Whereas resolution computes only the inconsistency degree of a set of necessity-valued clauses $\tilde{C}$, semantic evaluation, based on the Davis and Putnam procedure, computes also the best models (or one of the best models, if preferred) of $\tilde{C}$. It works only in the propositional case. It consists in building a (small) part of a semantic tree for $\tilde{C}$, by evaluating literals successively. Some techniques improving the efficiency of semantic evaluation by transforming it into the search in a min-max tree, and then pruning branches by two ways, one being the well-known alpha-beta pruning method, the other one being a generalisation of the "model partition theorem" [16,17] defined for (classical) propositional logic by Jeannicot, Oxusoff and Rauzy [15].

3.3. Generalized clauses

Generalized clauses were defined in the classical case by Hooker [14]; if λ is a positive number, the generalized clause $\lambda: x_1 \vee ... \vee x_n$ expresses that at least λ literals among $\{x_1, ..., x_n\}$ are satisfied. A necessity-valued generalized clause is then defined as a pair $(\lambda: x_1 \vee ... \vee x_n\ \alpha)$, expressing that it is necessary to the degree that at least λ literals among $\{x_1, ..., x_n\}$ be satisfied. Incorporating generalized clauses into the language does not imply any specific problem at the formal level, since a generalized clause is in fact equivalent to a finite set of (normal) clauses: for example, $(2: x_1 \vee x_2 \vee x_3 \vee x_4\ \alpha)$ is equivalent to $\{(x_1 \vee x_2 \vee x_3\ \alpha), (x_1 \vee x_2 \vee x_4\ \alpha), (x_1 \vee x_3 \vee x_4\ \alpha), (x_2 \vee x_3 \vee x_4\ \alpha)\}$. The purpose of incorporating generalized clauses is to make the language richer in order to represent some problems involving a lot of such constraints. The constraint "it is α-necessary that *at most* λ literals among $\{x_1, ..., x_n\}$ be satisfied" can be represented by the necessity-valued generalized clause

$$(n-\lambda: \neg x_1 \vee ... \vee \neg x_n\ \alpha)$$

and the constraint "it is α-necessary that *at most* λ literals among $\{x_1, ..., x_n\}$ be satisfied" can be represented by the two necessity-valued generalized clauses

$$\{(\lambda: x_1 \vee ... \vee x_n\ \alpha), (n-\lambda: \neg x_1 \vee ... \vee \neg x_n\ \alpha)\}$$

It can be easily seen that a generalized clause with $\lambda = 1$ reduces to a (normal) clause. Hooker shows in [14] that taking into account generalized clauses in the deduction process of propositional classical logic corresponds to a class of 0-1 linear programs.

In possibilistic logic we may have a collection of generalized clauses with the same "clause" part and without redundant information; for example, the set of necessity-valued generalized clauses

$\{(4: p \vee q \vee r \vee s \quad 0.2), (3: p \vee q \vee r \vee s \quad 0.5), (2: p \vee q \vee r \vee s \quad 0.8), (p \vee q \vee r \vee s \quad 1)\}$

expresses that it is somewhat (but weakly) necessary that all literals p, q, r and s be satisfied, rather necessary that at least 3 of them be satisfied, strongly necessary that at least 2 of them be, and absolutely necessary that at least 1 of them be. To avoid heavy notations induced by such collections of necessity-valued generalized clauses we introduce $\tilde{\lambda}$-*clauses*, where $\tilde{\lambda}$ is a fuzzy natural number (a *fuzzy natural number* is a fuzzy set on the set of natural numbers, i.e. it is defined by its membership function from $\mathbb{N}$ to [0,1]); see [7]; hence the $\tilde{\lambda}$-clause representing the previous collection of necessity-valued generalized clauses is

$$((4_{|0.2} ; 3_{|0.5} ; 2_{|0.8} ; 1_{|1}) : p \vee q \vee r \vee s)$$

Taking account of generalized and $\tilde{\lambda}$-clauses in the semantic evaluation procedure induces no problem (however it is less easy with resolution).

4. Possibilistic logic and soft-constrained problems

4.1. Introduction

In this section we deal with *propositional* necessity-valued logic. As shown in section 2, the inconsistency degree of a set of necessity-valued formulas $\mathcal{F} = \{(\varphi_1 \, \alpha_1), ..., (\varphi_n \, \alpha_n)\}$, verifies the equality (propositions 4 and 5) Incons $(\mathcal{F}) = 1 - \text{Max}_{\omega \in \Omega} \, \pi^*_{\mathcal{F}} (\omega)$ and computing the best model(s) of $\mathcal{F}$ comes down to find the interpretation(s) ω maximizing $\pi^*_{\mathcal{F}} (\omega)$, where $\pi^*_{\mathcal{F}} (\omega) = \text{Min} \{ 1- \alpha_i \mid \omega \vDash \neg \varphi_i, i = 1, ..., n\}$. In a more compact way, it reduces to the discrete optimisation problem

$$\text{Max}_{\omega \in \Omega} \, \text{Min} \{ 1 - \alpha_i \mid \omega \vDash \neg \varphi_i, i = 1, ..., n\}$$

or equivalently to this other one

$$\text{Min}_{\omega \in \Omega} \, \text{Max} \{ \alpha_i \mid \omega \vDash \neg \varphi_i, i = 1, ..., n\}$$

So, computing Incons $(\mathcal{F})$ and the best model(s) of $\mathcal{F}$ is a min-max discrete optimisation problem; hence, problems of the same nature, which have the general form

$$\text{Min}_{x \in X} \, \text{max}_{y \in Y} \, f(x,y)$$

where X and Y are finite, can be translated in necessity-valued logic and solved by resolution or semantic evaluation; moreover, if semantic evaluation is used, the set of best models of $\mathcal{F}$ will give the set of optimal solutions for the min-max discrete optimisation problem.

Of course, the problem of computing the inconsistency degree of $\mathcal{F}$ is NP-complete (since it is proved in [17] to be equivalent to at most n problems of satisfiability in propositional classical logic, where n is the number of formulas in $\mathcal{F}$); thus, resolution and semantic evaluation is (in the case where we use non-Horn clauses) exponential[1] and it is clear that for a given problem, there generally exists a specific algorithm whose complexity is at least as as good as (often better than) the complexity of possibilistic deduction procedures.

Thus, we do not claim to give, for the problems we shall deal with, a more efficient algorithm than those already existing; however, we think that translation in possibilistic logic is useful, for several reasons:
- the search method is independent of the problem;
- the pruning properties in the search tree can confer to the algorithm a good average complexity (even polynomial, in some cases);
- possibilistic logic enables a richer representation capability in the formulation of a problem (one can specifies complex satisfaction constraints not specifiable in the classical formulation).

Thus, necessity-valued logic appears to be a logical framework for expressing in a declarative way some min-max discrete optimisation problems. In the next subsection we give an example of such a problem, the bottleneck assignment problem. Lastly we shall deal with more general soft-constrained problems, or problems with *prioritized* constraints and we shall give a detailed example.

4.2. A min-max discrete optimisation problem: the "bottleneck" assignment

The min-max assignment problem (also called "bottleneck assignment problem") is formulated as follows: one has n tasks to assign to n machines (one and only one task per machine); if machine i is assigned to task j, the resulting cost is a_{ij}. Then the total cost of the global assignment is not the sum, but the *maximum* of the costs of the elementary assignments. Thus one looks for a permutation $\mathcal{P}$ of $\{1,2,...,n\}$ such that $\text{Max}_{\{i=1,2,...,n\}}[a_{i,\mathcal{P}(i)}]$ be minimum[2].

The min-max assignment problem associated to the n x n matrix $A = (a_{i,j})$ can be translated and solved in possibilistic logic: first, the coefficients of A are supposed to lay in [0,1] (if it is not the case, we normalize A); then, to A one associates the set of necessity-valued clauses and generalized clauses EC (A), whose atoms are $\{B_{i,j}, i=1...n, j=1...n\}$:

$EC(A) = EC_1(A) \cup EC_2(A)$, with

$EC_1(A) =$

$\{(B_{1,1}\vee...\vee B_{1,n}\ \ 1), ... , (B_{n,1}\vee...\vee B_{n,n}\ \ 1)\}$ (at least one task per machine)

$\cup\ \{(n-1:\neg B_{1,1}\vee...\vee\neg B_{1,n}\ \ 1),...,(n-1:\neg B_{n,1}\vee...\vee\neg B_{n,n}\ \ 1)\}$ (at most one task per machine)

$\cup\ \{(B_{1,1}\vee...\vee B_{n,1}\ \ 1), ... , (B_{1,n}\vee...\vee B_{n,n}\ \ 1)\}$ (at least one machine per task)

---

[1]However, their *average* complexity may be polynomial in some particular cases (see for example [19] for the case of classical logic).
[2]The min-max assignment problem can be solved in polynomial time using the Ford-Fulkerson algorithm for searching a maximal flow in a network.

$\cup$ {(n-1: $\neg B_{1,1} \vee ... \vee \neg B_{n,1}$ 1)),..., (n-1:$\neg B_{1,n} \vee ... \vee \neg B_{n,n}$ 1) } (at most one machine per task)[1];
$EC_2(A) = \{ (\neg B_{i,j} \; a_{i,j}) \; 1 \leq i \leq n, 1 \leq j \leq n \}$.

Intuitively, stating that $B_{i,j}$ is true if (i, j) belongs to the assignment, it can be seen that searching for the optimal assignment for A is equivalent to searching the best model of EC(A). This is expressed more formally by the following result :

Proposition 9 : If $\mathcal{P}$ (n) is the permutation set of {1,...,n}, then we have $\text{Min}_{P \in P(n)} \text{max}_{i=1,...,n}(a_{i,\mathcal{P}(i)}) = $ Incons (EC(A)), and the best model(s) give the optimal solution(s) for the assignment problem.
The proof is omitted. It is in [17].

The possibilistic formulation of a min-max assignment problem enables the formalization of extra *soft* constraints. For example, the constraint "exactly one task per machine" can be weakened; we give two examples of such weakenings (with n = 3):
    - "the machine 3 may, if necessary, can be assigned a second task but certainly not a third one" is translated by replacing the clauses of EC (A) corresponding to the 3rd range by {(2: $\neg B_{3,1} \vee \neg B_{3,2} \vee \neg B_{3,3}$ α) ; ($\neg B_{3,1} \vee \neg B_{3,2} \vee \neg B_{3,3}$ 1) }, knowing that the least α, the more permissive the expression "if necessary";
    - "the 2nd task may, if necessary, not be done" is translated by replacing the clause $(B_{1,2} \vee B_{2,2} \vee B_{3,2}$ 1) by $(B_{1,2} \vee B_{2,2} \vee B_{3,2}$ α) .
    Complex soft constraints can also be formalized; for example:
    - "if task 3 is assigned to machine 1 then task 2 should rather be assigned to machine 2", translated by $(\neg B_{1,3} \vee B_{2,2}$ α);
    - "it is absolutely forbidden to assign tasks 1 and 2 to the same machine, excepted maybe machine 3", translated by { ($\neg B_{1,1} \vee \neg B_{1,2}$ 1) ; ($\neg B_{2,1} \vee \neg B_{2,2}$ 1) ; ($\neg B_{3,1} \vee \neg B_{3,2}$ α) };
    - "if it rains, task 3 may not be done, if necessary; but if it does not rain, then it must be done", translated by {($\neg$rains $\vee B_{1,3} \vee B_{2,3} \vee B_{3,3}$ α); (rains $\vee B_{1,3} \vee B_{2,3} \vee B_{3,3}$ 1) }.

4.3. Problems with prioritized constraints

More generally, min-max discrete optimisation problems may come from constraint satisfaction problems, where the constraints are weighted by necessity degrees measuring their priority, and where the constraint set is "partially" inconsistent, in the sense of section 2. Solving such a "prioritized" constraint satisfaction problem consists in finding the solution minimizing the degree of the most important constraint among those which are violated. For the same reasons evoked in section 4.1., possibilistic logic offers a general logical framework for representing and solving these problems. The system GARI of Descotte and Latombe [2], which is an approach

---

[1]Any three of these four sets of clauses imply the fourth one, so it is useless, for example, translating that there is at one machine per task.

more oriented to production rules than to logical deduction, is very similar to ours, since it computes the solution establishing the best compromise under a set of ordered antagonistic constraints; it has been efficiently applied in the planning field. More recently, Borning et al. [1] have incorporated hierarchies in constraint logic programming, where priorities are represented by an atomic formula in the body of a Horn clause; Satoh [20] proposes a formalisation of soft constraints, based on circumscription; in his approach the most preferable solutions are the ones that satisfy *as many* prioritized constraints as possible, contrarily to ours or Descotte and Latombe's; he does not provide any algorithmic issues. Lastly, our work is also somewhat linked to Wrzos-Kaminski and Wrzos-Kaminska's [21] who define priorities upon justifications in an ATMS.

We now give an example of constraint satisfaction with priorities and partial inconsistencies and its translation into possibilistic logic. We plan to organize a meeting, and we have to decide whom telling to attend it. The set of potential participants is {Albert, Betty, Chris, David, Elizabeth, Florence, George, Harold}, and there are a lot of constraints restricting the possible choices:

(1) there must be around 5 participants to the meeting; the ideal number would be 5; however, 4 or 6 participants may be accepted (4 – priority 0.2 for "more than 4" – being more acceptable than 6 – priority 0.3 for "less than 6" –), or even 3 or 7 (weakly acceptable, and 3 – priority 0.5 – more acceptable than 7 – priority 0.6 –); lastly, this number *must* be between 3 and 7.

(2) the potential participants belong to two distinct clans, the first one composed by {Albert, Betty, Chris, David} and the second one by {Elizabeth, Florence, George, Harold} ; it would be preferable to have as many participants from each clan; a difference of one unit is tolerable – priority 0.6 for "not more than 0" –, but not a difference of two or more units.

(3) to these general constraints add some "individual" constraints:
- it is absolutely necessary (1) that Albert or Betty be present;
- it is almost necessary (0.9) that if Albert comes then either Chris or David comes;
- it is relatively necessary (0.8) that if Chris and David come then Albert comes too;
- it is quite preferable (0.7) that David and Elizabeth be not together at the meeting;
- it is preferable (0.6) that if David comes then Harold comes too and conversely;
- it is rather preferable (0.5) that if Florence comes then Harold comes too;
- it is moderately preferable (0.4) then Elizabeth, Chris and George be not simultaneously present;
- it is somewhat preferable (0.2) that if Florence and George come then so does Elizabeth.

This list of constraints is expressed by the following necessity-valued clauses and generalized clauses (where A means "Albert comes", etc.):

(1) $A+B+C+D+E+F+G+H \geq 5$ [priority 0.2] gives the generalized necessity-valued clause

$$(5 : A \vee B \vee C \vee D \vee E \vee F \vee G \vee H \quad 0.2)$$

and similarly we get the two following ones

$$(4 : A \lor B \lor C \lor D \lor E \lor F \lor G \lor H \quad 0.5)$$

$$(3 : A \lor B \lor C \lor D \lor E \lor F \lor G \lor H \quad 1)$$

$A+B+C+D+E+F+G+H \leq 5$ [priority 0.3] gives

$$(4 : \neg A \lor \neg B \lor \neg C \lor \neg D \lor \neg E \lor \neg F \lor \neg G \lor \neg H \quad 0.3)$$

and similarly we get the two following ones

$$(3 : \neg A \lor \neg B \lor \neg C \lor \neg D \lor \neg E \lor \neg F \lor \neg G \lor \neg H \quad 0.6)$$

$$(2 : \neg A \lor \neg B \lor \neg C \lor \neg D \lor \neg E \lor \neg F \lor \neg G \lor \neg H \quad 1)$$

These 6 generalized clauses can be synthesized by the two $\tilde{\lambda}$-clauses

$$((3|_1 ; 4|_{0.5} ; 5|_{0.2}) : A \lor B \lor C \lor D \lor E \lor F \lor G \lor H)$$

$$((2|_1 ; 3|_{0.6} ; 4|_{0.3}) : \neg A \lor \neg B \lor \neg C \lor \neg D \lor \neg E \lor \neg F \lor \neg G \lor \neg H)$$

(2) the constraint $|(A+B+C+D)-(E+F+G+H)| = 0$ [priority 0.6] gives

$$(4 : A \lor B \lor C \lor D \lor \neg E \lor \neg F \lor \neg G \lor \neg H \quad 0.6)$$

$$(4 : \neg A \lor \neg B \lor \neg C \lor \neg D \lor E \lor F \lor G \lor H \quad 0.6)$$

and the constraint $|(A+B+C+D)-(E+F+G+H)| \leq 1$ [priority 1] gives

$$(3 : A \lor B \lor C \lor D \lor \neg E \lor \neg F \lor \neg G \lor \neg H \quad 1)$$

$$(3 : \neg A \lor \neg B \lor \neg C \lor \neg D \lor E \lor F \lor G \lor H \quad 1)$$

which can be synthesized by the two $\tilde{\lambda}$-clauses

$$((3|_1 ; 4|_{0.6}) : A \lor B \lor C \lor D \lor \neg E \lor \neg F \lor \neg G \lor \neg H)$$

$$((3|_1 ; 4|_{0.6}) : \neg A \lor \neg B \lor \neg C \lor \neg D \lor E \lor F \lor G \lor H)$$

(2) the other constraints give the following necessity-valued clauses

$(A \lor B \quad 1)$; $(\neg A \lor C \lor D \quad 0.9)$; $(\neg C \lor \neg D \lor A \quad 0.8)$; $(\neg D \lor \neg E \quad 0.7)$; $(\neg D \lor H \quad 0.6)$; $(\neg H \lor D \quad 0.6)$; $(\neg F \lor H \quad 0.5)$; $(\neg E \lor \neg G \lor \neg C \quad 0.4)$; $(\neg F \lor \neg G \lor E \quad 0.2)$.

Applying semantic evaluation to this set $\mathcal{C}$ of necessity-valued clauses and generalized clauses give as inconsistency degree $\textbf{Incons}\ (\mathcal{C}) = 0.3$ and as the best model of $\mathcal{F}$

$$\bar{\omega} = \{A, \neg B, C, D, \neg E, F, G, H\}$$

i.e. the persons whom we should tell to attend the meeting are Albert, Chris, David, Florence, George and Harold. This interpretation satisfies all constraints except $(4: \neg A \lor \neg B \lor \neg C \lor \neg D \lor \neg E \lor \neg F \lor \neg G \lor \neg H \quad 0.3)$ and $(\neg F \lor \neg G \lor E \quad 0.2)$.

6. Conclusion

Thus, due to the original feature of (the necessity-valued fragment of) possibilistic logic, *the gradation of inconsistency*, it is a well-adapted logical framework for representing constraint satisfaction problems with priorities. The optimal solution(s) is (are) given by the best model(s) of the corresponding set of necessity-valued clauses by the possibilistic semantic evaluation algorithm, working in the propositional case. For some problems which are better expressible in first-order logic (for example, problems related to fuzzy graphs), possibilistic resolution works but only computes the inconsistency degree; however the search for the best model(s) by possibilistic resolution is not hopeless. Thus we might think of *using possibilistic logic as a programming language*, which would be useful as well in the field of constraint satisfaction with priorities as in the field of logic programming with uncertainty.

Several extensions of necessity-valued possibilistic logic should enable a more expressive language. We mention here three of them:

(1) the use of complete possibilistic logic [4] (incorporating in the language also lower bounds of possibility degrees) would enable the expression of "permission" constraints: $\Pi(\varphi) \geq \alpha$ expresses that φ is at least α-allowed, which is weaker than $N(\varphi) \geq \beta$ for any $\beta > 0$, since we have the property $N(\varphi) > 0 \Rightarrow \Pi(\varphi) = 1$, expressing that a formula must be completely allowed before being somewhat necessary.

(2) Possibility and necessity measure take place in a more general framework, called *decomposable measures* (see [6]), obtained by replacing the operators "min" and "max" of possibility theory by more general *triangular norms and conorms* [6]; these general measures include probabilities, and also a measure enabling the representation of discrete optimisation problems with additive costs (which are the most well-known ones).

(3) An other way to generalise possibilistic logic is to take as value set for possibility and necessity measures, not [0,1] but more generally a lattice, where order is not necessarily total. This would enable the representation of problems non-totally ordered priorities, for example problems with fuzzy priority degrees, or Boolean ones, etc.

Semantic aspects of these two types of generalised possibilistic logics are very similar to "basic" possibilistic logic; the main point to study is the extension of methods and results about automated deduction, which is not straightforward.

References

[1]   Borning A., Maher M., Martindale A., Wilson M. (1989) "Constraint hierarchies in logic programming", Proc. ICLP'89, 149-164.

[2]   Descotte Y., Latombe J.C. (1985) , "Making compromises among antagonistic constraints in a planner", Artificial Intelligence, 27, 183-217.

[3]   Dubois D., Lang J., Prade H. (1987) "Theorem proving under uncertainty — A possibility theory-based approach". Proc. of the 10th Inter. Joint Conf. on Artificial Intelligence (IJCAI 87), Milano, Italy, 984-986.

[4]   Dubois D., Lang J., Prade H. (1989) "Automated reasoning using possibilistic logic : semantics, belief revision and variable certainty weights". Proc. of the 5th Workshop on Uncertainty in Artificial Intelligence , Windsor, Ontario, 81-87.

[5]   Dubois D., Lang J., Prade H. (1991) "Fuzzy Sets in Approximate Reasoning. Part 2: Logical approaches", Fuzzy Sets and Systems 40 (1), 203-244.

[6]   Dubois D., Prade H. (1982) "A class of fuzzy measures based on triangular norms. A general framework for the combination of uncertain information", Int. J. of Intelligent Systems, 8(1), 43-61.

[7]   Dubois D., Prade H. (1988) (with the collaboration de Farreny H., Martin-Clouaire R., Testemale C.) "Possibility Theory : an Approach to Computerized Processing of Uncertainty". Plenum Press, New York.

[8]   Dubois D., Prade H. (1987) "Necessity measures and the resolution principle". IEEE Trans. on Systems, Man and Cybernetics, 17, 474-478.

[9]   Dubois D., Prade H. (1991) "Epistemic entrenchment and possibilistic logic", à paraître dans Artificial Intelligence.

[10]   Dubois D., Prade H., Testemale C. (1988) "In search of a modal system for possibility theory". Proc. of the Conf. on Artificial Intelligence (ECAI), Munich, Germany, 501–506.

[11]   Froidevaux C., Grossetête C. (1990) "Graded default theories for uncertainty", ECAI 90, Stockholm, 283-288.

[12]   Gärdenfors P., Makinson D. (1988) "Revision of knowledge systems using epistemic entrenchment", in M. Vardi ed., Proc. Second Conference on Theoretical Aspects of Reasoning about Knowledge (Morgan Kaufmann).

[13]   Ginsberg M.L. (1988) "Multi-valued logics : a uniform approach to reasoning in artificial intelligence". Computational Intelligence, 4, 265–316.

[14]   Hooker J.N. (1986) "A quantitative approach to logical inference", Decision Support Systems 4, 45-69.

[15]   Jeannicot S., Oxusoff L., Rauzy A. (1988) "Evaluation sémantique : une propriété de coupure pour rendre efficace la procédure de Davis et Putman". Revue d'Intelligence Artificielle, 2(1), 41-60.

[16]   Lang J. (1990) "Semantic evaluation in possibilistic logic", Proc. of the 3rd Inter. Conference on Information Processing and Management of Uncertainty in Knowledge-Based Systems, Paris, 51-55.

[17]   Lang J. (1991) "Logique possibiliste: aspects formels, déduction automatique, et applications", PhD thesis, University of Toulouse (France), January 1991

[18]   Nilsson N. "Probabilistic logic". Artificial Intelligence, 28, 71–87.

[19]   Purdom P.W., "Search rearrangement backtracking and polynomial average time", Artificial Intelligence 21, 117-133.

[20]   Satoh K. (1990), "Formalizing soft constraints by interpretation ordering", Proc. ECAI 90, Stockholm, 585-590.

[21]   Wrzos-Kaminski J., Wrzos-Kaminska A., "Explicit ordering of defaults in ATMS", Proc. ECAI 90, Stockholm, 714-719.

[22]   Zadeh L.A. "Fuzzy sets as a basis for a theory of possibility". Fuzzy Sets and Systems, 1, 3–28.

An Approach to Data-Driven Learning

Zdravko Markov

Institute of Informatics, Bulgarian Academy of Sciences
Acad.G.Bonchev st. Bl.29A, 1113 Sofia, Bulgaria

Abstract

In the present paper a data-driven approach to learning is described. The approach is discussed in the framework of the Net-Clause Language (NCL), which is also outlined. NCL is aimed at building network models and describes distributed computational schemes. It also exhibits sound semantics as a data-driven deductive system. The proposed learning scheme falls in the class of methods for *learning from examples* and the learning strategy used is *instance-to-class generalization*. Two basic examples are discussed giving the underlying ideas of using NCL for *inductive concept learning* and *learning semantic networks*.

1. Introduction

Machine Learning is one of the "hard" AI problems. Most of the achievements in this area have been done in narrow and well formalized domains. However even in these restricted cases Machine Learning is still a hard problem. Consider Concept Learning as an instance of Inductive Learning from Examples. This problem in a logical framework is described as follows [1]: a *concept is a predicate*. The *concept description* is a set of clauses explaining the predicate in logical sense. *Given a set of ground instances of a predicate (examples) the learning system is to find its description.* The practical concept learning systems impose various restrictions on the above generally stated problem. Almost all of these restrictions concern the description of the learned predicate . The description language is often limited to definite clauses (clauses without negated goals) [2,3]. This allows to avoid the problems with negation. In most systems the explanation of the hypothesis is based on deduction where the negation is defined as failure, i.e. the Closed World Assumption (CWA) holds. However CWA is inconsistent with the very idea of learning, i.e. the ability of the system to acquire new knowledge presumes that the system's world is open.

Most of the concept learning methods are based on classical deduction systems like Prolog, e.g. [2,3,4]. In these cases the learning algorithms are substantially different from the process of deduction, used for logical explanation of the concepts. Actually we have two separate systems - one for learning and another for proving (explaining) the learned predicate. When generating hypotheses a *global exhaustive search* through the clauses is always required. For example in the Prolog based learning systems, proving (explaining) a concept (predicate) is based on *goal-driven search* and building up the predicate description is achieved by analyzing all positive literals (facts and clause heads) in the database. The latter is in a sense a *data-driven search*. That is to say learning is data-driven and explanation - goal-driven.

The basic idea of our approach is *to implement concept learning methods in a data-driven deductive system*. This would provide for integration of both learning and explanation (which in the interactive systems is a part of the overall learning process) in an unified computational environment. The data-driven computation provides also for constructive definition of negation, avoiding in such way the conceptual contradiction between CWA and learning.

The main contribution of the present paper is embedding a learning mechanism in a data-driven computational environment, called Net-Clause Language (NCL). NCL is aimed at building network models and describes distributed computational schemes in connectionist sense [5,6]. It exhibits also a sound semantics as a data-driven deductive system in the framework of clausal and non-clausal logic. An almost complete description of NCL can be found in [7].

The next Section outlines the basic concepts of NCL. A description of the NCL built-in learning mechanism along with an example of its use is given in Section 3. Section 4 discusses an interpretation of the example from Section 3 in the framework of inductive concept learning. Another example of using NCL learning for building semantic networks is given in Section 5. Section 6 briefly discusses the advantages and drawbacks of the obtained results.

2. Overview of NCL

Syntactically the Net-Clause language (NCL) is an extension of the standard Prolog. Its semantics however is aimed at modeling graph like structures (networks), consisting of nodes and links. The nodes specify procedures unifying terms, and the links are channels along which the terms are propagated. The language is designed for *describing distributed computation schemes, without centralized control using unification as a basic processing mechanism*.

The basic constructors of NCL programs are the *net-clauses*. A net-clause is a sequence of *nodes*, syntactically represented as structures (complex terms), separated by the special delimiter ":". The *network links* are implicitly defined by shared variables among different nodes in a net-clause. The variables in NCL are called *net-variables*.

Generally there are two types of nodes - *free nodes* and *procedural nodes*. The free nodes are structures (in the form of Prolog facts) used to access net-variables, inside and outside the net-clause. The procedural nodes are the active elements in the network. Procedures unifying terms are associated to the procedural nodes. The procedures are activated under certain conditions, defined *locally* in each node. The control in NCL is based on the unification procedure. Generally when unifying net-variables two possible results can occur: *binding net-variables* to terms or *sharing net-variables*. These possibilities define the two activation schemes in NCL. Each one of them is specified by a particular type of procedural node. We restrict the further considerations to the *spreading activation scheme*, based on net-variable binding. The other activation scheme (activation-by-need) is discussed elsewhere, e.g. [8] and does not relate directly to the NCL learning.

The spreading activation control scheme is defined by procedural nodes written in the following syntax: $\mathbf{node}(X_1,...,X_n,M,<\mathbf{procedure}>)$. The purpose of the node procedure is to unify terms, particularly to bind variables, which in turn could further propagate both data (terms) and control (activation) among other nodes in the network. M is an integer number and its semantics is to define a *threshold*, determining the amount of data required to activate the procedure. X_i are net-variables which serve as channels for term propagating. They can be used both as *ex-*

citatory links and as *inhibitory links* for the activation of the procedure. The excitatory links are represented as *simple (ordinary) variables* and the inhibitory links are represented as *negated variables* (written as $\sim X_i$). The procedure is activated if the difference between the number of the bound simple variables and the number of the bound negated ones is equal to M. Here is an example:

input1(X): input2(Y): output(Z):
node(X, $\sim$Y,1,X1 = 1): node(Y, $\sim$X,1,Y1 = 1): node(X1,Y1,1,Z = 1).

The above NCL program implements the textbook neural network example realizing the XOR function. The following NCL queries illustrate its work (logical "1" is encoded as a bound net-variable and "0" - as a free one; the answers of the NCL interpreter are given in italics):

<- input1(1), input2($\_$), output(Z).
Z = 1

<- input1(1), input2(1), output(Z).
Z = _1

In [7] the logical semantics of NCL is defined by establishing a correspondence between Horn clauses and Net-clauses. Here we extend this correspondence to *general logic programs* defining the following transformation rules:

1. Each *program clause* is translated into a net-clause, where the clause head is represented by a spreading activation node and the clause body - by a collection of free nodes. $X_1,...,X_m$ are all variables occurring in the literals $A_1,...,A_p$. $Y_1,...,Y_n$ are all variables occurring in $B_1,...,B_q$. Such net-clauses we call *data-driven rules*.

$$p(Z_1,...,Z_n) <-- \qquad\qquad node(X_1,...,X_m, \sim Y_1,..., \sim Y_n, m, p(Z_1,...,Z_n)):$$
$$A_1,...,A_p, \qquad <=> \qquad A_1: ... A_p:$$
$$not\ B_1,...,not\ B_q \qquad\qquad B_1: ... B_q.$$

2. The *goal clause* is represented as a net-clause built out of free nodes, which can share variables, thus introducing means to share variables in the original Horn clause goal.

$$<-- B_1,...,B_n \qquad <=> \qquad B_1: ... B_n.$$

3. The *unit clauses* are represented as a net-clause query (data), which activates the net-clause program.

$$C_1 <--$$
$$... \qquad <=> \qquad <- C_1,...,C_n$$
$$C_n <--$$

The spreading activation scheme implements *data-driven inference (forward chaining)* in Horn clauses. Furthermore unlike the standard Logic Programming, in NCL the set of allowable formulae can be extended to a class of formulae in non-clausal form. From logical point of view a net-clause in general is a conjunction of Horn clauses, where the scope of the universal quantifiers is extended to all Horn clauses represented by a net-clause. Thus a net-clause allows communication links to be established between several Horn clauses through the shared variables. Therefore the procedural semantics of NCL can be expressed in terms of *non-clausal resolution*, as it is shown in [7].

The following example illustrates the NCL implementation of general logic programs. It shows also a special case (not shown in the transformation rule 1) when positive and negated goals share variables. Then additional variables (T1,T2) are introduced to propagate the truth values of the goals from the clause body to the clause head (the node procedure).

a(X) :- not b(X), c(X)   node(T1, ~ T2,1,a(X)): b(X,T2): c(X,T1).
b(1).   < = >   <- b(1,true),c(2,true).
c(2).

The NCL data-driven inference has also a connectionist interpretation. The data-driven rule (the right-hand side of transformation rule 1) can be viewed as a threshold element, fired when the predicate **p** succeeds. *The success of a predicate in terms of NCL data-driven inference means that all its variables are instantiated.* So the net-variables $X_1,...X_m$ are excitatory links when bound indicating the success of the predicates in the clause body. $Y_1,...,Y_n$ are inhibitory links suppressing the success of **p** if some of the negated predicates in the body have succeeded (some of Y_i's have been bound). Furthermore we can introduce partial success of **p** by using a threshold less than **m**. In this case **p** will succeed even when not all A_i's have succeeded or some B_i's have succeeded.

3. Learning in NCL

When programming in NCL the structure of the net-clauses should be programmed explicitly. However this task is quite difficult, especially in the case of large number of nodes and links and irregular structure of the network. A natural approach to solve this problem is to extend the data-driven paradigm to the process of NCL programming. This means to allow *data themselves to determine the proper network structure for their processing*.

The NCL network structure is determined by the network links specified by the shared net-variables appearing in the nodes. Thus to implement a learning procedure in NCL a mechanism for automatic sharing of the net-variables is required. The basic idea is to use *term generalization*. This operation is defined as follows: t_1 is *more general* than t_2, if there exists a substitution s, such that $t_1 s = t_2$, i.e. it defines a partial order on terms. In NCL we introduce slightly modified version of term generalization, called NCL generalization. Consider the term t and the substitution g = $\{t_1/X_1, t_2/X_2, ...,t_n/X_n\}$, where t_i are non-variable subterms, appearing in t ($t_i \neq t_j$, $i \neq j$), and X_i are variables ($X_i \neq X_j$, $i \neq j$). Here is an example of applying g. Let S be a set of ground terms:

$S = \{t(a,b,f(a,b)),t(b,c,f(b,c)),t(c,d,f(c,d)),...\}$. The terms in S are ground instances of the term $t(X,Y,f(X,Y))$. The term $t(X,Y,f(X,Y))$ is a result of applying generalization to any of the terms from S, i.e it is the most general of the terms in S. Hence $Sg = \{t(X,Y,f(X,Y))\}$, i.e. g reduces S to one element (similarly to the **mgu** of S).

A problem in the above definition is how to find a unique set of subterms t_i. For that purpose we introduce another term, which is used as a pattern for building the generalized term. So we come the following definition.

<u>Definition 1</u>: Let p and t be terms and $s = \{V_1/t_1,V_2/t_2,...,V_n/t_n\}$ - their unifier, i.e. ps = t. NCL generalization **g** of the term t, using the pattern **ι** is defined as the substitution $g = \{t_1/W_1,t_2/W_2,...,t_n/W_n\}$, where W_i are variables and $W_i = W_j$ iff t_i is unified with t_j $(i \neq j)$; $i,j = 1,...,n$. W_i are called *variables obtained by NCL generalization*.

The implementation of NCL generalization is based on the following idea: *during the activation of the net-clauses, the net-variables bound to unifiable ground terms are replaced with shared variables*. Thus the net-clause plays the role of the pattern p, and t is a conjunction of procedures, unifying free nodes in the net-clause. The substitution s is a set of variable bindings, obtained by the unification of p and t. To specify this set two built-in procedures are used: **top(Marker)** and **gen(Marker)**. They define the scope, where the generalization is performed. Procedure **top** fixes the beginning of creation of s returning a marker. Executing **gen** with the same marker performs the NCL generalization with the currently obtained net-variable bindings (the unifier s). The result of the generalization is the modified net-clause, i.e. p is used to store the result of applying g to t (p = tg).

Further we illustrate the learning scheme in NCL by solving a particular learning task, defined as follows: *Given an instance of a geometric figure to build a net-clause, such that it can recognize all figures, belonging to the corresponding class*. The geometric figures are represented as sequences of edges. Each edge is a structure in the form **edge(M,N,S,L)**, where **M** and **N** are the edge vertices, and S and L - its slope and length. (This representation is explained in details in [5,6,7].)

To solve the above stated problem we need:

1. *A learning environment*. This a collection of unconnected free nodes, represented by a net-clause with unique net-variables in the following form:

```
edge(_1,_2,_3,_4):              ...
edge(_5,_6,_7,_8):              fig(_37)
edge(_9,_10,_11,_12):          ...
edge(_13,_14,_15,_16):
edge(_17,_18,_19,_20):
edge(_21,_22,_23,_24):
...
```

2. A unified algorithm for changing the environment in accordance with the learning strategy used. This algorithm is specified by NCL queries of the following type:

<- top(M),edge(a,b,0,20),edge(b,c,45,20),edge(c,d,0,20),edge(d,a,45,20),   (1)
    get_bindings(M,L), add_node(L,fig(rhombus)), gen(M).

Query 1 specifies an instance of a rhombus as a sequence of its edges. **get_bindings** is a procedure returning a list of all variable bindings, obtained after the marker M. This is actually the substitution s, mentioned above. The procedure **add_node** adds a procedural node, containing **length**(L) net-variables, to the net-clause and binds them to the terms from L. Finally **gen** performs the NCL generalization, thus structuring the net-clause in the following way:

edge(_1,_2,_3,_4):   ...
edge(_2,_5,_6,_4):   node(_1,_2,_3,_4,_5,_6,_7,7,fig(rhombus)):
edge(_5,_7,_3,_4):   ...
edge(_7,_1,_6,_4):   fig(_37):
edge(_8,_9,_10,_11):   ...
edge(_12,_13,_14,_15):
...

Now instances of rhombuses can be recognized by queries of the following type:

<- edge(1,2,0,20),edge(2,3,50,20),edge(3,4,0,20),edge(4,1,50,20),fig(X).   (2)
X=rhombus

This is achieved by the built-in spreading activation mechanism, invoked when the net-variables in the free nodes are bound by the entered data (the sequence of edges). Then the corresponding procedural node is activated and its procedure executed, thus binding the output net-variable (_37) to "rhombus", which is indicated by the answer of the NCL interpreter, through the last goal in the query.

    Entering another pattern (an instance of a parallelogram) in the same way we can further specify the network structure.

<- top(M), edge(a,b,0,20),edge(b,c,45,30),edge(c,d,0,20),edge(d,a,45,30),   (3)
    get_bindings(M,L),add_node(L,fig(parallelogram)), gen(M).

Thus we obtain the following net-clause:

edge(_1,_2,_3,_4):   ...
edge(_2,_5,_6,_4):   node(_1,_2,_3,_4,_5,_6,_7,7,fig(rhombus)):
edge(_5,_7,_3,_4):   node(_1,_2,_3,_4,_8,_9,_10,_11,8,fig(parallelogram)):
edge(_7,_1,_6,_4):   ...
edge(_2,_8,_9,_10):   fig(_12):
edge(_8,_11,_3,_4):   ...
edge(_11,_1,_9,_10):

Note that the above net-clause cannot be represented as a set of Horn clauses (using the correspondence defined in the previous Section). Actually this is a conjunction of Horn clauses, which share variables.

Repeating the above learning-from-example procedure we can train the network to recognize more patterns. For that purpose we should extend the number of unconnected nodes providing in such way enough space for new patterns. Furthermore during the learning process the required nodes can be generated automatically *when necessary* (see procedure "object" in Section 5).

It is important to note that the obtained on the basis of above described learning algorithm NCL program is a *minimal* one in a sense that common parts of the figures (edges) appear only once. This is due to the sequential strategy of unifying the free nodes. When a new pattern is entered it may fit completely (if it is an instance of the learned class) or partially the connected free nodes in the network. The learning takes place in the latter case. Then some edges are unified with the free nodes of the already learned class and the remaining ones unify the existing unconnected f generalization.

The example illustrating the NCL learning algorithm refers to *one-layer networks*, i.e. networks with no more than one procedural node between the inputs and the outputs. Generally this scheme of learning can be represented by the following query:

$$<- top(M), <data>, get\_bindings(M,L), add\_node(L, <solution>), gen(M)., \qquad (4)$$

where < data > and < solution> are conjunctions of ground literals. This query represents an instance of a data-driven inference rule:

$$<data> => <solution> \qquad (5)$$

By applying generalization (procedure "gen") to rule (5) its scope of application is extended to all literals, which are ground instances of the literals appearing in < data >.

Queries (1) and (3) are examples of how to build (or extend) one-layer networks by generalization of rules of type (5). Furthermore using this technique we can build also multi-layer networks including hierarchical and disjunctive rules.

4. Inductive Concept Learning

In the present Section we shall interpret the last example in the framework of inductive concept learning. We modify the originally sated problem [1] and come to the following definition.

Definition 2: *NCL inductive concept learning*.

Given: 1. Background knowledge - a set of data-driven rules.

2. Learning environment - a rich enough set of free nodes with unique net-variables (unconnected free nodes).

3. An example of a predicate - ground literal P.

4. Positive data - ground literals $PD_1,...,PD_m$.

5. Negative data - ground literals $ND_1,...,ND_n$.

<u>Find:</u> *An updated background knowledge such that P is inferred by data-driven inference from the background knowledge and the positive data, but not from the background knowledge and the negative data.*

Note that compared to the standard definition of concept learning in our case the roles of the examples and the data are changed. This is due to the fact that the data in NCL are the invariant part of the knowledge, which is a subject of generalization.

The solution of the above stated problem is based only on the NCL generalization. Two basic operations are considered for this purpose:

1. *Processing positive data.* Net-clause query $< - PD_1,...,PD_m$ is executed. Then the NCL generalization is performed and the spreading activation node $node(X_1,...,X_k,k,P')$ is added to the background knowledge. $X_1,...,X_k$ are the variables obtained by NCL generalization and P' is a NCL generalization of P, in which the unique terms are preserved. It is important to note that the set of variables obtained by generalization contains variables not only from the free nodes directly matched the data. Since the data can activate some data-driven rules from the background knowledge other variables can appear in the generated hypothesis. This is the case when *hierarchical predicates* are learned. Further, since the bindings already caused activation of some nodes are excluded from the set of variables obtained by NCL generalization, the rules in a hierarchy cannot share variables, i.e. only disjunctive rules can share variables.

2. *Processing negative data.* Net-clause query $< - ND_1,...,ND_n$ is executed. Then the NCL generalization is performed and the net-variables obtained $(Y_1,...,Y_q)$ are included as inhibitory links in the corresponding spreading activation node, i.e. $node(X_1,...,X_k,{}^\sim Y_1,..., {}^\sim Y_q,k,P')$ is obtained.

The outlined scheme for NCL concept learning exhibits some specific properties:

1. Since different data-driven rules can share variables (or free nodes) the background knowledge is not just a conjunction of program clauses, rather it is a special kind of a first order formula. Thus at each induction step the current formula is extended with a new conjunct.

2. Each newly established hypothesis (a data-driven rule) is added to the existing background knowledge. Thus the set of unconnected free nodes (the learning environment) is structured and further transformed into background knowledge. Since the data-driven rules can share variables, during NCL learning the existing background knowledge might be changed. This may happen in the case when a new hypothesis shares some variables belonging to a data-driven rule from the background knowledge. To avoid such effects we should follow a particular learning strategy. Generally two NCL *learning strategies* could be used:

(a) *using positive data only.* In this case the data should be specified in a special order - *from more specific to more general*, i.e. at each induction step in the background knowledge there should be no data-driven rule, which subsumes the currently induced hypothesis.

(b) *using positive and negative data.* The learning examples can be processed in arbitrary order. However after inducing a hypothesis an update of the background knowledge is required. The positive data for the newly generated hypothesis should be used as negative ones for modifying all wrongly activated data-driven rules.

3. The NCL learning network is *minimal* in a sense that all literals belonging to several data-driven rules appear only ones. (This property was explained in the previous Section.)

5. Learning Semantic Networks

Semantic networks (SNs) are often used as a notation for a subset of predicate logic. Thus the problem of learning SNs could be considered in the framework of concept learning. However, there is a difference in the way predicates are learned and explained. In the predicate logic, predicates are proved to be true either by goal-driven or by data-driven computation. That is to say, we always have a hierarchy of predicates. Such a hierarchy is used when a predicate is being learned - a set of lower level predicates are searched to be used as its explanation. In SNs there is no hierarchy between the relations (links). Therefore if we want to create an SN using its logical equivalent form we should build all possible combinations of predicates, i.e. we need an explanation of each predicate by all the others. In this section we show how NCL can be used for learning Semantic Networks directly dealing with nodes and links instead of predicates.

A semantic network can be easily represented in NCL. An SN fragment, consisting of two objects and a relation between them is represented as follows:

$$r(O1,O2): node(O1,O2,2, < \text{indicate that } r(O1,O2) \text{ is true} >):$$
$$o1(O1): node(O1,1, < O1 \text{ belongs to class } o1 >)): \qquad (1)$$
$$o2(O2): node(O2,1, < O2 \text{ belongs to class } o2 >)).$$

Various inferences can be done using this representation. We can check whether a ground instance of a relation is true by the query "<- r(o1,o2).". By "<- o1(a),o2(b)." we can obtain all relations, which hold between the objects "a" and "b". Furthermore using a threshold 1 in the first node we can search a path between two objects (classes).

The above representation is particularly suitable for the purposes of learning. To show this we shall consider an example, referring to the problem of learning family relationships, discussed in two different frameworks by Hinton [9] and Quinlan [10].

There are two stylized isomorphic families as it is shown in Figure 1. 12 relationship types are considered - *wife, husband, mother, father, daughter, son, sister, brother, aunt, uncle, niece* and *nephew*.

Hinton uses a connectionist network with 36 input units (12 relationships types + 24 individuals), three hidden layers and 24 output units, corresponding to the 24 individuals. Given both a relationship R and an individual A as an input, the network produces an output vector {B}, where the active units correspond to the individuals, each of which has relationship R to A. Thus there are 104 such input-output pairs. The network is trained by randomly chosen 100 of them and the remaining 4 are used as a test set.

Quinlan specifies the input-output vectors as tuples in the form <A,B> for each relationship R. This gives 24 tuples for each input-output pair, where each tuple is either positive (A has relationship R to B) or negative (A has no relationship R to B). The system FOIL he described generates clauses for each relation R. The experiments are based on the same random division of the input-output pairs - 100 for training and the remaining 4 - for testing the obtained clauses.

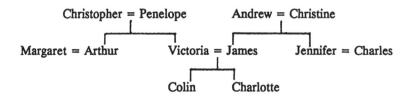

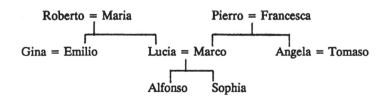

Figure 1. Two family trees ("=" denotes "married to")

Our representation of the problem is simpler. We specify only the positive examples of relationships. Thus instead of 104X24 = 2496 tuples, we have 112 examples in the predicate form:

< relationship > (< individual1 > , < individual2 >)

The learning system is to find a net-clause built out of fragments of the following type:

relationship(X,Y): node(X,Y,2,write(relationship(X,Y))):   (2)
node(X,1,X = individual1): node(Y,1,Y = individual2).

The process of building up this net-clause is quite straightforward and it is based only on NCL generalization. Unlike the Quinlan's system FOIL our system is incremental, i.e. it processes the training examples one by one, refining incrementally the currently built net-clause. At each step it processes a positive example of a relationship given as a an NCL query. Here are two instances of such queries:

<- top(M), relation(son(colin,james)), object(colin(colin)),   (3)
    object(james(james)), gen(M).

<- top(M), relation(wife(victoria,james)), object(victoria(victoria)),   (4)
    object(james(james)), gen(M).

"relation" and "object" are the only two procedures the learning system is built of. Their definitions (in Prolog) are the following:

relation(X) :- call(X),!.
relation(X) :- get_top_layer(P), asserta(P),call(X).

object(X) :- call(X),!.
object(X) :- functor(X,F,1),functor(Y,F,1),arg(1,Y,V),assertz((Y:node(V,1,F=V))),call(X).

Initially the net-clause consists of unconnected free nodes with attached procedural nodes, representing relationships (as shown in the SN fragment 2). This net-clause consists of all types of relationships, where for each type there are several free and attached procedural nodes. Actually these are all potential relationships between the individuals in the family. This is the only *a priori information* supplied to the system before the learning takes place.

To simplify the considerations we shall use only three relationship types, which are sufficient to describe the underlying ideas of the learning algorithm. So, consider the following initial net-clause (the procedural nodes are also skipped for simplicity):

wife(_1,_2) : husband(_3,_4) :
son(_5,_6) : daughter(_7,_8) :   (5)
brother(_9,_10) : sister(_11,_12).

Both procedures "relation" and "object" are based on the idea that when a call to a free node fails (the first clause fails) then new free nodes are added to the current net-clause (done by the second clause of the procedures). These newly created nodes are then unified with the example object or relationship (procedure "call" is executed) and finally NCL generalization is performed.

The algorithm for processing relationships is similar to this of processing objects. However when the call fails instead of adding only the failed node (representing an individual), a whole layer of nodes is added to the current net-clause. Actually this is exactly the initial set of nodes, but taken from the top of the net-clause. The process of copying is performed by the sequence "get_top_layer(P),asserta(P)" in the second clause of "relation". "get_top_layer(P)" unifies P with the top layer of the current net-clause, taking a *pattern of nodes in the form of the initial net-clause (5)*. Then these nodes are added *in the beginning* of the current net-clause (using the built-in procedure "asserta").

The above sequence of actions ensures that at each step the top layer of the current net-clause contains all variables already shared by NCL generalization. This is actually a fragment of the semantic network representing the already processed and generalized examples of relationships. After execution of queries 2 and 3 the initial net-clause 4 is structured in the following way:

wife(_1,_2) : husband(_3,_4): son(_5,_2):
daughter(_7,_8) : brother(_9,_10) : sister(_11,_12):   (6)
node(_1,1,_1=victoria) : node(_2,1,_2=james) : node(_5,1,_5=colin).

Note that the relationships "wife" and "son" are already connected by a common individual, represented by the net-variable _2. The three individuals appeared in the examples are represented by procedural NCL nodes (nodes in the SN), created by the procedure "object".

Because of the sequential *top-down* access to the free nodes, when the next training example appears, it always uses the top layer free nodes *first* to be unified with, thus using all already generalized relationships (in the above case these are connected nodes "wife" and "son").

When enough examples are processed the top layer of the current net-clause contains all potential relationships between the individuals in the family along with all connections between the different relationship types, represented as shared net-variables. So, the current net-clause looks like the following one:

wife(_1,_2) : husband(_2,_1): son(_3,_2):
daughter(_4,_1) : brother(_3,_4) : sister(_4,_3):

...

wife(_1,_2) : husband(_3,_4): son(_5,_2):
daughter(_7,_8) : brother(_9,_10) : sister(_11,_12):
node(_1,1,_1=victoria) : node(_2,1,_2=james) : node(_5,1,_5=colin).

Note that the initial net-clause is at the bottom of the current net-clause and at the top we have the generalized description of the family.

Using the obtained net-clause, queries of the following types can be answered (the NCL answers are given in italics):

<- aunt(margaret,colin).
aunt(margaret,colin)

<- colin(colin),jennifer(jennifer).
aunt(jennifer,colin)
nephew(colin,jennifer)

<- uncle(colin,charlotte)
no

<- emilio(emilio),lucia(lucia),sophia(sophia).
brother(emilio,lucia)
sister(lucia,emilio)
daughter(sophia,lucia)
mother(lucia,sophia)
uncle(emilio,sophia)
niece(sophia,emilio)

The described algorithm builds a complete description of the two family relationships even when some of the examples are missing. If the system is first given a full set of training examples for one of the families and the missing examples are from the other one, then it builds a complete description of the family relationships. If the missing examples are spread among the families then the algorithm requires a second pass through the same sequence of examples, where the top layer of the net-clause built after the first pass is used as an initial net-clause for the second one.

Because of the high connectivity of the network it is quite difficult to estimate the minimal number of training examples, required to build a complete description of the families. Furthermore this number depends on the types of the relationships in the skipped examples. Obviously these are the reasons Hinton and Quinlan used statistical experiments to test the performance of their systems in learning family relationships.

However there is a simple criterion, which can be applied in our case. The described learning algorithm can be summarized as generalization of patterns of connected relations. In this scheme learning a relation means including it in the already generalized pattern of relations (the top layer of the current net-clause). So, if we have the relation $R(O_1,O_n)$ and a path of other already learned relations (links) connecting O_1 and O_n, e.g $R_1(O_1,O_2)$, $R_2(O_2,O_3)$,...,$R_{n-1}(O_{n-1},O_n)$, then to learn R we need at least one learning example. This criterion holds not only for the discussed example of learning family relationships. Obviously it is applicable in the general case of learning semantic networks in the form (1).

6. Discussion

The NCL learning mechanism was originally aimed at supporting the process of programming in NCL, since the conventional explicit programming was rather unsuitable for building distributed computational models. The developed algorithm for generalization of ground terms (NCL generalization) is a convenient and quite efficient tool for creating complex net-clauses by using small patterns of ground instances of their nodes. Furthermore NCL generalization exhibits some good features as a building block for more complicated learning algorithms. Here we discussed two of them - inductive concept learning and learning semantic networks. Though this discussion was based on examples, some more general observations could be made.

The advantages of the discussed NCL learning schemes can be summarized as follows:

1. The process of learning is inherently connected with the basic mechanisms of the Net-clause language. Learning a predicate and the process of its explanation are both based on data-driven computation. Since learning and explanation are both needed for any learning system, NCL can be viewed as an unified environment for implementing such systems.

2. The concept description language in NCL learning is not restricted to Horn clauses or general logic programs as in most of the learning systems ([3,4,10]), rather it is a subset of a first order language, where clauses can share variables.

3. Using NCL as an implementation language simplifies drastically the learning systems. (For example the program code for learning semantic networks, given in Section 4 was simply two two-line Prolog predicates.) This is because some major implementation problems of the learning algorithms as the global search through clauses, the heuristics for restricting the number of generated hypotheses etc. are actually built-in features of the Net-clause language.

4. Instead of a complete learning system (such as the described in [4,10]) NCL provides a set of building blocks for implementing various learning algorithms. Thus a learning algorithm can be easily built or adapted to the particular task to be solved.

We can point out some negative features, too:

1. Recursive concepts can not be learned. Though recursion does not concern connectionist

models, it is a basic notion for the symbolic ones. As NCL exhibits features of both of them, it would be an advantage to extend NCL concept learning to recursive predicates. Actually recursive net-clauses are allowed in NCL. However they use a special lazy unification mechanism (described in [7]), which is not compatible with the NCL generalization.

2. To ensure the consistency of the learned net-clause a special order of the training examples is required (as mentioned in Section 4). Otherwise a modification of the existing nodes is required (adding new variables or changing the thresholds), which in turn changes the overall structure of the net-clause. Unfortunately this is not an immediate feature of NCL and requires copying of the whole net-clause, which is very inefficient.

The future work will address the two above mentioned drawbacks of the discussed NCL learning schemes. Another direction for further research will be the proposed algorithm for learning semantic networks. It seems very promising, particularly because it can be viewed as a natural extension of the concept learners.

References

1. Genesereth, M.R, N.J.Nilsson, *Logical foundations of Artificial Intelligence*, Morgan Kaufmann, Los Altos, 1987.

2. Shapiro, E.Y., Inductive inference of theories form facts, Tech. Rept.192, Department of Computer Science, Yale University, New Haven, CT (1981).

3. Shapiro, E.Y., *Algorithmic Program Debugging* (MIT Press, Cambridge, MA, 1983).

4. De Raedt, L. and M. Bruynooghe, On Negation and Three-valued Logic in Interactive Concept-Learning, in: *Proceedings of ECAI-90*, Stockholm, Sweden, August 6-10, 1990, pp.207-212.

5. Markov, Z., A framework for network modeling in Prolog, in: *Proceedings of IJCAI-89*, Detroit, U.S.A (1989), 78-83, Morgan Kaufmann.

6. Markov, Z. and C. Dichev and L. Sinapova, The Net-Clause Language - a tool for describing network models, in: *Proceedings of the Eighth Canadian Conference on AI*, Ottawa, Canada, 23-25 May, 1990, 33-39.

7. Markov, Z. & Ch. Dichev. The Net-Clause Language - A Tool for Data-Driven Inference, In: *Logics in AI*, Proceedings of European Workshop JELIA'90, Amsterdam, The Netherlands, September 1990, pp. 366-385 (Lecture Notes in Computer Science, No.478, Springer-Verlag, 1991).

8. Markov, Z., L. Sinapova and Ch. Dichev. Default reasoning in a network environment, in: *Proceedings of ECAI-90*, Stockholm, Sweden, August 6-10, 1990, pp.431-436.

9. Hinton, G.E. Learning distributed representations of concepts. In: *Proceedings of the Eight Annual Conference of the Cognitive Science Society*, Amherst, MA: Lawrence Erlbaum, 1986.

10. Quinlan, J.R. Learning logical definitions from relations. *Machine Learning*, Vol 5, No.3, August 1990, pp.239-266. Kluwer Academic Publishers, Boston.

Extending Abduction
from Propositional to First-Order Logic

Pierre Marquis

CRIN / INRIA-Lorraine
Campus Scientifique - B.P. 239
54506 Vandœuvre–lès–Nancy Cedex
FRANCE
e-mail: marquis@loria.crin.fr

Abstract

Abduction is often considered as inference to the best explanation. In this paper, we appeal to abduction as a way to generate all but only explanations that have "some reasonable prospect" of being valid.
We first provide a logical definition of abduction. We next study how abductive reasoning can be mechanized in propositional logic. We finally propose the generalization of this study to first–order logic and highlight main obstacles for mechanizing abduction in that frame.

1 Introduction

Abduction is a basic form of logical inference pioneered by Aristotle (Aristotle, ?) and further investigated by philosophers, logicians (in particular, C.S. Peirce (Peirce, 1931)) and AI researchers. The essence of abductive inference is the generation of the best factual explanations as to why a phenomenon is observed given what is already known (Stickel, 1988).

In this paper, we appeal to abduction as a way to generate all but only explanations that have "some reasonable prospect" of being valid. Section 2 defines abductive reasoning in a logical frame and characterizes the explanations to be preferred in a formal way. Section 3 is dedicated to the mechanization of abduction in propositional logic. We first describe the explanations to be preferred in terms of prime implicant / implicate. We next present some methods for implementing abduction. We also precise the role of the preferred explanations among all the explanations before providing some computational issues. Section 4 proposes the generalization of this study to first–order logic and highlights main obstacles for mechanizing abduction in that frame. In particular, we prove that a first–order formula is not always equivalent to the conjunction (even infinite) of its prime implicates. We finally present two classes of first–order formulas in which abductive reasoning can be performed. Section 5 concludes this paper.

2 Abduction

In this section, we first provide a logical definition of abduction. We next describe the explanations to be preferred in formal terms. We then propose an alternative characterization of abduction with regards to deduction.

From now on, we assume the reader familiary with the syntax and the semantics of propositional and first-order logics. We consider a first-order language FORML built on non-logical symbols from L in the usual way. As in (Gallier, 1986), ⊨ denotes logical entailment and ▪ denotes logical equivalence.

Let $\mathcal{Th}$, f and h be three closed formulas of FORML, which represent (respectively) the given knowledge about the considered domain, an observed event of this domain and an explanation of this one. Let us assume that f is compatible with $\mathcal{Th}$ (i.e. ($\mathcal{Th} \wedge f$) is satisfiable) but f is not a logical consequence of $\mathcal{Th}$. In this situation, $\mathcal{Th}$ does not explain f. Hence, it is necessary to complete $\mathcal{Th}$, that means to derive additional facts h explaining f in the intended interpretation described by $\mathcal{Th}$.

2.1 A Formal Definition of Abduction

> *Definition 2.1*[1]   An existentially quantified conjunction h of literals is an *abductive explanation* of f w.r.t. $\mathcal{Th}$ if $\mathcal{Th}$, $h \models f$.

Example 2.1

Let $\mathcal{Th}$ be the formula $\forall x$ (elephant(x) $\Rightarrow$ grey(x)) and f be the formula grey(Babar).

Then elephant(Babar), (elephant(Celeste) $\wedge$ ¬grey(Celeste)),

(elephant(Babar) $\wedge$ woman(The_old_lady)) are three abductive explanations of f w.r.t. $\mathcal{Th}$.

Note that we require abductive explanations to be conjunctions of facts. This syntactical restriction allows to distinguish abduction between other explanation inference models, *e.g.* inductive generalization, which can be considered as a way to derive general laws in implicative forms. Moreover, as we shall see in the next section, such a syntactical restriction is necessary to ensure that important semantical restrictions on explanations are significant.

2.2 Preferred Explanations

A key issue in abductive reasoning is to pick up the best explanations. But it seems so subjective and task–dependant to characterize them that there is no hope of devising a general algorithm that will compute only the best explanations. In addition, the validity of abductive explanations in the intended model cannot be deductively established in the general case since abduction is a purely inductive model of reasoning (Kugel, 1977). Thus, *there is no truth–preserving way to generate abductive explanations.*

For supporting the essence of abductive inference, some restrictions are needed in order to guarantee that the explanations are interesting in some sense. Various abductive schemes have been developed for discriminating such explanations (Stickel, 1988).

In the following sections, we only consider explanations that have "some reasonable prospect" of being valid. Computing such explanations is required, for instance, in truth maintenance (Reiter and de Kleer, 1987) and diagnostic problems (El Ayeb *et al.*, 1990a). These explanations can be formally characterized by two semantical properties, *consistency* and *minimality* (Cox and

---

[1] This definition is the classical one (Pople, 1973) (Morgan, 1975). An alternative and more general definition is due to P.T. Cox and T. Pietrzykowski who do not restrict the prefixes of abductive explanations to purely existential ones (Cox and Pietrzykowski, 1986).

Pietrzykowski, 1986). Intuitively, minimal and consistent abductive explanations are interesting in the sense that they are not too general and do not trivially imply the considered phenomenon by producing an inconsistency with the given knowledge.

Definition 2.2   An abductive explanation h of f w.r.t. Th is consistent (with Th) if ($\mathit{Th} \wedge h$) is a satisfiable formula.

Definition 2.3   An abductive explanation h of f w.r.t. Th is minimal if every abductive explanation of f w.r.t. Th being a logical consequence of h is equivalent to h.

The minimality property eliminates explanations that are unnecessarily general. In other words it prevents from over–restricting the models of the abductive explanations.

Example 2.1 (continued)
    elephant(Babar) is a minimal abductive explanation of grey(Babar) w.r.t. $\forall x$ (elephant(x) $\Rightarrow$ grey(x)). (elephant(Babar) $\wedge$ woman(The_old_lady)) is not a minimal abductive explanation of grey(Babar) w.r.t. $\forall x$ (elephant(x) $\Rightarrow$ grey(x)).

Consistency eliminates explanations that are not related to f but, because they produce unsatisfiability with Th, are explanations nevertheless. Such explanations have to be removed since they are not valid in the intended interpretation.

Example 2.1 (continued)
    elephant(Babar) is a consistent abductive explanation of grey(Babar) w.r.t. $\forall x$ (elephant(x) $\Rightarrow$ grey(x)). (elephant(Celeste) $\wedge$ ¬grey(Celeste)) is an inconsistent abductive explanation of grey(Babar) w.r.t. $\forall x$ (elephant(x) $\Rightarrow$ grey(x)).

Note that, if the abductive explanations of a phenomenon w.r.t. a theory were not syntactically constrained, then the set of all the minimal and consistent explanations of f w.r.t. Th would be reduced to $\{(\mathit{Th} \Rightarrow f)\}$ (up to logical equivalence) (Marquis, 1989). But ($\mathit{Th} \Rightarrow f$) is surely not an interesting explanation of f since it is a logical consequence of f.

2.3 Abduction vs. Deduction

As has been already pointed out by H.E. Pople, the process of abductive inference can be accomplished by means of a procedure based on the machinery already developed for deductive logic. Resolution-based abductive systems (Pople, 1973) (Cox and Pietrzykowski, 1986) and other proof–theoretic methods for mechanizing abduction (Morgan, 1971) (Morgan, 1975) are built upon the following model–theoretic simple result:

$$\mathit{Th}, h \models f \text{ iff } \mathit{Th}, \neg f \models \neg h.$$

This duality fully explains why abductive reasoning can be performed in a deductive way: *Inferring explanations can be viewed, for the main part, as inferring theorems* (Plotkin, 1971). The following sections will show such a duality to be of the utmost value in mechanizing abduction.

3 Mechanizing Abduction in Propositional Logic

In this section, we first define the minimal explanations of a phenomenon w.r.t. a theory in terms of prime implicant /implicate. We next provide some methods for computing consistent and minimal explanations. We finally show that the minimal explanations of a phenomenon characterize all its explanations.

3.1 Characterizing Minimal Explanations

Let us consider a propositional language PROPps freely generated by a countably infinite set $PS = \{P, Q, R ...\}$ of atoms. A *cube* C is a finite conjunction $(L_1 \wedge L_2 \wedge ... \wedge L_n)$ of literals with no literal repeated. A *clause* C is a finite disjunction $(L_1 \vee L_2 \vee ... \vee L_n)$ of literals with no literal repeated. Cubes and clauses are considered as sets of literals when it is convenient. CUBE is the subset of PROPps which contains every cube. CLAUSE is the subset of PROPps which contains every clause. CUBE / ≡ (resp. CLAUSE / ≡) denotes the quotient set of CUBE (resp. CLAUSE) by the equivalence relation ≡.

According to Definition 2.1, the abductive explanations of a phenomenon f w.r.t. a theory Th are the cubes h such that $Th, h \models f$. But $Th, h \models f$ iff $h \models (Th \Rightarrow f)$. Consequently, the abductive explanations h of f w.r.t. Th are the implicants of the formula $(Th \Rightarrow f)$. According to Definition 2.3, the minimal ones are the maximal implicants for the ordering induced by $\models$ in CUBE / ≡. These implicants are exactly the prime implicants of $(Th \Rightarrow f)$:

Definition 3.1   A *prime implicant* of a formula A of PROPps is a cube C such that:
  - $C \models A$;
  - if C' is a cube such that $C' \models A$ and $C \models C'$ then $C' \models C$.

The concept of prime implicant (due to W.V. Quine (Quine, 1959)) appears to be the key for characterizing the minimal explanations of a phenomenon. The prominent part taken by such a notion in propositional abductive reasoning has been already pointed out in dual terms in (Reiter and de Kleer, 1987).

Since $h \models (Th \Rightarrow f)$ iff $(Th \wedge \neg f) \models \neg h$, the abductive explanations h of f w.r.t. Th are the negations of the implicates of $(Th \wedge \neg f)$ and the minimal ones are the negations of the prime implicates of $(Th \wedge \neg f)$ where:

Definition 3.1'   A *prime implicate* of a formula A of PROPps is a clause C such that:
  - $A \models C$;
  - if C' is a clause such that $A \models C'$ and $C' \models C$ then $C \models C'$.

Fortunately, Reiter-de Kleer's and our own characterization of minimal abductive explanations are equivalent, as shown by the next theorem:

Theorem 3.1   C is a prime implicant of A iff $\neg C$ is a prime implicate of $\neg A$.

3.2 Computing Consistent and Minimal Explanations

In this section, we first provide some methods for computing the minimal explanations of a phenomenon w.r.t. a theory. We next cope with the problem of selecting the consistent explanations among all the minimal ones and we show that such a selection can be accomplished "on-the-fly" during the computation of the minimal explanations. We finally present some computational issues.

3.2.1 Computing Minimal Explanations

We review hereafter some classical *variable–elimination* methods for generating the prime implicants and the prime implicates of propositional formulas.

Let us consider two sets of literals C_1 and C_2 that contain respectively the literal L_1 and the literal $\neg L_1$. The following inference rule between sets of literals:

$$\frac{C_1 \quad C_2}{C_1 \cup C_2 \setminus \{L_1 \, \neg L_1\}}$$

is referred to as the *consensus rule* when C_1 and C_2 represent cubes (Kuntzmann and Naslin, 1967) and referred to as the *resolution rule* when C_1 and C_2 represent clauses (Robinson, 1965). The resulting set is a consensus of C_1 and C_2 in the first case and a resolvent of C_1 and C_2 otherwise.

A consensus derivation of C_n from a set of cubes S (considered as the disjunction of its elements) is a finite (and non empty) sequence $C_1, ..., C_n$ of cubes such that every C_i is an element of S or is a consensus of two cubes C_j, C_k with $j < i$ and $k < i$.

We note $S \vdash_C C$ if there exists a consensus derivation of C from S. We define the general consensus strategy C by $C(S) = S \cup \{C_3 \mid C_1, C_2 \in S, C_3$ is a consensus of C_1 and $C_2\}$. We define also $C^0(S) = S$, $C^1(S) = C(S), ..., C^{i+1}(S) = C(C^i(S)), ...$

Applying the consensus method requires the transformation of the considered formula A into disjunctive normal form DNF(A).

Let $max_\models(C^i(DNF(A)))$ be the set of the maximal elements of $C^i(DNF(A))$ for $\models$. Applying the consensus method for generating prime implicants consists in determining the first integer k such that
$max_\models(C^k(DNF(A))) = max_\models(C^{k+1}(DNF(A)))$:

A resolution derivation of C_n from a set of clauses S (considered as the conjunction of its elements) is a finite (and non empty) sequence $C_1, ..., C_n$ of clauses such that every C_i is an element of S or is a resolvent of two clauses C_j, C_k with $j < i$ and $k < i$.

We note $S \vdash_R C$ if there exists a resolution derivation of C from S. We define the general resolution strategy R by $R(S) = S \cup \{C_3 \mid C_1, C_2 \in S, C_3$ is a resolvent of C_1 and $C_2\}$. We define also $R^0(S) = S$, $R^1(S) = R(S), ..., R^{i+1}(S) = R(R^i(S)), ...$

Applying the resolution method requires the transformation of the considered formula A into conjunctive normal form CNF(A).

Let $min_\models(R^i(CNF(A)))$ be the set of the minimal elements of $R^i(CNF(A))$ for $\models$. Applying the resolution method for generating prime implicates consists in determining the first integer k such that
$min_\models(R^k(CNF(A))) = min_\models(R^{k+1}(CNF(A)))$:

$max_\models(C^k(DNF(A)))$ *is the set of prime implicants of A, noted* $PI(A)$.

$min_\models(R^k(CNF(A)))$ *is the set of prime implicates of A, noted* $IP(A)$.

Since resolution is used to prove contradiction, consensus can be used to prove validity. This rule is *falsity–preserving*: If an interpretation I is not a model of C_1 and not a model of C_2, then I is not a model of any consensus C_3 of C_1 and C_2. In other words every model of C_3 is a model of $(C_1 \lor C_2)$. Further, the empty cube is valid. So, if DNF(A) $\vdash_C$ { }, then A is a valid formula. The converse is also true: The general consensus strategy is *complete for validity*. This property can be proved as a trivial consequence of the duality between consensus and resolution, shown in the following theorem:

Theorem 3.2 $C_1, ..., C_n$ is a consensus derivation from DNF(A) iff $\neg C_1, ..., \neg C_n$ is a resolution derivation from CNF($\neg$A).

From Theorems 3.1 and 3.2, the general consensus (resp. resolution) method can be used to generate prime implicates (resp. implicants). These simple methods are clearly brute-force ones. However, they can be refined since, as we shall see in Section 4, every resolution strategy which is *complete in consequence-finding* can be used to compute prime implicates (or implicants). Finally, some other methods for deriving prime implicants (and implicates), such as the *multiplication methods*, exist (Slagle *et al.*, 1970) (Jackson and Pais, 1990) (Marquis, 1991). Multiplication methods are based on the following theorem:

Theorem 3.3   Let A and B be two propositional formulas.
$$\mathcal{PI}((A \land B)) = \max_{\vdash}(\{(C_A \land C_B) \mid C_A \in \mathcal{PI}(A), C_B \in \mathcal{PI}(B)\})$$
$$\mathcal{IP}((A \lor B)) = \min_{\vdash}(\{(C_A \lor C_B) \mid C_A \in \mathcal{IP}(A), C_B \in \mathcal{IP}(B)\}).$$

In contrast to consensus-based and resolution-based methods, these techniques cannot be easily extended from propositional to first-order logic. This is the reason why we shall not detail them further in this paper.

3.2.2 Computing Consistent Explanations

When the minimal explanations of f w.r.t. $\mathcal{T}h$ are computed, we have to remove those which are not consistent with $\mathcal{T}h$. It is obvious that such an elimination can be performed since propositional satisfiability is decidable.

A much better method would avoid the generation of inconsistent explanations when computing the minimal ones. Unfortunately, such a method does not exist: As pointed out by M.E. Stickel (Stickel, 1988), the consistency of explanations must be checked outside the abductive reasoning inference system. However, the next theorem shows that there is a very simple elimination technique in propositional logic:

Theorem 3.4 $\mathcal{PI}((\mathcal{T}h \Rightarrow f)) \setminus \mathcal{PI}(\neg\mathcal{T}h)$ is the set of all the minimal and consistent abductive explanations of f w.r.t. $\mathcal{T}h$.

Using an *incremental algorithm* for computing prime implicants, such as IPIA (Kean and Tsiknis, 1990), allows to compute the prime implicants of $(\mathcal{T}h \Rightarrow f)$ incrementally from the prime implicants of $\neg\mathcal{T}h$. Therefore, implementing abduction requires only to store the prime implicants of $\neg\mathcal{T}h$ when they are generated and to (eventually) remove them afterwards from the set of prime implicants of $(\mathcal{T}h \Rightarrow f)$. If $\mathcal{T}h$ is not frequently modified, storing the prime implicants of $\neg\mathcal{T}h$ is a good way to improve (in practice "only") both the computation of $\mathcal{PI}((\mathcal{T}h \Rightarrow f))$ and the elimination step when several phenomena f are to be explained. Additionally, we know that the compiling of knowledge bases as their sets of prime implicates allows to significantly enhance the performance of numerous AI systems (Mathieu and Delahaye, 1990). It is then worthwhile noting that, when a knowledge base is logically compiled, much of the effort for performing abduction is done: From Theorem 3.1, the prime implicants of $\neg Th$ are exactly the negations of the prime implicates of Th.

3.2.3 Some Computational Issues

Though effective in propositional logic, *abduction cannot be mechanized efficiently in general*. There are two reasons for that: On one hand, the number of prime implicants (or prime implicates) of a propositional formula can be exponential in the number of its atoms in the worst case (see, for instance, the formula $(... (L_1 \Leftrightarrow L_2) \Leftrightarrow L_3) \Leftrightarrow L_4) ... \Leftrightarrow L_i) ...)$ where L_i are literals). Moreover, G.M. Provan has recently proved that this worst case is reached for almost all propositional formulas (Provan, 1990).

On the other hand, the complexity of mechanizing abduction is not only due to the number of prime implicants. As shown in (Selman and Levesque, 1990), computing a consistent and minimal explanation h of a propositional atom w.r.t. a propositional Horn theory, such that h is built on a given subset of PS, is NP-hard. Note that this complexity does not come from the subproblem of checking consistency since it can be solved in polynomial time in this situation (Dowling and Gallier, 1984).

3.3. Minimal Explanations vs. Explanations

As pointed out in (Reiter and de Kleer, 1987), the sets $PI(A)$ and $IP(A)$ can be viewed as bases of A in CUBE / $\equiv$ and in CLAUSE / $\equiv$ (respectively). There is no loss of information in representing A by its prime implicants or by its prime implicates.

Theorem 3.5   Let C be a cube and A, B be two formulas:
1. $C \models A$ iff there exists a prime implicant C' of A such that $C \models C'$;
2. $A \equiv \vee (C \mid C \in PI(A)) \, C$;

Theorem 3.5'   Let C be a clause and A, B be two formulas:
1'. $A \models C$ iff there exists a prime implicate C' of A such that $C' \models C$;
2'. $A \equiv \wedge (C \mid C \in IP(A)) \, C$.

Properties 3.5.1 and 3.5.1', which are known as *covering properties*, highlight the central role taken by the minimal abductive explanations of a phenomenon among all its abductive explanations. To be more specific, they show that *minimal abductive explanations describe all abductive explanations*: If h' is an abductive explanation of f w.r.t. Th which is valid in the intended model, then there exists at least one minimal abductive explanation h of f w.r.t. Th such that h is a logical consequence of h'. Consequently, h is also valid in the intended model. Moreover, assuming f to be valid in the intended interpretation allows to state that $(Th \Rightarrow f)$ is valid. Hence, at least one minimal abductive explanation h is valid in the intended interpretation. Note that it does not mean that abduction is a valid form of reasoning: There is generally no deductive way to choose the valid explanations of f.

4 Mechanizing Abduction in First-Order Logic

In this section, we first generalize the propositional definitions of consistent and minimal explanations to first-order logic. We next investigate the mechanization of abduction in that frame. We then show that the minimal explanations of a phenomenon do not generally characterize all its explanations. In addition, we propose two classes of first–order formulas for which the covering properties remain true and in which abduction is mechanizable.

4.1 Characterizing Minimal Explanations

The propositional notions of cube and clause can be extended to first–order logic: A (first–order) cube (resp. clause) C is a formula of FORML the prefix of which is only composed of existential (resp. universal) quantification and the matrix is a finite conjunction (resp. disjunction) of literals with no literal repeated. As usual, cubes and clauses are represented as sets of literals when it is convenient.

In a first-order frame, the minimal explanations of f w.r.t. Th can still be considered as the prime implicants of $(Th \Rightarrow f)$ (or the negations of the prime implicates of $(Th \land \neg f)$) where:

Definition 4.1 A (first–order) *prime implicant* of a formula A of FORML is a cube C such that: - $C \vDash A$; - if C' is a cube such that $C' \vDash A$ and $C \vDash C'$ then $C' \vDash C$.	**Definition 4.1'** A (first–order) *prime implicate* of a formula A of FORML is a clause C such that: - $A \vDash C$; - if C' is a clause such that $A \vDash C'$ and $C' \vDash C$ then $C \vDash C'$.

As a consequence of the duality between existential and universal quantification[1], Theorem 3.1 is still valid in first-order logic. Hence, prime implicants and prime implicates can be generated by the same methods in that frame. Note that, though much research has been devoted to the notion of prime implicant in the propositional case (see, for instance, (Kuntzmann and Naslin, 1967)), this concept has never been tackled in first-order logic. The forthcoming sections attempt to fill this gap.

4.2 Computing Consistent and Minimal Explanations

In this section, we show to what extent the propositional methods for computing minimal and consistent explanations can be generalized to first-order logic.

4.2.1 Quantifier-Free Normal Forms

Like resolution, the consensus method, also known as *inductive resolution*, has been extended from a propositional frame to a first-order one (Morgan, 1975). As clausal transformation is necessary for resolution, the (first-order) consensus method requires a transformation step into *cubic normal form*. This transformation proceeds in two steps. The considered formula is first transformed into prenex disjunctive normal form in the usual way. Then, universal quantifiers must be removed. This last step is closed to skolemization but must preserve validity (since consensus is complete for validity). Let A be a formula in prenex disjunctive normal form and let A_{skc} be its universal quantifier–free normal

---

[1] Quantification $\exists x$ can be viewed as a notation for $\neg \forall x \neg$.

form. Askc is defined as $\neg((\neg A)skr)$ where $(\neg A)skr$ is the simplest usual Skolem normal form corresponding to the prenex conjunctive normal form $\neg A$. Obviously, skc is a transformation which preserves validity since skr preserves satisfiability: A is valid iff $\neg A$ is unsatisfiable iff $(\neg A)skr$ is unsatisfiable iff $\neg((\neg A)skr)$ is valid. From the duality between skc and skr, a skc transformation of A consists in replacing in A any occurrence of a universally quantified variable x in the scope of existential quantifications $\exists x_1, ..., \exists x_n$ by the term $\alpha(x_1, ..., x_n)$ where α is a Skolem function.

The Skolem transformations skr and skc generally modify the set of prime implicants (and prime implicates) of the formula A on which they are applied. For instance, let A be the formula $\forall x\ P(x)$. Then Askc is $P(\alpha)$. There is no satisfiable prime implicants of A but the set of prime implicants of Askc is $\{P(\alpha)\}$. We shall not be concerned with this problem in this paper (see (Cox and Pietrzykowski, 1984) for more details) but we only consider the generation of prime implicants for purely existential formulas and of prime implicates for purely universal formulas. Note that, $(Th \Rightarrow f)$ is purely existential (and $(Th \wedge \neg f)$ is purely universal) when Th is purely universal and f is purely existential.

4.2.2 Generating Prime Implicants/Implicates

We shall show in the following that the consensus-based and the resolution-based methods can be extended to derive first-order prime implicants (implicates).

Let us first indicate how these methods are generalized: A consensus (resolvent) of two cubes (clauses) C_1 and C_2 is a binary consensus (resolvent) of C_1 and C_2, or a binary consensus (resolvent) of C_1 and a factor of C_2, or a binary consensus (resolvent) of a factor of C_1 and C_2, or a binary consensus (resolvent) of a factor of C_1 and a factor of C_2. Factor and binary consensus (resolvent) are defined by:

• If a substitution σ is a most general unifier of two (or more) literals from C, then Cσ is a factor of C ;

• If there exists two literals L_1 and L_2 with opposite signs in two cubes (clauses) C_1 and C_2 that do not share common variables, and if σ is a most general unifier of L_1 and L_2, then the cube (clause) $(C_1\sigma \setminus \{L_1\sigma\}) \cup (C_2\sigma \setminus \{L_2\sigma\})$ is a binary consensus (resolvent) of C_1 and C_2.

From these definitions, it clearly follows that Theorem 3.2 still holds in first-order logic. More generally, every property related to resolution is closed to a dual one about consensus. Since a great deal of study has been devoted to resolution since J.A. Robinson's work, we shall refer to this notion for describing how to generate prime implicants (and implicates). To be more specific, we shall compute the minimal explanations of f w.r.t. Th as negations of the prime implicates of $(Th \wedge \neg f)$.

As evoked above, the general resolution strategy introduced in Section 3 can be used to derive the prime implicates of any set of first-order clauses S (in particular, when S represents the clausal normal form of $(Th \wedge \neg f)$). This property is a consequence of *the subsumption theorem* (Lee, 1967):

Theorem 4.1   Let S be a set of clauses and C a non-tautological clause. If C is a logical consequence of S then there exists a resolution derivation of a clause C' from S such that C' subsumes C.

Note that this theorem also holds for some refinements[1] of resolution. For example, it remains true for linear resolution (Minocozzi and Reiter, 1972) but it is not valid in general for semantic resolution or ordered resolution[2]. For instance, let us consider $S = \{(\neg P \vee Q), (\neg Q \vee R)\}$ and the prime implicant $(\neg P \vee R)$ of S. On one hand, S admits no I–resolution step for any interpretation I that contains $\{\neg P, R\}$, due to lack of electron. In fact, only the prime implicates of S which are false in I can be obtained by I–resolution from S (Slagle *et al.*, 1969). On the other hand, S admits no O–resolution step with the ordering O defined by $Q < P < R$ which excludes resolution on the literal Q.

When the subsumption theorem holds for R, R is (by definition) complete in consequence–finding. The interest of this notion for generating prime implicates relies on the following theorem:

Theorem 4.2   If C is a prime implicate of a set of clauses S and if R is complete in consequence–finding, then $S \vdash_R C'$ with $C' \equiv C$.

4.2.3 Some Computational Issues

From Theorem 4.2, it is sufficient to remove the implied clauses from the set of all resolvents from S to generate the prime implicates of S. Unfortunately, it is not possible to decide logical entailment between clauses in the general case (Schmidt–Schauβ, 1986).

More generally, there is no way to test whether a clause is a prime implicate of a given set of clauses: *Computing prime implicates is not semi–decidable in first–order logic* since S is satisfiable iff there exists a prime implicate of S which is not the empty clause. For the same reason, there is no effective way to check the consistency of explanations in the general case. Finally, as the next section will show, it is generally impossible to perform abduction in first-order logic, even for classes of formulas in which the logical entailment relation (hence, the consistency one) is decidable.

4.3 Minimal Explanations vs. Explanations

In this section, we first show that the minimal abductive explanations of a phenomenon are not sufficient for characterizing all its explanations in the general case. We next propose some classes of first-order formulas for which the covering properties remain true and in which abduction is mechanizable.

4.3.1 The General Case

In contrast to propositional logic, the existence of a basis of prime implicates (resp. prime implicants) for a set of first-order clauses (resp. cubes) S is not ensured. Intuitively, there are eventually *some holes in the prime implicates covering* of S, as the next theorem shows:

Theorem 4.3   A set of first-order clauses S is not always equivalent to the conjunction (even infinite) of its prime implicates: If C is a logical consequence of S, a prime implicate C' of S such that $C' \vDash C$ does not always exist.

---

[1] By definition, a resolution strategy R is a mapping from a set of clauses S to a subset of the union of S with the set of all the instances of all the resolvents of S.
[2] However, ordered resolution can be used to generate only prime implicates built on a prespecified set of predicates. This is helpful in some problems, such as diagnostic, where predicate specific abduction is required (El Ayeb *et al.*, 1990b).

For instance, let us consider the set:

$$S = \{\forall x\, \forall y\, \forall z\, ((P(x,y) \wedge P(y, z)) \Rightarrow P(x, z)),\ \forall x\, \neg P(x, x)\}.$$

$C = \forall x\, \forall y\, (P(x,y) \Rightarrow \neg P(y, x))$ is a non-tautological consequence of S. Assume that there is a prime implicate C' of S such that C' $\models$ C. From Theorem 4.2, a clause equivalent to C' can be reached by applying any resolution strategy complete in consequence-finding on S. We shall now prove that the previous assumption leads to a contradiction by analyzing the set of implicates of S which implies C. Let us consider the general resolution strategy R introduced in Section 3 and the set $R^n(S)$, $n \geq 0$. By induction on n:

- There is no positive clause in $R^n(S)$;
- Negative clauses in $R^n(S)$ are variants of
$C^-_i = \forall\, (\neg P(x_0, x_1) \vee \neg P(x_1, x_2) \vee \ldots \vee \neg P(x_i, x_0))$ with $0 \leq i \leq n$;
- Other clauses (mixed ones) in $R^n(S)$ are variants of
$C_j = \forall\, (\neg P(x_{-2}, x_{-1}) \vee \neg P(x_{-1}, x_0) \vee \ldots \vee \neg P(x_{j-1}, x_j)) \vee P(x_{-2}, x_0)$ with $0 \leq j \leq n$.

For n = 0, this property obviously holds. Let us prove it for n = k + 1. For the resolution step k + 1, the positive resolvents can only be derived by resolution between the positive clauses and the negative or mixed ones, computed during the previous steps. But this conflicts with the fact that there is no positive clause in $R^k(S)$ (induction hypothesis). In the same vein, the negative resolvents can only be obtained by resolving between the negative clauses and the positive or mixed ones, computed during the previous steps. Hence, it is sufficient to consider how resolution applies between C^-_i and C_j with $0 \leq i, j \leq k$. It is easy to see that any resolution step between C^-_i and C_j provides a variant of C^-_{i+j-2}. Finally, the mixed resolvents in $R^{k+1}(S)$ can be generated by resolution between:

- The mixed clauses and the positive, negative or mixed ones ;
- The positive clauses and the negative ones

computed during the previous steps. From the previous case, we know that resolving between mixed clauses and negative ones does not provide any mixed clause. Consequently, it is sufficient to consider how resolution applies between C_i and C_j with $0 \leq i, j \leq k$. It is easy to see that any resolution step between C_i and C_j provides a variant of C_{i+j-2}.

Let us now prove that the implicates C' of S which implies C are necessarily negative resolvents. If C' implies C then $(C' \wedge (\neg C)skr)$ is unsatisfiable. But $(\neg C)skr$ is a conjunction of positive literals. As a consequence, C' is a negative clause:

- If C' is a positive clause, there is no resolvent from $(C' \wedge (\neg C)skr)$;
- If C' is a mixed (and eventually self-resolving clause), there is at least one positive literal in each resolvent from $(C' \wedge (\neg C)skr)$.

In these cases, the empty clause cannot be obtained by resolution from $(C' \wedge (\neg C)skr)$. Hence, since R is refutationally complete, $(C' \wedge (\neg C)skr)$ is satisfiable if C' is not a negative clause.

Finally, the implicates of S which imply $C = C^-_1$ are not prime ones.

For all $i \geq 1$, $C^-_{2i-1} = \forall\, (\neg P(x_0, x_1) \vee \neg P(x_1, x_2) \vee \ldots \vee \neg P(x_{i-2}, x_{i-1}) \vee \neg P(x_{i-1}, x_i) \vee \ldots \vee \neg P(x_{2(i-1)}, x_{2i-1}) \vee \neg P(x_{2i-1}, x_0))$ subsumes $C^-_{i-1} = \forall\, (\neg P(x_0, x_1) \vee \neg P(x_1, x_2) \vee \ldots \vee \neg P(x_{i-2}, x_{i-1}) \vee \neg P(x_{i-1}, x_0))$ $(i \geq 1)$ with $\sigma_i = \{(x_{i-1}, x_0), (x_i, x_1), \ldots, (x_{2i-1}, x_{i-2}), (x_0, x_{i-1})\}$ but C^-_{i-1} does not subsume C^-_{2i-1}. Since subsumption and implication are equivalent on negative clauses (Gottlob, 1987), no prime implicates of S implies C.

From a theoretical point of view, note that the previous counter-example is closed to formulas from the *Bernays–Schönfinkel class*, which is *solvable* (satisfiability for this class is decidable (Dreben and Goldfarb, 1979)). So solvability is not sufficient to ensure that the covering properties hold.

From an applied point of view, it is worthwhile noting that such a problematical set of clauses S is not out of the ordinary since S just describes a binary relation P which is transitive and irreflexive, hence asymmetric. There are numerous axiomatizations of AI problems in which a transitive and irreflexive relation occurs. Let us mention, for instance, the relation < of D. McDermott (McDermott, 1982) and the relation before of J.F. Allen (Allen, 1983) in the field of temporal reified logics.

Another important problem which concerns the mechanization of abduction in first–order logic lies in the possible infinitude of the prime implicants/implicates of a formula. Considering only constant function symbols and requiring solvability are not sufficient to cope with this problem in the general case. For example, with:

$$\mathcal{Th} = \forall x\ \forall y\ \forall z\ ((P(x,\ y) \wedge P(y,\ z)) \Rightarrow P(x,\ z))$$
$$f = P(a,\ b),$$

each of the following conjunctions:

$$h_0 = f = P(a,\ b)$$
$$h_1 = \exists x_1\ (P(a,\ x_1) \wedge P(x_1,\ b))$$
$$h_2 = \exists x_1\ \exists x_2\ (P(a,\ x_1) \wedge P(x_1,\ x_2) \wedge P(x_2,\ b)) \dots$$

is a minimal and consistent explanation of f w.r.t. $\mathcal{Th}$.

4.3.2 Some Effective Subcases

Fortunately, there exist some classes of first-order formulas for which the covering properties remain true. As a consequence, we propose hereafter two classes of first-order formulas in which abduction is mechanizable.

We first need the following result. Let the *depth* of a clause be the maximal depth of its literals, where the depth of a literal is inductively defined in the usual way (see, for instance, (Gallier, 1986)). Then, for every clause C and C', if the depth of C' is strictly greater than the depth of C and C is not a tautology, then C is not a logical consequence of C' (Schmidt–Schauß, 1986). Consequently, the maximal depth of the clauses C' which imply C is *bounded*.

A/ 2-Clauses

Let us consider a set S of 2–clauses (*i.e.* clauses the matrix of which are composed of at most two literals). The *size* of the resolvents from S (that is, the number of literals occurring in them) is (at most) two. Therefore, the set of the resolvents from S which imply C is finite (up to logical equivalence). Hence, the set of the prime implicates of S which imply C is finite.

Since the logical entailment between 2-clauses is decidable (Schmidt–Schauß, 1986), abduction is mechanizable in this situation when no function symbol occurs (this last constraint on S is necessary to prevent from generating an infinite set of prime implicates $I\mathcal{P}(S)$: If

$S = \{\forall x \ (P(x) \ \Rightarrow \ P(f(x)))$, $P(a)\}$ then $I\mathcal{P}(S) = \{\forall x \ (P(x) \ \Rightarrow \ P(f(x)))$, $P(a)$, $P(f(a))$, $P(f(f(a)))$, $...\}$ is infinite).

B/ Monadic logic

Similar arguments allow us to conclude that the prime implicates covering property holds for any set of clauses S and any non-tautological clause C which are built solely on unary and constant symbols.

In this case, the size of any depth-bounded clause C' cannot grow infinitely without providing a subsuming factor of C', that is, a clause C" which is a factor of C' and which subsumes C". Such a clause C" is obviously equivalent to C'. Hence, the set of the resolvents from S which imply C is finite (up to logical equivalence) and, as a consequence, the set of the prime implicates of S which imply C is finite.

Since the logical entailment between such clauses is decidable, abduction is mechanizable in the first-order monadic logic (that is, the logic the language of which is built on unary predicate and constant function symbols only).

This last result is considerable for some predicate specific abduction problems, such as circuit diagnostic, in which the assumable predicates (which represent the diagnostic states of the components) are unary and can be instantiated by constant symbols (which represent the considered components) only (El Ayeb et al., 1990a).

5 Conclusion

This paper has proposed abduction as a way to generate all the explanations of a phenomenon that have "some reasonable prospect" of being valid. In a first time, we have defined abduction in a logical frame and characterized the explanations to be preferred. We then have described these explanations in a propositional frame. Next, we have provided some methods to compute them and precised the role of the preferred explanations among all the explanations. The second part of this paper was dedicated to the generalization of this study to first–order logic. Main obstacles in mechanizing first–order abduction were presented. In particular, we have shown that a first-order formula is not necessarily equivalent to the conjunction (even infinite) of its prime implicates. Nevertheless, two classes of first–order formulas in which abductive reasoning can be performed were finally exhibited.

Bibliography

Allen, J.F. [1983]. «Maintaining Knowledge about Temporal Intervals», *Communications of the Association for Computing Machinery* 26, pp. 832-843.

Aristotle [?]. *Organon III : les premiers analytiques*, traduction et notes de J. Tricot, librairie philosphique J. Vrin, Paris, 1971.

Chang, C.L. and Lee, R.C.T. [1973]. *Symbolic Logic and Mechanical Theorem Proving*, Academic Press, New York.

Cox, P.T. and Pietrzykowski, T. [1984]. «A Complete, Nonredundant Algorithm for Reversed Skolemization», *Theoretical Computer Science* 28, pp. 239-261.

Cox, P.T. and Pietrzykowski, T. [1986]. «Causes for Events: Their Computation and Applications», Proc. *International Conference on Automated Deduction*, pp. 608–621.

Dowling, W.F. and Gallier, J.H. [1984]. «Linear-time Algorithms for Testing the Satisfiability of Propositional Horn Formulae», *Journal of Logic Programming* 3, pp. 267-284.

Dreben, B. and Goldfarb, W.D. [1979]. *The Decision Problem: Solvable Classes of Quantificational Formulas*, Addison-Wesley Publishing Company, Reading.

El Ayeb, B., Marquis, P. and Rusinowitch, M. [1990a]. «Deductive / Abductive Diagnosis: The DA–Principles», Proc. *European Conference on Artificial Intelligence*, pp. 47-52.

El Ayeb, B., Marquis, P. and Rusinowitch, M. [1990b]. «A New Diagnosis Approach by Deduction and Abduction», Proc. *International Workshop on Expert Systems in Engineering*, Lecture Notes in Artificial Intelligence 462, pp. 32-46.

Gallier, J.H. [1986]. *Logic for Computer Science - Foundations of Automatic Theorem Proving*, Harper & Row, New York.

Gottlob, G. [1987]. «Subsumption and Implication», *Information Processing Letters* 24, pp. 109–111.

Jackson, P. and Pais, J. [1990]. «Computing Prime Implicants», Proc. *International Conference on Automated Deduction* , Lecture Notes in Artificial Intelligence 449, pp. 543-557.

Kean, A. and Tsiknis, G. [1990]. «An Incremental Method for Generating Prime Implicants/Implicates», *Journal of Symbolic Computation* 9, pp. 185-206.

Kugel, P. [1977]. «Induction, Pure and Simple», *Information and Control* 35, pp. 276–336.

Kuntzmann, J. and Naslin, P. [1967]. *Algèbre de Boole et machines logiques*, Dunod, Paris.

Lee, R.C.T. [1967]. *A Completeness Theorem and a Computer Program for Finding Theorems Derivable from Given Axioms*, Ph.D. thesis, University of California.

Marquis, P. [1989]. «Computing Most Specific Generalisations in Propositional Calculus», Proc. *European Working Session on Learning*, pp. 135-138.

Marquis, P. [1991]. *Contribution à l'étude des méthodes de construction d'hypothèses en intelligence artificielle*, Ph.D. thesis, Université de Nancy I.

Mathieu, P. and Delahaye, J.P. [1990]. «The Logical Compilation of Knowledge Bases», Proc. *European Workshop JELIA'90*, Lecture Notes in Artificial Intelligence 478, pp. 386-398.

McDermott, D. [1982]. «A Temporal Logic for Reasoning About Processes and Plans», *Cognitive Science* 6, pp. 101-155.

Minicozzi, E. and Reiter, R. [1972]. «A Note on Linear Resolution Strategies in Consequence–Finding», *Artificial Intelligence* 3, pp. 175–180.

Morgan, C.G. [1971]. «Hypothesis Generation by Machine», *Artificial Intelligence* 2, pp. 179-187.

Morgan, C.G. [1975]. «Automated Hypothesis Generation Using Extended Inductive Resolution», Proc. *International Joint Conference on Artificial Intelligence*, pp. 351-356.

Peirce, C.S. [1931]. *Collected Papers of Charles Sanders Peirce*, vol. 2, C. Hartsthorne and P. Weiss (ed.), Harvard University Press, Cambridge.

Plotkin, G.D. [1971]. «A Further Note on Inductive Generalisation», *Machine Intelligence* 6, pp. 101–124.

Pople, H.E. Jr [1973]. «On the Mechanization of Abductive Logic», Proc. *International Joint Conference on Artificial Intelligence*, pp. 147-152.

Provan, G.M. [1990]. «The Computational Complexity of Multiple-Context Truth Maintenance Systems», Proc. *European Conference on Artificial Intelligence*, pp. 523-527.

Quine, W.V. [1959]. «On Cores and Prime Implicants of Truth Functions», *American Mathematical Monthly* 66, pp. 755-760.

Reiter, R. and de Kleer, J. [1987]. «Foundations of Assumption-Based Truth Maintenance Systems: Preliminary Report», Proc. *National Conference on Artificial Intelligence (AAAI 87)*, pp. 183-188.

Robinson, J.A. [1965]. «A Machine-oriented Logic based on the Resolution Principle», *Journal of the Association for Computing Machinery* 12:1, pp. 23-41.

Schmidt-Schauβ, M. [1986]. «Some Undecidable Classes of Clause Sets», SEKI–Report SR-86-08, Fachbereich Informatik, Kaiserslautern Universität.

Selman, B. and Levesque, H.J. [1990]. «Abductive and Default Reasoning: A Computational Core», Proc. *National Conference on Artificial Intelligence (AAAI 90)*, pp. 343-348.

Slagle, J.R., Chang, C.L. and Lee, R.C.T. [1969]. «Completeness Theorems for Semantic Resolution in Consequence-Finding», Proc. *International Joint Conference on Artificial Intelligence*, pp. 281-285.

Slagle, J.R., Chang, C.L. and Lee, R.C.T. [1970]. «A New Algorithm for Generating Prime Implicants», *IEEE Transactions on Computers* C-19(4), pp. 304-310.

Stickel, M.E. [1988]. «A Prolog–like Inference System for Computing Minimum–Cost Abductive explanations in Natural–Language Interpretation», Proc. *International Computer Science Conference*, Hong–Kong.

Building in Equational Theories into the Connection Method

Uwe Petermann

Leipzig University, Dept. of Informatics
Augustusplatz 10-11, O-7010 Leipzig, Germany

Abstract

We consider possibilities of building in equational theories into a proof procedure based on the connection method.

The connection method has been enhanced by the following rules for equality reasoning: paramodulation, RUE-resolution and demodulation (all modulo a built in equational theory). Completeness results are proved for the two former rules. The considered procedures accept arbitrary first order formulas. The translation of improvements from one class of proof calculi (resolution) to another (connection method) contributes to a better understanding of the relations between different proof calculi.

1 Introduction

Building in theories is a powerful method for the reduction of the search space encountered by reasoning procedures. The idea is rather simple. The inferences related to axioms and rules of the built in theory are treated by a specialized reasoner. This one is more efficient because it takes advantage of its knowlegde about the theory.

This technique has been used already at the early beginning of practical experiences with theorem proving. One has observed quickly that the uniform treatment of the axioms of the equality relation cause an enormous growth of the search space of the resolution process. The countermeasure were specialized rules for the treatment of the equality sign. Rules like paramodulation [RW69] and demodulation [WRCS67] allow a search space reduction. But they are still insufficient. Further improvements like J. B. Morris' E-resolution [Mo69] and V. J. Digricoli's RUE-resolution [DH86] have been invented in order to guide the paramodulation steps. The paramodulation and the demodulation principle have been refined giving equations a direction and including Knuth-Bendix like completion techniques (cf. [BG90], [Prae90].

A general view on building in theories has been given by M. Stickel with his concept of theory resolution [Sti85]. Special cases of this approach others than the already mentioned are for instance taxonometric theories.

The mentioned rules for equality reasoning have first been developed as improvements of the resolution method. For a consideration of the connection graph procedure with built in theories see [BB]. Theory resolution for the matrix method has been considered only for the ground case (cf. [Sti85]).

The aim of the present paper is to envolve the mentioned rules for equality reasoning into the connection method and to reconstruct completeness results within this framework. This should be helpful for a better understanding of the relations between different proof calculi.

The lifting of M. Stickel's result concerning the matrix method to full first-order predicate calculus has been done by the present author in [Pe90a,Pe90b] (cf. subsection 2.4 below). The connection procedure presented there accepts formulas which are neither skolemized nor in clausal form. To avoid Skolemization seems to be important because Skolemization may destroy syntactical properties of formulas which are important for the unification. For an example see [DEL90].

A general characterization of equational theory connections (cf. subsection 2.3 below) has been given in [Pe90b]. This characterization enables us in the present paper to prove completeness results for the connection method enhanced by the following rules for equality reasoning: paramodulation (see section 3) and RUE-resolution modulo an equational theory (see section 4). The latter result is obtained in a strong form and verifies a conjecture of Digricoli [DH86]. Moreover his rule has been generalized in the sense that it works now modulo a built in equational theory. For demodulation modulo an equational theory the soundness has been proved (see section 5).

2 The connection method with a built in equational theory

2.1 First order equational theories

A *first order theory* is a triplet $T = (\mathcal{L}, \vdash, \mathcal{A})$ where $\mathcal{L}$ is a first order language with equality, $\vdash$ is the first order inference relation with equality and $\mathcal{A}$ a set of formulas from $\mathcal{L}$ called the axioms of T. A theory will be called *open* if all of its axioms are open, i.e. quantifier free, formulas. It will be called *empty* if its axiom set is empty. If all axioms of a theory are equations then it will be called *equational.* By a T-unification algorithm we mean an algorithm computing a complete set of T-unifiers for each set of sets of terms. Given a theory T we are interested in a procedure which gives for a formula α from $\mathcal{L}$ an answer to the question whether α is T-derivable or not. Such a procedure will be called a *proof procedure with built in theory T.*

2.2 The ground case: T-derivability of open formulas

For the empty theory exists a simple criterion for the derivability of an open formula α:

> in every disjunction of a conjunctive normal form of α must exist a complementary pair of literals.

Such pairs are called *connections* [Bi82]. This notion may be generalized for the case of an arbitrary open theory $T = (\mathcal{L}, \vdash, \mathcal{A})$. Arbitrary T-derivable disjunctions of literals may be considered instead of pairs of complementary literals. Such formulas are called T-*connections.* For a given theory T it is now necessary to characterize a set $\mathcal{U}$ of T-connections such that

> for an arbitrary open formula α:
>
> α is T-derivable iff
>
> every disjunction of a conjunctive normal form of α contains an element of $\mathcal{U}$ as a subformula.

Such a set will be called T-*complete.* Given a set of T-connections $\mathcal{U}$ a formula will be called $\mathcal{U}$-*complementary* if every disjunction of its conjunctive normal form contains a T-connection belonging to $\mathcal{U}$ as a subformula [1].

For example the formula α having the following form

$$(P(b) \wedge (\neg ab = 1 \vee \neg Q(b))) \vee (\neg P(b) \vee (a = b^{-1} \wedge Q(a^{-1})))$$

---

[1] Building the conjunctive normal form of a non-open formula quantified subformulas are treated as they were atomic.

is derivable in the theory of groups. This may be seen easily considering its matrix representation

$$(2.1) \qquad \left[\left[\begin{bmatrix} P(b) \\ [\neg ab = 1 \quad \neg Q(b)] \end{bmatrix}\right] \begin{bmatrix} \neg P(b) & \begin{bmatrix} a = b^{-1} \\ Q(a^{-1}) \end{bmatrix} \end{bmatrix}\right].$$

The following three paths (sets of submatrices) through this matrix

$$\left[\begin{bmatrix} [P(b)] & [\neg P(b)] & \begin{bmatrix} a = b^{-1} \\ Q(a^{-1}) \end{bmatrix} \end{bmatrix}\right]$$

$$[[\neg ab = 1 \quad \neg Q(b)] \quad [\neg P(b)] \quad [Q(a^{-1})]]$$

$$[[\neg ab = 1 \quad \neg Q(b)] \quad [\neg P(b)] \quad [a = b^{-1}]]$$

represent a conjunctive (non-normal) form equivalent to α. Every of those paths contains one of the following group theory connections:

$$P(b) \vee \neg P(b), \qquad \neg ab = 1 \vee \neg Q(b) \vee Q(a^{-1}), \qquad \neg ab = 1 \vee a = b^{-1}.$$

This makes sure that α is complementary.

2.3 Complete sets of equational theory connections

Now we give a characterization of complete sets of equational theory connections.

Proposition 2.1 *Let* $T = (\mathcal{L}, \vdash, \mathcal{A})$ *be an equational theory. Let* $\mathcal{U}$ *be the set of all formulas which are either of the form*

$$\bigvee_{i=1}^{n} \neg s_i = s_i' \vee \neg P(t_1 \dots t_m) \vee P(t_1' \dots t_m')$$

where P *is a* m-*ary predicate symbol and*

$$\bigvee_{i=1}^{n} \neg s_i = s_i' \vee \bigwedge_{j=1}^{m} t_j = t_j'$$

is T-*derivable or*
which are T-*derivable formulas of the form*

$$\bigvee_{i=1}^{n} \neg s_i = s_i' \vee t = t'.$$

Then $\mathcal{U}$ *is a* T-*complete set of* T-*connections.*

Proof: Suppose a given open formula α is not $\mathcal{U}$-complementary. Then there exists a disjunction β of its conjunctive normal form which has no subformula of one of the forms given above. Then consider the $\mathcal{A}$-algebra A freely generated by the negated equations in β. (cf. [Re87] definition 3.2.1., p. 114). A is a term algebra factorized by a minimal congruence relation determined by $\mathcal{A}$ and the set of generating equations. It may be extended to a model of T which falsifies β and thereby α.

q.e.d.

Remark: The construction of the free $\mathcal{A}$-algebra A is more general than used in the previous proof. It applies to arbitrary theories consisting of conditional equations interpreted in partial algebras.

2.4 Lifting to the full first-order calculus

The lifting of the connection procedure with built in theory from the ground to the full first-order case relies on a Herbrand theorem formulated at the end of this section. The main idea of the lifting will be discussed by the help of an example. Let $T = (\mathcal{L}, \vdash, \mathcal{A})$ be the theory of groups. We consider the T-derivability of the following formula α

$$\exists x(P(x) \wedge \forall a(\neg ax = 1 \vee \neg Q(x))) \vee \forall b(\neg P(b) \vee \exists y(y = b^{-1} \wedge Q(y^{-1})))$$

which may be represented by the matrix

$$\left[\exists x \begin{bmatrix} P(x) \\ \forall a\,[\neg ax = 1 & \neg Q(x)] \end{bmatrix} \quad \forall b\left[\neg P(b) \quad \exists y\begin{bmatrix} y = b^{-1} \\ Q(y^{-1}) \end{bmatrix}\right]\right].$$

The basic idea is

(1) to transform the formula α by instantiations of existential and eliminations of universal quantifiers and

(2) to find a substitution σ such that after the application of σ to the formula β obtained by a sequence of transformations holds

   (a) $\beta[\sigma]$ is $\mathcal{U}$-complementary and

   (b) $\beta[\sigma]$ might be obtained from α by a sequence of T-derivability invariant transformations.

Condition (1) and (2a) assure that $\beta[\sigma]$ is T-derivable whereas condition (2b) implies that α is T-derivable too.

In our example after the elimination of the quantification of b, the instantiation of the variable x and the elimination of the quantification of a one obtains the following matrix

$$\left[\exists x'A \quad \begin{bmatrix} P(x) \\ [\neg ax = 1 & \neg Q(x)] \end{bmatrix} \quad \begin{bmatrix} \neg P(b) & \exists y\begin{bmatrix} y = b^{-1} \\ Q(y^{-1}) \end{bmatrix} \end{bmatrix}\right]$$

where A denotes the submatrix

$$\begin{bmatrix} P(x') \\ \forall a'\,[\neg a'x' = 1 & \neg Q(x')] \end{bmatrix}.$$

Finally, instantiating y we obtain the matrix

$$(2.2) \quad \left[\exists x'A \quad \begin{bmatrix} P(x) \\ [\neg ax = 1 & \neg Q(x)] \end{bmatrix} \quad \begin{bmatrix} \neg P(b) & \begin{bmatrix} y = b^{-1} \\ Q(y^{-1}) \end{bmatrix} & \exists y'B \end{bmatrix}\right]$$

where B denotes the matrix

$$\begin{bmatrix} y' = b^{-1} \\ Q(y'^{-1}) \end{bmatrix}.$$

Now, lets compare matrix (2.2) with matrix (2.1) in subsection 2.2. Every path through matrix (2.2) contains one of the following subformulas

$$P(x) \vee \neg P(b), \qquad \neg ax = 1 \vee \neg Q(x) \vee Q(y^{-1}), \qquad \neg ax = 1 \vee y = b^{-1}$$

which become T-connections after the application of the substitution

$$\sigma = [x\backslash b \quad y\backslash a].$$

Hence, applying σ to (2.2) we obtain the $\mathcal{U}$-complementary matrix

$$(2.3) \quad \left[\exists x' A \begin{bmatrix} P(b) \\ [\neg ab = 1 \quad \neg Q(b)] \end{bmatrix} \quad \begin{bmatrix} \neg P(b) & \begin{bmatrix} a = b^{-1} \\ Q(a^{-1}) \end{bmatrix} & \exists y' B \end{bmatrix}\right].$$

Therefore matrix (2.2) together with the substitution σ satifies the conditions (1) and (2a) mentioned above. For proving condition (2b) we observe that matrix (2.3) may be obtained from the original matrix by the following sequence of three $\mathcal{T}$-derivability invariant transformations:

(1) elimination of the universal quantification of b,

(2) instantiation of x by b and elimination of the universal quantification of a and

(3) instantiation of y by a.

Formulas obtained by transformations of the latter kind will be called *compound instances* of the original formula. Formulas obtained by the restricted form of transformations (like those used constructing (2.2) from α) will be called *simple compound instances*. If β is a simple compound instance of a formula α and σ is a substitution such that $\beta[\sigma]$ is a compound instance of α then σ will be called *acceptable*. Not all substitutions making a compound instance of a formula $\mathcal{U}$-complementary are acceptable. Acceptability of substitutions may be tested by a purely syntactic criterion. This way we avoid Skolemization.

The complexity of the acceptability test seems to be comparable with that of the occurs check for substitutions after Skolemization if a Martelli-Montanari-like unification is used (cf. [Pe90b]). The benefit of avoiding Skolemization is that Skolemization may destroy syntactical properties needed for the termination of unification algorithms. For an example see [DEL90].

Now we may formulate the Herbrand theorem ([Pe90a], for the proof see [Pe90b]).

Theorem 2.2 *If $\mathcal{T} = (\mathcal{L}, \vdash, \mathcal{A})$ is an open theory and $\mathcal{U}$ is a $\mathcal{T}$-complete set of $\mathcal{T}$-connections then for every formula α holds: α is $\mathcal{T}$-derivable if and only if there exists a simple compound instance β of α and an acceptable for β substitution σ such that $\beta[\sigma]$ is $\mathcal{U}$-complementary.*

2.5 The systematic complementarity test

Because of lack of space we can give only a rough idea of the control and data structure for the systematic complementary test. For details the reader is referred to [Bi82] or [Pe90b].

We assume that $\mathcal{T}$ is an open theory and that $\mathcal{U}$ is a $\mathcal{T}$-complete set of $\mathcal{T}$-connections.

A *proof state* consists of a matrix and an acceptable substitution[2]. An *initial proof state* has the form $(M, [])$ where M is the matrix representing the formula to be proved and $[]$ the empty substitution, a *final proof state* has the form $([], \sigma)$ where $[]$ denotes the empty matrix. A final proof state should be reached from an initial one by a sequence of proof steps. In a proof state

$$\left(\left[\left[\begin{bmatrix} L_1 \\ \dots \end{bmatrix} \ \dots \ \begin{bmatrix} L_n \\ \dots \end{bmatrix} \right] \ \begin{bmatrix} L_0 \\ \dots \end{bmatrix} \ R \right], \sigma \right)$$

$L_1, \dots, L_n$ form the *actual path*, denoted by p, and L_0 is the *actual goal* of the *actual clause* $\begin{bmatrix} L_0 \\ \dots \end{bmatrix}$ [3].

By p^+ will be denoted the set $p \cup \{L_0\}$. The matrix R having the form

$$\left[\begin{bmatrix} K_1 \\ \dots \end{bmatrix} \dots \begin{bmatrix} K_m \\ \dots \end{bmatrix} \ R' \right]$$

will be called the *remainder*[4]. The aim is to find a substitution σ' such that for some indices $1 \leq i_1, \dots, i_l \leq n$

---

[2]Nevertheless, the features for the acceptability test will be neglected in the following.

[3]For simplicity $L_0, \dots, L_n$ and $K_1, \dots, K_m$ are assumed to be literals.

[4]m may be 0 and R' may be the empty matrix.

$$(L_{i_1} \vee \ldots \vee L_{i_l} \vee L_0 \vee K_1 \vee \ldots \vee K_m)\,[\sigma' \circ \sigma]$$

is a $\mathcal{T}$-connection and $\sigma' \circ \sigma$ is acceptable. Hence, a unification procedure modulo $\mathcal{U}$ is necessary. This is more complex than $\mathcal{T}$-unification.

If $m = 0$ then the actual goal has been solved. If a goal has been solved then it may be deleted from the matrix. Then the remaining goals of the actual clause or of those on the left of the actual clause have to be solved. If there are no such goals then the resulting proof state is a final one.

If $m \geq 0$ then the proof step is called an *extension* and the resulting proof state looks as follows:

$$\left(\left[\left[\left[\begin{matrix}L_1\\\ldots\end{matrix}\right]\ldots\begin{bmatrix}L_n\\\ldots\end{bmatrix}\begin{bmatrix}L_0\\\ldots\end{bmatrix}\begin{bmatrix}K_1\\\ldots\end{bmatrix}\ldots\begin{bmatrix}K_{m-1}\\\ldots\end{bmatrix}\right]\begin{bmatrix}K_m\\\ldots\end{bmatrix}\;R'\;\right],\sigma' \circ \sigma\right)$$

with the new actual goal K_m.

In the following both proof steps will be considered in more detail.

If a step of these kinds is not possible then a simple instantiation of a subformula from the remainder may be tried.

3 Connection method with paramodulation

Though it is nice to have the general characterization of the set $\mathcal{U}$ of equational theory connections given in proposition 2.1 unification modulo $\mathcal{U}$ seems to be a too complex step. The well known paramodulation rule [RW69] may be viewed as a possibility to split the unification modulo $\mathcal{U}$ into less complex steps. Namely, it will be sufficient to have $\mathcal{T}$-unification.

First we give the definition of this rule for an equational theory $\mathcal{T}$ built in into the connection method. At next an example will be considered and the completeness will be proved.

Definition 3.1 *(The paramodulation rule for the connection method.)*

(1) *Suppose the literals K and $L(s)$ belong to the set p^+ of a proof state and that K is either $\neg s' = t$ or $\neg t = s'$. If σ' (universally) unifies $s'[\sigma]$ and $s[\sigma]$ then from the proof state*

$$\left(\left[\ldots\begin{bmatrix}K\\\ldots\end{bmatrix}\ldots\begin{bmatrix}L(s)\\\ldots\end{bmatrix}\ldots\right],\;\sigma\right)$$

leads a paramodulation step to the proof state

$$\left(\left[\ldots\begin{bmatrix}K\\\ldots\end{bmatrix}\ldots\begin{bmatrix}L(s)\\\ldots\end{bmatrix}\;L(t)\;\ldots\right],\;\sigma'\circ\sigma\right).$$

(2) *Suppose $L(s)$ belongs to the actual path or is the actual goal of a proof state. If σ' $\mathcal{T}$-unifies $s[\sigma]$ and $t[\sigma]$ then a paramodulation step leads from*

$$\left(\left[\ldots\begin{bmatrix}L(s)\\\ldots\end{bmatrix}\ldots\right],\;\sigma\right)\;\text{to}\;\left(\left[\ldots\begin{bmatrix}L(s)\\\ldots\end{bmatrix}\;L(t)\ldots\right],\;\sigma'\circ\sigma\right).$$

(3) *The actual goal in a proof state*

$$(M,\sigma)$$

is solved if the set p^+ defined accordingly to subsection 2.5 contains either

- *literals K and $\neg L$ such that σ unifies (universally) K and L (i.e. a connection in the usual sense) or*

- *a non-negated equation $s = t$ such that σ unifies (universally) s and t.*

Example: Let $\mathcal{T}$ be the theory of monoids. We consider the matrix

$$\left(\left[\left[\begin{matrix}[\neg ax = 1 & \neg Q(a(xc))] \\ & P(x)\end{matrix}\right]\quad \left[\begin{matrix}Q(c(yb)) \\ yb = 1\end{matrix}\right]\quad \neg P(b)\quad \exists x'A\quad \exists y'B\right],[\,]\right).$$

The actual path is $\neg ax = 1 \vee \neg Q(a(xc))$. The actual goal $Q(c(yb))$ may be solved by five paramodulation steps. These steps simulate the computation of the $\mathcal{T}$-connection which may be obtained applying the substitution

$$\sigma = [x\backslash b\quad y\backslash a] \quad\text{to}\quad \neg ax = 1 \vee \neg Q(a(xc)) \vee Q(c(yb)).$$

1: Because $a(xc) = (ax)c$ is derivable in the theory of monoids the following proof state may be obtained accordingly to definition 3.1.(2)[5].

$$\left(\left[\left[\begin{matrix}[\neg ax = 1 & \neg Q(a(xc))] \\ & \ldots\end{matrix}\right]\quad \neg Q((ax)c)\quad \left[\begin{matrix}Q(c(yb)) \\ \ldots\end{matrix}\right]\quad \ldots\right],[\,]\right).$$

2: The formula $\neg ax = 1 \vee \neg Q((ax)c) \vee Q(1c)$ is a monoid theory connection and the matrices

$$[\neg Q((ax)c)]\quad\text{and}\quad\left[\neg Q((ax)c)\quad\left[\begin{matrix}Q(1c) \\ \neg Q(1c)\end{matrix}\right]\right]$$

represent equivalent formulas. This justifies a paramodulation step accordingly to definition 3.1.(1) giving

$$\left(\left[\left[\begin{matrix}[\neg ax = 1 & \neg Q(a(xc))] \\ & \ldots\end{matrix}\right]\quad \neg Q((ax)c)\quad \neg Q(1c)\quad \left[\begin{matrix}Q(c(yb)) \\ \ldots\end{matrix}\right]\ldots\right],[]\right).$$

3: The formula $\neg ax = 1 \vee \neg Q(1c) \vee Q(c)$ is a monoid theory connection. This justifies the paramodulation step accordingly to definition 3.1.(2) giving

$$\left(\left[\left[\begin{matrix}[\neg ax = 1 & \neg Q(a(xc))] \\ & \ldots\end{matrix}\right]\quad \neg Q((ax)c)\quad \neg Q(1c)\quad \neg Q(c)\quad \left[\begin{matrix}Q(c(yb)) \\ \ldots\end{matrix}\right]\ldots\right],[]\right).$$

4,5: Subterm yb of $Q(c(yb))$ may be unified with the left side of the negated equation $\neg ax = 1$. Two paramodulation steps accordingly to definition 3.1.(1) and (2) give:

$$\left(\left[\left[\begin{matrix}[\neg ax = 1 & \neg Q(a(xc))] \\ & \ldots\end{matrix}\right]\quad \neg Q((ax)c)\quad \neg Q(1c)\quad \neg Q(c)\quad \left[\begin{matrix}Q(c(yb)) \\ \ldots\end{matrix}\right]\quad Q(c1)\quad Q(c)\ldots\right],\sigma\right).$$

Since we obtained a usual connection the new subgoal $Q(c)$ has been solved and also the original subgoal $Q(c(yb))$. The wanted unifier has been found.

Proposition 3.2 *Let $\mathcal{T}$ be an equational theory such that there exists a $\mathcal{T}$-unification algorithm. Then the connection method with the paramodulation rule is a sound and complete proof procedure with built in theory $\mathcal{T}$.*

Proof: The soundness of the rule is obvious (cf. the remarks in the previous example).

For the completeness we show that every step establishing a $\mathcal{T}$-connection according to subsection 2.5 may be simulated by a sequence of paramodulation steps producing the same or a more general unifier and the same new subgoals.

---

[5]Dots symbolize unchanged parts of the matrix.

For this purpose we consider a proof state

$$\left(\left[\left[\left[\begin{smallmatrix}\neg s_1 = s_1'\\ \dots\end{smallmatrix}\right] \dots \left[\begin{smallmatrix}\neg s_n = s_n'\\ \dots\end{smallmatrix}\right]\right] \left[\begin{smallmatrix}t = t'\\ \dots\end{smallmatrix}\right] \left[\ R\ \right]\right], \sigma\right)$$

and a substitution σ' such that

$$\left(\bigvee_{i=1}^{n} \neg s_i = s_i' \vee t = t'\right)[\sigma' \circ \sigma]$$

is an equational theory connection and $\sigma' \circ \sigma$ is acceptable. Then the term $t[\sigma' \circ \sigma]$ may be transformed stepwise into $t'[\sigma' \circ \sigma]$. In each step a subterm $r'[\sigma' \circ \sigma]$ is substituted by a subterm $r''[\sigma' \circ \sigma]$ where the equation $r'[\sigma' \circ \sigma] = r''[\sigma' \circ \sigma]$ is either an axiom instance or one of the negated equational literals of the $\mathcal{T}$-connection. By induction on the length of this sequence may be constructed a sequence of paramodulation steps giving a unifier σ'' such that $\sigma'' \circ \sigma$ is more general than $\sigma' \circ \sigma$ and acceptable too. The paramodulation step does not destroy subgoals. Therefore no problem appears if such literals occur in their original form in other $\mathcal{T}$-connections needed for the $\mathcal{U}$-complementarity.

q.e.d.

Possible improvements of the paramodulation rule are the combination with term orderings and completion techniques i.g. [BG90].

4 Connection method with RUE-resolution

While finding a $\mathcal{U}$-unifier immediately seems to be a too complex step paramodulation has another disadvantage. It gives no hint how to find the subsequent paramodulation steps leading to a unifier. This also has been recognized early.

In the present section we discuss Digricoli's RUE-resolution [DH86] which is an improvement of Morris' E-resolution [Mo69]. The aim of this rule is to guide the search for the appropriate paramodulation steps. Suppose the prover tries to extend a disjunction $\neg P(t_1 \dots t_m) \vee P(t_1' \dots t_m')$ or an equation $t = t'$ to an equational theory connection. Then it should try to construct systematically a set of equations

$$\{\{s_1, s_1'\}, \dots, \{s_n, s_n'\}\}$$

such that

$$\bigvee_{i=1}^{n} \neg s_i = s_i' \vee \neg P(t_1 \dots t_m) \vee P(t_1' \dots t_m') \quad \text{or} \quad \bigvee_{i=1}^{n} \neg s_i = s_i' \vee t = t'$$

are $\mathcal{T}$-connections respectively. Such sets will be called disagreement sets.

Example: Let us again assume that $\mathcal{T} = (\mathcal{L}, \vdash, \mathcal{A})$ is the equational theory of monoids and let us consider the proof state

$$\left(\left[[\dots] \left[\begin{smallmatrix}\neg ax = 1\\ \dots\end{smallmatrix}\right] \left[\left[\begin{smallmatrix}a(xc) = c(yb)\\ yb = 1\end{smallmatrix}\right] [R']\right]\right], []\right).$$

In the connection method with built in theory $\mathcal{T}$ the actual goal $\neg ax = 1$ may be solved by an extension step establishing the $\mathcal{T}$- and the usual connection

$$(\neg ax = 1 \vee a(xc) = c(yb))[x\backslash b\ \ y\backslash a] \quad \text{and} \quad (\neg ax = 1 \vee yb = 1)[x\backslash b\ \ y\backslash a].$$

The connection prover enhanced by the RUE-rule will solve the goal $\neg ax = 1$ by the following steps $1 - 5$.

1. Extension step to $a(xc) = c(yb)$. $a(xc) = (ax)c$ is an axiom instance. The prover tries with the disagreement set $\{\{(ax)c, c(yb)\}\}$ and reaches the proof state

$$\left(\left[\left[\cdots\quad\begin{bmatrix}\neg ax=1\\ \cdots\end{bmatrix}\quad\begin{bmatrix}a(xc)=c(yb)\\ yb=1\end{bmatrix}\quad\right]\quad(ax)c=c(yb)\quad[R']\right],[]\right).$$

2. $1\cdot c = c$ and $c\cdot 1 = c$ are axioms and $\{\{ax = 1\},\{1 = yb\}\}$ is a disagreement set for the actual goal $(ax)c = c(yb)$. The next proof state is:

$$\left(\left[\left[\cdots\begin{bmatrix}\neg ax=1\\ \cdots\end{bmatrix}\begin{bmatrix}a(xc)=c(yb)\\ yb=1\end{bmatrix}(ax)c=c(yb)\right]\quad\begin{bmatrix}ax=1\\ 1=yb\end{bmatrix}\quad[R']\right],[]\right).$$

3.,4. The connection $\neg ax = 1 \lor ax = 1$ solves the new goal $ax = 1$. Substitution $[x\backslash b\quad y\backslash a]$ unifies the right side of the other new goal $1 = yb$ with the left side of the negated equational literal $\neg ax = 1$. The new proof state is:

$$\left(\left[\left[\cdots\quad\begin{bmatrix}\neg ax=1\\ \cdots\end{bmatrix}\quad\right]\quad[yb=1]\quad[R']\right],[x\backslash b\quad y\backslash a]\right).$$

5. The obtained substitution solves the goal $yb = 1$ and thereby the goal $\neg ax = 1$ too. Finally we reach the proof state

$$([[\ldots]\quad[\ldots]\quad[R']],[x\backslash b\quad y\backslash a]).$$

Below the concepts applied in the example will be defined more precisely.

Definition 4.1 *(T-disagreement set)*

(1) Disagreement sets of a pair of terms

    (a) If s and t are non-identical terms then the set $\{\{s,t\}\}$ is the origin disagreement set of this pair.

    (b) If s and t have the form $f(s_1,\ldots,s_n)$ and $f(t_1,\ldots,t_n)$ then

$$\{\{s_1,t_1\},\ldots,\{s_n,t_n\}\}$$

is the topmost disagreement set of this pair.

    (c) If D is a disagreement set of a pair of terms s and t, $\{s',t'\}$ belongs to D and has the disagreement set D' then

$$D'' = (D\setminus\{\{s',t'\}\})\cup D'$$

is a disagreement set of s and t. D'' is said to be below of D (symbolized by $D\prec D''$).

(2) The positive literal $s = s'$ has the topmost disagreement sets

$$\{\{s,s'\}\}$$

as well as

$$\{\{s,t\},\{s',t'\}\}\quad\text{and}\quad\{\{s,t'\},\{s',t\}\}$$

if $t = t'$ is an instance of an axiom from $\mathcal{A}$.

(3) Both $\{\{s,t\},\{s',t'\}\}$ and $\{\{s',t\},\{s,t'\}\}$ are topmost disagreement sets of the pair of equational literals $\neg s = s'$ and $t = t'$.

(4) If D_i is a disagreement set of s_i and t_i for $i = 1,\ldots,n$ then $\bigcup_{i=1}^{n}D_i$ is a disagreement set of the pair of literals $P(s_1,\ldots,s_n)$ and $\neg P(t_1,\ldots,t_n)$.

It is easy to prove that the disagreement sets defined above meet the motivation given at the beginning of this section. For a more detailed discussion let us return to the sample computation.

Case (2) of the previous definition justifies the choice of the set

$$\{\{(ax)c, c(yb)\}\}$$

as a disagreement set applied in the first step in that computation.

Not all disagreement sets which might be constructed are of real interest for the proof because they contain equations which are neither related to axioms of the built in theory nor to negated equational literals on the actual path or which might be used for an extension step. In the first step of the previous example could be formed the disagreement set $\{\{a, c\}, \{xc, yb\}\}$ which would not contribute to the proof. This is an example for a disagreement set which is not viable in the sense of the following definition. During the proof search only viable disagreement sets should be taken in acount. The property of being viable depends on the presence of unifiers. For this reason a substitution appears as a parameter of the following definition.

Definition 4.2 *(T-viable disagreement set)*
Let the equations $s_i = t_i$ for $i = 1, \ldots, n$ form a disagreement set D and let σ be a substitution and S a set of equations. Then D will be called T-viable with σ and S if and only if for every $i = 1, \ldots, n$ holds either

(1) *for every side r of the i-th equation exists an equation $a = b$ which belongs to S or which is an instance of an axiom from A such that either*

    (a) *σ unifies r either with a or with b or*

    (b) *for one side c of $a = b$ holds that r and c have the same leading function symbol and there exists a disagreement set below of $\{\{r, c\}\}$ which is T-viable with σ and S*

or

(2) *σ T-unifies the terms of the i-th equation.*

Let us examine again our sample computation. The disagreement set

$$\{\{(ax)c, c(yb)\}\}$$

from the 1st step is T-viable with the empty substitution and $S = \{\neg ax = 1\}$ because condition (1) is satisfied: $(ax)c$ matches on the leading function symbol with one side of the axiom $1 \cdot c = c$. The disagreement set $\{\{ax, 1\}\}$ is below the origin disagreement set of $(ax)c$ and $1 \cdot c$ and it is viable because $\neg ax = 1$ belongs to S (condition (1b)). Moreover $c(yb)$ unifies with one side of the associativity axiom (condition (1a)).

Substitutions like those appearing in definition 4.2 have to be found by a unification procedure. In the present case we use a modification of the qualified RUE-unification rule introduced by Digricoli. Using the terminology of [JK90] a general unification algorithm relies on transformations of multisets of multiequations which might be considered as a generalization of the disagreement sets. We explain the basic transformations rules decomposition, merging and mutation by the following examples in the theory of monoids[6].

(1) **decomposition:** The multiequation $a(xc) == c(yb)$ will be substituted by the multiequations $a == c$ and $xc == yb$.

(2) **merging:** The multiequations $x == t$ and $x == t'$ will be merged into one multiequation $x == t == t'$.

---

[6]The merging rule may have various forms depending on the built in theory.

(3) mutation: In the theory of monoids a multiequation $a(xc) == c(yb)$ may be substituted by a multiequation $(ax)c == c(yb)$.

If after a sequence of transformations applied to an origin disagreement set[7] all obtained multiequations are in the *fully decomposed form* $x_1 == \ldots == x_n == t$ and the occurs check does not fail then a $\mathcal{T}$-*unifier* has been computed.

We obtain a *partial* $\mathcal{T}$-*unifier* if the occurs check does not fail when applied to the fully decomposed multiequations only and every not fully decomposed multiequation of the form $f(\ldots t \ldots) == f(\ldots t' \ldots)$ (where t and t' are on corresponding positions) satisfies the viability condition (when treated as an element of an disagreement set).

If the occurs check does not fail when applied to the fully decomposed multiequations only and the built in theory is empty and to none of the not fully decomposed multiequations may be applied the decomposition rule then the obtained substitution is called the *most general partial unifier* (mgpu).

Morris' E-resolution [Mo69] uses the mgpu together with the topmost disagreement set. Anderson [Ann70] observed that E-resolution is not complete. Digricoli [DH86] uses the topmost viable disagreement set together with the so called qualified mgpu: The application of the decomposition rule is not forced for a multiequation $f(\ldots x \ldots) == f(\ldots t \ldots)$ (where x and t are on corresponding positions) if the pair $\{f(\ldots x \ldots), f(\ldots t \ldots)\}$ satisfies the viability condition.

The following example motivates our decision to use the partial $\mathcal{T}$-unifier instead of the qualified mgpu. In order to obtain an equational connection from the following formula

$$\neg f(a, h(c)) = f(k(a), h(b)) \vee \neg a = d \vee \neg k(a) = b \vee f(a, h(y)) = f(k(a), h(b))$$

one has to apply the substitution $\sigma = [y \backslash c]$. In this case Digricoli's condition does not prevent the prover from forming the mgpu $[y \backslash b]$. Therefore the appropriate unifier cannot be found.

Definition 4.3 *(The RUE-resolution rule for the connection method)*
   Let $\mathcal{T}$ be an equational theory.
   Then from the proof state with actual goal G

$$\left(\left[\left[\ldots \begin{bmatrix} L_j \\ \ldots \end{bmatrix} \ldots \right] \; \begin{bmatrix} G \\ \ldots \end{bmatrix} \; \left[\ldots \begin{bmatrix} M_k \\ \ldots \end{bmatrix} \ldots [R] \; \right] \; \right] \; \right], \; \sigma \right)$$

leads a RUE-resolution step to the proof state

$$\left(\left[\left[\ldots \begin{bmatrix} L_j \\ \ldots \end{bmatrix} \ldots \begin{bmatrix} G \\ \ldots \end{bmatrix} \ldots \begin{bmatrix} M_k \\ \ldots \end{bmatrix} \ldots \right] \; \begin{bmatrix} s_1 = s_1' \\ \ldots \\ s_n = s_n' \end{bmatrix} \; [R] \right], \; \sigma' \circ \sigma \right)$$

if one of the following conditions (1) *or* (2) *is satisfied:*

(1) There are

   (a) a negated and a non-negated literal K and K' having the same predicate symbol and

   (b) sets

      i. $S_{path} = \{L_j\}_{j=1}^m$ of literals belonging to the actual path or being the actual goal,

      ii. $S_{rest} = \{M_k\}_{k=1}^l$ of literals in clauses of the remainder and

      iii. S_Θ of negated equality literals different from K

      such that $\{K, K'\} \cup S_\Theta \subset S_{path} \cup S_{rest}$ and S_{rest} is subset of a path through the considered matrix.

---

[7]taken as a set of multiequations

(c) Moreover, σ' is a partial T-unifier for the origin disagreement set of K and K' and the equations $s_i = s'_i$ for $i = 1, \ldots, n$ form the topmost disagreement set of the pair K and K' which is T-viable with σ' and S_Θ and

(d) if K and K' are equality literals then one side r of the former must either

    i. be T-unified by σ' with one side of the latter literal or

    ii. match with one side r' of the latter literal on the leading function symbol.

In the latter case there must exist a disagreement set below $\{\{r, r'\}\}$ which is T-viable with σ' and S_Θ.

(2) There are

(a) a non-negated equational literal K and

(b) sets

    i. $S_{path} = \{L_j\}_{j=1}^m$ of literals belonging to the actual path or being the actual goal,

    ii. $S_{rest} = \{M_k\}_{k=1}^l$ of literals in clauses of the remainder and

    iii. S_Θ of negated equality literals

such that $\{K\} \cup S_\Theta \subset S_{path} \cup S_{rest}$ and S_{rest} is subset of a path through the considered matrix.

(c) Substitution σ' is a partial T-unifier for the equation K and moreover, the equations $s_i = s'_i$ for $i = 1, \ldots, n$ form the topmost disagreement set of the literal K which is T-viable with σ' and S_Θ.

(d) If the considered disagreement is the origin for K and has been constructed using an axiom instance then one side r of K must either

    i. be T-unified by σ' with one side of the axiom instance or

    ii. match with one side r' of the axiom instance on the leading function symbol.

In the latter case there must exist a disagreement set below $\{\{r, r'\}\}$ which is T-viable with σ' and S_Θ.

A goal in a proof state is solved if the set p^+ contains a complementary pair (in the usual sense) or a positive equality literal unified by the substitution of the proof state.

Remark: The first case of the previous definition accords to Digricoli's RUE-rule whereas the second case corresponds to his NRF-rule. Both rules have been formulated in their strong form i.e. with the restriction concerning the unifier and the so called equality restriction (conditions (1.d) and (2.d)).

The reader might prove that all steps in the sample computation in this section meet the definition above. It is interesting that condition (1.d) of definition 4.3 prevents the prover from trying a RUE-step taking $\neg ax = 1$ and $a(xc) = c(yb)$ as complementary literals in step 1 of this computation. While condition (1.d.ii) is satisfied there is no T-viable disagreement set below $\{\{ax, a(xc)\}\}$ or $\{\{ax, c(yb)\}\}$. The situation is different when solving the goal $1 = yb$ in step 4. Then this literal together with $\neg ax = 1$ satisfies condition (1.d.ii) of definition 4.3 because the disagreement set $\{\{a, y)\}, \{x, b\}\}$ is T-viable with the substitution $[x \backslash b \quad y \backslash a]$. The same is true in step 5.

Proposition 4.4 Let T be an equational theory such that there exists a T-unification algorithm. Then the connection method with the RUE-resolution rule is a sound and complete proof procedure with built in theory T.

Proof: Soundness of the rule follows from the construction of disagreement sets and partial $\mathcal{T}$-unifiers. The completeness proof has a similar structure like the completeness proof for the paramodulation rule.

q.e.d.

With the completeness of the connection method with the restricted RUE-resolution has been proved Digricoli's conjecture expressed for the resolution calculus enhanced by the RUE-rule in strong form with equality restriction.

A subject to further work is the exploration of the construction of the partial $\mathcal{T}$-unifers.

5   Connection method with demodulation

The motivation for the demodulation rule [WRCS67] is similar to that of paramodulation. The former is in some sense a special case of the latter.

Definition 5.1 *(The demodulation rule for the connection method.)*
Suppose the literal $\neg s' = t'$ belongs to the actual path and occurs within a unit clause and $L(s)$ is the actual goal of a proof state. If σ' $\mathcal{A}$-unifies $s'[\sigma]$ with $s[\sigma]$ and $t'[\sigma]$ with $t[\sigma]$ then from the proof state

$$\left(\left[\ldots \quad [\neg s' = t'] \quad \ldots \quad \begin{bmatrix} L(s) \\ \ldots \end{bmatrix} \quad \ldots \right] \ , \ \sigma \right)$$

leads a paramodulation step to the proof state

$$\left(\left[\ldots \quad [\neg s' = t'] \quad \ldots \quad \begin{bmatrix} L(t) \\ \ldots \end{bmatrix} \quad \ldots \right] \ , \ \sigma' \circ \sigma \right).$$

Proposition 5.2 *Soundness of demodulation*
The connection procedure enhanced by the demodulation rule is sound.

Proof: The negated equality literal occurs in a unit clause. It may be instantiated repeatedly in order to get the wished instance. This instance is element of every path through the matrix. Therefore in every path the literals $L(s)$ and $L(t)$ may be considered equivalent. Hence, it does not matter that the actual goal has been destroyed.

q.e.d.

Interesting improvements of the demodulation rule are again related with the application of term orderings and the combination with completion techniques.

6   Summary

Paramodulation, RUE-resolution and demodulation are well known rules for reasoning with equality.

It's the value of the presented results that they give a structured approach for carrying over various improvements developed within one family of proof calculi (resolution, connection graph resolution) to another family (variants of the connection method). This is possible because of the generality of the Herbrand theorem 2.2. and of the characterization of equational theory connections. The work to be done for obtaining completeness results for proof procedures with refined rules is simply to prove that it is possible to find all equational theory connections which are necessary during the construction of a proof. Complete proof calculi for equality reasoning accepting non-normal form formulas has been obtained this way.

The characterization of equational theory connections covers even the case of conditional equational theories interpreted in structures with partial operations. Nevertheless, the algorithmic features of this generalization have to be elaborated.

Further interesting topics are approaches including term orderings and completion techniques. Experimental work has to be done in the future too.

References

[Ann70]   Anderson R., Completeness results for E-resolution, In Proc. AFIP 70, Spring Joint Comp. Conf., AFIPS Press, Reston VA, pp 653-656, 1970.

[Ans81]   Andrews P., Theorem proving via general matings, J.ACM, 28, 193-214, 1981.

[BB]   K. H. Blaesius, H. J. Buerckert (Eds.), Deduction systems (in german), Oldenbourg-Verlag, Muenchen, Wien, 1987.

[BG90]   Bachmair L., Ganzinger H., On restrictions of ordered paramodulation with simplification, Proc. CADE 1990, pp. 427-441, 1990.

[Bi82]   Bibel W., Automated theorem proving, Verlag Vieweg, 1982, 2nd ed. 1987.

[DEL90]   Debart F., Enjalbert P., Lescot M., Multi modal logic programming using equational and order-sorted logic, Report, Lab. d'Informatique, Univ. de Caen, 1990.

[DH86]   Digricoli V. J., Harrison M. C., Equality-based binary resolution, J. ACM, Vol. 33, Nr. 2, April 1986, pp. 253-289.

[JK90]   Jouannaud J. P., Kirchner C., Solving equations in abstract algebras: a rule-based survey of unification, Report 561/1990, L.R.I., Univ. de Paris-Sud, 1990.

[Mo69]   Morris J. B., E-resolution: An extension of resolution to include the equality relation, In Proc. IJCAI 1969 (Washington D.C.), Walker D.E., Norton L.M., Eds., pp. 287-294, 1969.

[Pe90a]   Petermann U., Towards a connection procedure with built in theories, in Proc. JELIA 90, European Workshop on Logic in AI, Amsterdam Sept. 90, LNCS, Springer Publ., 1990.

[Pe90b]   Petermann U., Building in Theories into a first-order proof procedure based on the connection method, Preprint NTZ-Nr. 16/90, Naturwissenschaftlich - Theoretisches Zentrum and Department of Informatics, Leipzig University, 1990.

[Prae90] Praecklein A., Solving equality reasoning problems with a connection graph theorem prover, SEKI-Report, SR-90-07, FB Informatik, Univ. Kaiserslautern, 1990.

[Re87]   Reichel H., Initial computability, algebraic specification and partial algbras, Oxford Univ. Press, 1987.

[RW69]   Robinson G., Wos L., Paramodulation and theorem proving in first-order theories with equality, Machine Intelligence, 4, 1969.

[Sti85]   Stickel M., Automated deduction by theory resolution, J. Autom. Reasoning, 1, 4 (1985), 333-356.

[WRCS67] L.T. Wos, G.A. Robinson, D.F. Carson, L. Shalla, The concept of demodulation in theorem proving, J.ACM, Vol. 14, Nr. 4, Oct. 1967.

Logical Fiberings and Polycontextural Systems

J.Pfalzgraf *
RISC-Linz
Johannes Kepler University
A-4040 Linz, Austria
email K311576@AEARN.BITNET

Abstract

Based on the notion of abstract fiber spaces the concept of a logical fibering is developed. This was motivated by a project where so-called polycontextural logics were discussed. The fiber space approach provides a rather general framework for the modeling of such non classical logics. It gives the possibility to construct a variety of new logical spaces from a given (indexed) system of logics. These spaces are in some sense parallel (inference) systems. We can give a straight forward definition and classification of the so-called transjunctions arising in polycontextural logics. These are bivariate operations having values distributed over different logical subsystems. Univariate, bivariate operations are introduced in functional notation. The group generated by the generalized negation operations and system changes is investigated. We make some remarks on aspects of applicability and links to other work.

1 Introduction

The following work was initiated by a joint project of two university groups and an industrial company on so-called 'Polycontextural Logic', abbreviated PCL. It is of importance for the whole understanding to give some motivating background information, we do this subsequently and in the next section when dealing with basics from PCL.

The PCL approach to a nonclassical generalization of two valued logics in form of a whole system of classical logical spaces distributed over an indexing set of values is heavily influenced by the philosophical work of Gotthard Günther (cf. references to PCL) and it cannot be seen without these roots. G.Günther's work has been extensively studied and partly continued by R.Kaehr and coworkers (we call them 'PCL group', for short).

One of the main arguments of the PCL group was that this so-called 'transclassical logic' should be suitable as a logical basis for modeling of (living) communicating systems. In fact, parts of that theory had been discussed and developed at the Biological Computing Laboratory (BCL), Urbana Ill., in the sixties, in the realm of research done to establish a new 2nd order cybernetics which requires a new logical basis (as was argued).

Many unconventional philosophical and metaphysical considerations can be found around the whole PCL theoretic approach. And it is not always easy to follow or adopt these thoughts. Unfortunately, much of the literature on the subject is not easy to access or available.

Therefore, one aspect of this article is to draw attention to some of these ideas and results and also to our own formal mathematical approach in the field of abstract fiberings which shows, among others, that PCL systems can be derived as a special class of logical fiberings. It should be emphasized here that I am a non-specialist in PCL and not a member of the PCL group.

Of further interest would be possible links to other work (e.g. in the field of the project MEDLAR (ESPRIT BRA 3125) on methods of practical reasoning), in particular to labelled deductive systems (cf. [GA]).

In many discussions during that project with the PCL group, intuitively, I always had the impression that a mathematical formulation of such distributed logical systems can be given naturally in a general framework using categories, fiberings, indexed systems (and related fields) as a formal mathematical language.

For example, in this way it is possible to give a simple definition of the notion of *transjunctions* (this is a typical nonclassical bivariate operation in PCL systems) and their classification. On the basis of the

*sponsored by the Austrian Ministry of Science and Research (BWF), ESPRIT BRA 3125 "MEDLAR"

notion of an abstract fibering we establish a method to construct (many valued) logical spaces from a given (indexed) system of 2-valued logics.

This way, it is possible to derive a variety of new logical spaces systematically and it is easy to examine situations where formulas are consistent locally (in each subsystem) but not globally.

We apply the construction method to show that PCL systems can be derived as a special class of logical fiberings. A particular example in a 3-valued PCL motivates, more general, that the fibering approach leads to a method for decomposing (parallelizing) a given multiple-valued logic into 2-valued components (this is subject for further study).

The following presentation is not as rigorous as it could be since there are natural links to disciplines like indexed categories, toposes and sheaves which provide a more general framework for all such considerations (cf. selected titles in the literature list). A more general treatment can be subject of future work.

Further comments follow in the subsequent text. We also make a remark on intended possible applications in robot multitasking problems and possible links to other projects.

More details of the material presented in this article can be found in [PF3].

2 Remarks on PCL

We give some preparatory comments which, of course, can only represent a very limited perspective.

As previously remarked, PCL arose from particular philosophical considerations and has its individual understanding of communication and interaction. For a thorough understanding of the arguments of the PCL group it is necessery to have some insight in the written work.

A lot of material (case studies, reports) exists where ideas of polycontexturality are applied to nonformal, descriptive modeling of processes and scenarios (cf. e.g. [R1, 2]).

Basic principles are among others:
Distribution of several classical (2-valued) logics ('loci'). At least 3 loci are involved; the individual spaces are pairwise isomorphic ('locally'). For two classical spaces (components) of the whole system a third one has the function of *mediation*. Thus in certain respect, a pair of local components of the polycontextural system needs a third classical space for 'reflection' (for mediation) in the general communication process and in interaction. Although two components are isomorphic as 2-valued logical spaces, their placement ('index') plays a role in the common context of the whole system - the whole PCL system is multivalued. The 'transitions' ('communication') between the particular subsystems is of essential structural importance. Each individual subobject conceives the world through the same logic, but from a different place in the system; locally the results are all the same since the same 2-valued logic is placed there; but globally there may be differences in the results since reasoning is performed at different ontological places, the places being enumerated (labelled) by an index set; the role of the latter is twofold also constituting a global set of values for the PCL as a multivalued logic. This involvement is somewhat subtle (we refer to the literature on PCL).

The development of PCL as initiated by G. Günther is deeply influenced and based on philosophical considerations and it cannot be seen without these roots (we refer to the literature; cf. also some selected quotations from PCL literature in [PF3]).

We point out that we are neither specialists in these philosophical foundations nor experts in PCL as represented by the PCL group.

We find some aspects of this work quite interesting and were motivated to model such PCL-systems as certain logical fiberings.

In this very general framework the PCL systems form a special class of multiple- valued logical spaces representable as certain fiberings.

3 Some Basic Notions from PCL

We use L or L_i (if an index is necessary) to denote a classical 2-valued logical space (a 1st order language w.r.t. a symbol set which we do not explicitly specify if there is no need to do so).

3.1 Local and Global Systems

A PCL system is an m-valued logical system consisting of n classical 2-valued subsystems denoted by L_i, $i = 1, 2, \ldots, n$, where $n = \binom{m}{2}$. The (global) truth values are denoted by $1, 2, \ldots, m$. (Thus each subsystem L_i can be associated with a 2-element subset of $\{1, 2, \ldots, m\}$.) The two (local) truth values within the (classical) subsystem L_i are defined by $\Omega_i = \{T_i, F_i\}$.

The total (global) system is denoted by $\mathcal{L}^{(m)}$, where $\mathcal{L}^{(m)}$ can be seen as the disjoint union (coproduct) of sets: $L_1 \amalg \ldots \amalg L_n$.

In addition to these basic constituents of a PCL the following so-called *mediation scheme* - we write MS for short - is an essential data for the definition of a PCL.

We show such a scheme for the case $m = 3$ (hence $n = 3$): Notation $MS3$

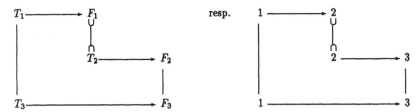

It contains the following information:

The arrow $T_i \rightarrow F_i$ expresses an ordering of the two values within the subsystem L_i, and

$\supset\!\!-\!\!\subset$ expresses the fact that an F-value in one system (L_1) becomes a T-value in another (L_2) (a "change" of truth values when changing the corresponding subsystems)- i.e. a 'semantical change'.

The vertical lines have to be interpreted as identifications.

The right diagram is a short notation where the global values are inserted (indicating the relations(identifications) between the corresponding local values).

Thus, the MS describes the global relations between the local values and contains informations about what happens if one passes from one subsystem to another. It expresses how the collection of the value sets $\Omega_i = \{T_i, F_i\}$ of the individual L_i becomes the global set of values $\{1, 2, 3\}$, respectively.

In logical fibering notation (cf. section 4) we shall express this by an equivalence relation on the union Ω^3 of all the local value sets $\Omega_i = \{T_i, F_i\}$. From the set of all local values the global value set is then obtained as a set of residue classes: $T_1 \equiv T_3$, $F_1 \equiv T_2$, $F_2 \equiv F_3$.

If we denote the equivalence class of T_1 by $[T_1]$, etc., then the three 'global' values are $1 = [T_1] = [T_3]$, $2 = [F_1] = [T_2]$, $3 = [F_2] = [F_3]$, corresponding to the foregoing mediation scheme.

REMARK. In PCL a certain enumeration convention for the subsystems is defined. For a motivation of this kind of indexing and enumeration as well as the mediation scheme we refer to the particular literature on PCL. We do not go into these details here (cf. also $[PF3]$).

NOTATIONAL CONVENTION. In PCL 'vector-like' formulas are studied. Hence, for terms, formulas (expressions) in a PCL vector notation is used: $X = \begin{pmatrix} x_1 \\ \vdots \\ x_m \end{pmatrix}$, $Y = \begin{pmatrix} y_1 \\ \vdots \\ y_m \end{pmatrix}$, etc., x_i corresponds to expressions in subsystem L_i , respectively. We shall make no dictinction between column and row notation of expressions X,Y, ...

3.2 Negations in a PCL

Particular negation operations are introduced in PCL via tableaux. These univariate operations consist of negations in particular subsystems combined with system changes. For more details we refer to the PCL literature (cf. also $[PF3, \text{section } 3]$). With our approach we give a general investigation of negations and system changes and describe the group they generate (section 5).

3.3 Bivariate Operations, Transjunctions

For notational simplicity, again we restrict the considerations to $\mathcal{L}^{(3)}$.

The tableau method is also used to introduce bivariate operations. Since every subsystem is a classical first order system we are led to bivariate operations which are defined componentwise, as for example:

$X \wedge \vee \wedge Y$ is to be interpreted in $\mathcal{L}^{(3)}$ as the operation where a conjunction is performed in the subsystem L_1 a disjunction in L_2 and a conjunction in L_3. Analogously the operation $X \wedge \vee \rightarrow Y$ has to be understood.

In vector notation: $X \wedge \vee \wedge Y = \begin{pmatrix} x_1 \wedge y_1 \\ x_2 \vee y_2 \\ x_3 \wedge y_3 \end{pmatrix}, X \wedge \vee \rightarrow Y = \begin{pmatrix} x_1 \wedge y_1 \\ x_2 \vee y_2 \\ x_3 \rightarrow y_3 \end{pmatrix}.$

Question: can all such operations be formed consistently ?

Answer: it turns out that this is not the case in general, e.g. $X \wedge \vee \wedge Y$ can be formed but $X \wedge \vee \rightarrow Y$ cannot be defined consistently in $\mathcal{L}^{(3)}$. This will become clear when we consider semantical aspects and evaluation of such expressions.

The identification of certain truth values plays an important role from a semantical point of view and it impacts the introduction of bivariate operations. We will come back to this later.

We give an example of a formula: $N_1(N_1 X \wedge \wedge \wedge Y) = X \vee \wedge \wedge N_1 Y.$

For a typical PCL like proof by tableaux we refer to the literature (cf. also $[PF3, 3.11]$).

In section 6 we shall give a short proof of that formula in an operational way using our notation.

REMARK. In a PCL system a new type of binary operation arises: an operation where the four output values w.r.t. the four inputs in a local subsystem L_i are distributed over other subsystems $L_j, j \neq i$. These (non classical) operations are also introduced via tableaux in the PCL literature, they are called *transjunctions*.

Dealing with logical fiberings, it is easy to see how to define and classify such operations, cf. section 6, 7.

3.4 Remark on Implementations

Some rules for forming PCL formulas of the above type have been implemented in *Prolog* during the previously mentioned project and were used to verify some formulas.

4 Logical Fiberings

All categories are assumed to be small categories (object and morphism classes are sets).

Subsequently, we introduce the concept of abstract fiber spaces in great generality. There is much material in the literature showing that fiberings provide a powerful tool - a 'formal mathematical language' where local-global relations of objects and data are expressible.

(For example, in $[PF2]$ we made practical experiences with that concept when we applied geometric fiberings to solve some open problems which formerly existed in a category of geometric spaces).

4.1 Preliminary Remarks on Indexed Systems

Fiberings and indexed systems are closely related from a formal point of view. As pointed out by P. Taylor in $[LNCS, p.\ 449ff]$, all consistency problems in forming families of sets w.r.t. a given indexing set I can be avoided when we interpret an indexed system $(A_i)_{i \in I}$ in terms of an abstract fibering ξ with a canonical projection (so-called *display map*) π in the following sense:

the "total space" A of the fibering $\xi = (A, \pi, I)$ is the coproduct (i.e. disjoint union) of the A_i, hence $A = \coprod_{i \in I} A_i$, and $\pi(a) = i$ for all $a \in A_i$, defines the projection map $\pi : A \rightarrow I$ from A to the "base space" I. Then $A_i = \pi^{-1}(i)$ is exactly the fiber over i.

After these remarks we come to our general definition.

4.2 Fiber Spaces

We define a *fiber space (fibering, bundle)* $\xi = (E, \pi, B, F)$ in a very general way for objects of a category (not only for topological spaces).
The map $\pi : E \to B$ is sometimes called *projection*, E is called *total space*, B the *base space*, the set of all preimages of an element $b \in B$, i.e. $\pi^{-1}(b)$ is called *fiber over b*.
F denotes the *typical fiber* with which each fiber $\pi^{-1}(b)$, $b \in B$, of the bundle is modeled.
A *covering* $\{U_i\}_{i \in I}$ of the base space B consists in general of subsets of B (whose union is B); depending on the category additional properties and conditions can be required (e.g. that they are open sets).
Typically, a fibering ξ is locally trivial (w.r.t. $\{U_i\}_{i \in I}$), i.e. the following diagram is commutative:

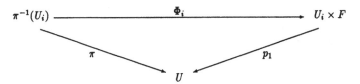

Φ_i is an isomorphism in the corresponding category, where

$$\Phi_i = (\pi, \phi_i), \qquad \phi_i : \pi^{-1}(U_i) \to F \qquad \text{(a morphism)}.$$

For $b \in U_i$, $\phi_{i,b} : \pi^{-1}(b) \overset{\cong}{\to} F$ is the fiber isomorphism induced by Φ_i (through this π^{-1} obtains its fiber structure).

The remaining properties concerning structure group, cocycle condition, etc. (cf. literature, e.g. [S]) can be formulated analogously here. We do not go into further details here since we only need the elementary features of the fiber space concept.

Of particular importance for our purposes is the case where the covering consists of one-point sets, i.e. the U_i are $1-$point sets (so every point of the base set B is an individual covering set).
The structure isomorphisms of the fibers are then given by:

$$\phi_b : \pi^{-1}(b) \overset{\cong}{\longrightarrow} F$$

and for $b \neq c$ the fibers over b resp. c can be compared with each other (*fiber transition* with F as "mediator") using

$$\phi_c^{-1} \circ \phi_b : \pi^{-1}(b) \to F \to \pi^{-1}(c).$$

4.3 Logical Fiberings

A *logical fibering* is an abstract fiber space ξ as defined above where the typical fiber is $F = L$, in our considerations L will be a classical first order logical space.
The base space will often be denoted by $B = I$, the "indexing set".
We can also think of the fibers as modeled via a *boolean algebra* as typical fiber. Then ξ would be an *abstract bundle of boolean algebras.*
In this article we shall deal only with coverings where each U_i is a one point set, i.e. $U_i = \{i\}$ for $B = I$, the base space B of the fibering being the indexing set I.
A *morphism* between two logical fiberings is defined in a similar way as this is done for bundles (c.f. e.g. [GO, Ch.4.5], [LS] and others). In some sense we tend to interpret a morphism as a process of transporting information between spaces.
In this way we obtain the *category of logical fiberings*. Logical fiberings over a certain constant base space I (index set) then form a 'comma category' $\mathcal{L} \downarrow I$ in the usual sense (cf. e.g. [GO]).

REMARK. Although we are working here with classical logics as fibers, we point out that all considerations can be done for more general objects in the fibers (e.g. different logics or algebras, etc.), based on a modified, generalized definition of an abstract fibering.

In a logical fibering the map $\pi : E \to I$ is always a morphism in the category of sets. The base set I can carry an additional own structure (e.g. partial order, graph, net, semigroup, algebra, topology, etc.). If E, I belong to the same category, then it is reasonable to require that π is a morphism in that category. In particular, it might be of interest in this framework to study logical fiberings which are bundles or sheaves of e.g. boolean algebras over a topological base space.

We want to mention here that the fibering approach reflects certain internal parallelism.

NOTE. Although we introduced the notions (fiberings, etc.) in great generality our interest in this article concentrates only on a certain class of logical fiberings w.r.t. one-point coverings. We also do not discuss specific structures of the base space (index set). But nevertheless, we wanted to introduce the notions in a certain generality for later use. (These aspects are of interest for further study).

4.4 Free Parallel Systems

The simplest form of a fibering or bundle is the "trivial fibering" $\xi = (E, \pi, B, F)$ with $E = B \times F$, π the first projection; the fiber over $i \in B$ is: $\pi^{-1}(i) = \{i\} \times F$.

In the context of our logical fiberings such a trivial fibering is a *parallel system of logics L_i* over an index set I as base space B and $F = L$ a classical first order logic.

We can think of reasoning processes running in parallel and independently within each fiber $L_i = \pi^{-1}(i)$. Transition ("communication") between fibers (loci) is described via the $\phi_i, i \in I$.

We call such a logical fibering a *"free parallel system"* $\mathcal{L}^I$. Its total space is denoted by E^I, note $E^I = \coprod_{i \in I} L_i$.

Simplest case (trivial fibering): all $\phi_i = id_L$. We shall make a difference between *local truth values* $\Omega_i = \{T_i, F_i\}$ in each 2-valued subsystem $L_i, i \in I$, and the set of *global values* Ω^I of the whole fibering. Parallel systems are characterized by the fact that there are no relations between different local values, i.e. the global value set is the mere coproduct (disjoint union)

$$\Omega^I = \coprod_{i \in I} \Omega_i$$

("free" parallel system). For $I = \{1, \ldots, n\}$, n a natural number, we define $\mathcal{L}^n := \mathcal{L}^I$. The corresponding total space is denoted by E^n.

NOTATION. A logical fibering which is derived from a free parallel fibering $\mathcal{L}^I$ by the introduction of an equivalence relation $\equiv$ on Ω^I will be denoted by $\mathcal{L}^{(I)}$, with $\Omega^{(I)} := \Omega^I / \equiv$. For $I = \{1, \ldots, n\}$ we use the notation $\mathcal{L}^{(n)}$. Accordingly, the total space of such a fibering is $E^{(I)}$ or $E^{(n)}$, respectively.

All the logical operations we are considering in a logical space $\mathcal{L}^{(I)}$ are induced by corresponding operations in $\mathcal{L}^I$. In a certain sense, $\mathcal{L}^{(I)}$ is a logical fibering with constraints (cf. section 7 for more details). For example, the PCL system $\mathcal{L}^{(3)}$ arises from the free parallel system $\mathcal{L}^3$ by introducing the equivalence relation $\equiv$ (given by $MS3$, cf. section 3) on the global values Ω^3 yielding $\Omega^{(3)} = \{1, 2, 3\}$.

REMARK. In the free parallel logical systems $\mathcal{L}^I$ there are no restrictions on the value sets Ω^I (e.g., in $\mathcal{L}^n$ there is a total amount of $2n$ global truth values). A variety of logical systems can be derived from $\mathcal{L}^I$ by introducing various equivalence relations on Ω^I. We can vary freely all data, i.e. base space, the fibers. In principle, we can combine different logics in a fibering when we allow different types of logical spaces for the fibers (by a corresponding generalization of the definition of an abstract fibering.)

If we consider (in the category of sets) the total space E^I of the fibering $\mathcal{L}^I$ as coproduct of the sets $L_i (i \in I)$, then $\mathcal{L}^I$ is a bundle over I, i.e. an object of the comma category Set$\downarrow I$, also denoted Bn(I), bundles over I, cf. $[GO, Ch.4.5]$.

This category actually is a *topos*, $[Goldblatt, loc.cit.]$. We do not go into these details here.

4.5 Notation for Logical Expressions

With the notion of a logical fibering we express the coexistence of various logical loci residing over an indexed system (base space) which itself can have an own structure (object of certain category). Our objective is to give a suitable formalization of the logical expressions in a fibering $\mathcal{L}^I$ which form the corresponding language of $\mathcal{L}^I$. This should be constructed from the local languages of the $L_i, i \in I$. In a free parallel fibering it is possible without restrictions to form global expressions consisting of a family of arbitrary local expressions (formed in parallel in each subsystem).

Canonically, this leads to the following formal *definition of a global expression* x in the language of $\mathcal{L}^I$, namely $x = (x_i)_{i \in I}$.

The language of a logical fibering is therefore obtained as the collection of all families of expressions from the local languages of the subsystems L_i.

Formally, all such families $x = (x_i)_{i \in I}$ form the *direct product* (in the categorical sense) of the sets of all local expressions, we denote this by $\prod_{i \in I} L_i$.

Logical connectives are introduced componentwise.

If I is a finite set then we use vector notation for expressions x, as already done previously.

We mention here that, alternatively, the set of all global expressions can be expressed as the set of all *sections* $s : I \to \coprod_{i \in I} L_i$, a section has the property $\pi \circ s = id_I$, π being the projection of the fibering. This is more compatible with our notion of a fibering (cf. P.Taylor in $[LNCS, p.451]$ and also $[PF3]$).

NOTATION. We use the symbol E^I also to denote the language corresponding to the logical fibering (keeping in mind that it is a direct product or all sections, respectively).

5 Univariate Operations, Negations

5.1 Preliminary Remark

The negation operation N_1 in the PCL $\mathcal{L}^{(3)}$ is defined (cf. e.g. $[PF3]$) in such a way that N_1 realizes a classical negation in L_1 and swaps the contents of the places L_2 and L_3 (system change) — i.e. realizes the transposition $2 \mapsto 3, 3 \mapsto 2$ (with cycle description for permutations, briefly written as (23)). This can be represented as follows:

$N_1 : \mathcal{L}^{(3)} \to \mathcal{L}^{(3)}$:

$$
\begin{pmatrix} x_1 \\ x_2 \\ x_3 \end{pmatrix} \xmapsto{N_1} \begin{pmatrix} \overline{x_1} \\ x_3 \\ x_2 \end{pmatrix} , i.e. \qquad N_1 X = N_1 \begin{pmatrix} x_1 \\ x_2 \\ x_3 \end{pmatrix} = \begin{pmatrix} \overline{x_1} \\ x_3 \\ x_2 \end{pmatrix}
$$

Such operations can be canonically extended to general logical fiberings.

5.2 Transpositions (System Changes)

Let $\mathcal{L} = \mathcal{L}^n$ (the following considerations can easily be generalized to $\mathcal{L}^I$).

Permutations play an important role in the definition of negation operations in PCL.

We recall that every *permutation group* (i.e. the group of all bijections of a set onto itself — here we consider mainly finite sets) can be generated by *transpositions*.

A transposition swaps two elements (numbers) and leaves the rest fixed, e.g.

$$
\begin{pmatrix} 1 \ldots i \ldots j \ldots n \\ 1 \ldots j \ldots i \ldots n \end{pmatrix} = (ij)
$$

(ij) is the cycle notation for permutations. We let permutations operate on the indices of the subsystems $L_1, \ldots, L_n$ and can then describe *system changes* as follows:

we denote by $\tau_{ij} : \mathcal{L} \to \mathcal{L}$ the transposition (system change)

$$
\tau_{ij}(X) = (x_1, \ldots, x_{i-1}, \phi_{ij}(x_j), \ldots, \phi_{ji}(x_i), x_{j+1}, \ldots, x_n),
$$

which means: the expression x_j in the place (fiber) over j is transferred to L_i by ϕ_{ij} (fiber transition isomorphism) and, conversely, the content of position i is brought to the fiber (logical place) L_j by means of ϕ_{ji}.

This corresponds to a "system change" by means of $\phi_{ij}, \phi_{ij}^{-1}$.

Note that for evaluations $w_i : L_i \to \Omega_i$ and $w_j : L_j \to \Omega_j$ it holds that: $\qquad w_i(\phi_{ij}(x_j)) = w_j(x_j)$.

To shorten notation we shall omit the ϕ_{ij} in our formulas and abbreviate $\tau_{ij} = (ij)$, if no confusion arises.

5.3   Negations in Subsystems

By a "*local*" or "*inner*" *negation* we mean one of the following operators:
$$n_i : \mathcal{L} \to \mathcal{L}, \quad n_i(x_1, \ldots, x_i, \ldots, x_n) = (x_1, \ldots, \overline{x_i}, \ldots, x_n), \text{ for } i = 1, \ldots, n.$$
So, only in the system L_i the negation is applied, all other places remain unchanged.
It holds that $n_i \circ n_j = n_j \circ n_i$, $for \ i \neq j$ and $n_i \circ n_i = Id_{\mathcal{L}} = $ the identity on $\mathcal{L}$.
Now we can compose negation operations on $\mathcal{L}$:
The *negation operator* $N_{ij}^k : \mathcal{L} \to \mathcal{L}$ is defined as the composition of the operators n_k and τ_{ij}, namely
$N_{ij}^k = \tau_{ij} \circ n_k$.
So, first a negation is carried out in the subsystem L_k and then the expressions on the positions i and j are interchanged (by τ_{ij}).

EXAMPLE. Negation operations in $\mathcal{L}^3$:
With the above notation we obtain the following operators ("global" negations) on the fibering $\mathcal{L} := \mathcal{L}^3$ by composition of elementary operations.
(The same can be done by passing to $\mathcal{L}^{(3)}$, but well-definedness problems have to be handled with care).
In particular we obtain the negation operations $N_1, \ldots, N_5$ using the notational convention of PCL (cf. $[R1], [R2], [PF3]$):

$$N_1 = n_1 \circ (23), \quad N_2 = n_2 \circ (13), \quad N_3 := N_2 \circ N_1, \quad N_4 := N_1 \circ N_2, \quad N_5 = N_1 \circ N_2 \circ N_1$$

Furthermore, $n := n_1 \circ n_2 \circ n_3 : \mathcal{L} \to \mathcal{L}$ (negation in each corresponding subsystem).
For every transposition $\tau \in \{(12), (13), (23)\}$ it holds: $\tau \circ n_i \circ \tau^{-1} = n_{\tau(i)}$ $\quad$ *where* $\tau = \tau^{-1}$, (i.e. τ operates on the elementary (inner) negations n_1, n_2, n_3 by conjugation).

Summarizing the previous considerations, we see that the negation operations and system changes generate a group. We do this here only for the particular example $\mathcal{L}^3$ and remark that this result also holds in the general case.

$\quad$ Let $\mathbf{N} = < n_1, n_2, n_3 >$ be the group of operators on $\mathcal{L}$ generated by the ("inner", "local") negations n_1, n_2, n_3.
$\mathbf{N}$ contains 8 elements and is isomorphic to the "elementary abelian 2-group"

$$\mathbf{N} \cong (\mathbb{Z}/2\mathbb{Z}) \times (\mathbb{Z}/2\mathbb{Z}) \times (\mathbb{Z}/2\mathbb{Z})$$

The transpositions τ_{ij} generate the group of "system changes", S_3, which is isomorphic to the full group of all permutations of 3 elements.
Combining the two groups we can define the group of *"global" negation operators* $\mathcal{N}$, which is generated by $\mathbf{N}$ and S_3: $\quad \mathcal{N} := < \mathbf{N}, S_3 >$.
Using the previous results it can be shown that $\mathcal{N}$ can be represented as a *semidirect product*:

$$\mathcal{N} := \mathbf{N} \rtimes S_3$$

(since it is possible to represent $\mathcal{N}$ as a product of groups $\mathcal{N} = \mathbf{N}.S_3$, $\mathbf{N} \lhd \mathcal{N}$ is a normal subgroup of $\mathcal{N}$ and S_3 operates on $\mathbf{N}$ by conjugation).
These considerations can be directly generalized and describe all the univariate operations we want to have in a (free parallel) logical fibering.

6   Bivariate Operations

6.1   Domain of Definition

Given a logical fibering $\mathcal{L}^I$ the question arises how to define bivariate logical operations, more precisely how does a suitable domain look like on which we can define an operation in a natural way, fitting to our fiber space concept.
In a natural way this leads us to the family of products $(L_i \times L_i)_{i \in I}$ where we can make a componentwise definition of a bivariate operation.
In a formally correct way and compatible with our fiber space notation such a family can be expressed by

the *pullback* (or fibered product) denoted by $E^I \times_I E^I$, cf. Taylor, in $[LNCS, p.449ff]$. See also $[PF3, 6.1]$ for more details.

A bivariate operation on $\mathcal{L}^I$ should map each pair of expressions $(x_i)_{i \in I}, (y_i)_{i \in I}$ in $\mathcal{L}^I$ to a new expression of $\mathcal{L}^I$ (an image of this mapping). Equivalently, such a pair corresponds to the familiy of pairs $((x_i, y_i))_{i \in I}$ which is exactly an element of the pullback $E^I \times_I E^I$.

REMARK. In a free parallel system $\mathcal{L}^I$ many bivariate operations can be introduced componentwise combining various bivariate operations, defined independently on each component (subsystem) L_i, $i \in I$.

6.2 Examples

For the sake of brevity we consider examples in $\mathcal{L}^3$ combining negations and bivariate operations. We want to demonstrate the operational way in which the formalism works. As mentioned in the beginning, all operations are expressed by mappings (functional notation).

(a) We show $N_1(N_1 X \wedge \wedge \wedge Y) = X \vee \wedge \wedge N_1 Y$.
Proof:

$$N_1(N_1 X \wedge \wedge \wedge Y) = N_1 \left(\begin{pmatrix} \overline{x_1} \\ x_3 \\ x_2 \end{pmatrix} \wedge \wedge \wedge \begin{pmatrix} y_1 \\ y_2 \\ y_3 \end{pmatrix} \right) =$$

$$= N_1 \begin{pmatrix} \overline{x_1} \wedge y_1 \\ x_3 \wedge y_2 \\ x_2 \wedge y_3 \end{pmatrix} = \begin{pmatrix} \overline{\overline{x_1} \wedge y_1} \\ x_2 \wedge y_3 \\ x_3 \wedge y_2 \end{pmatrix} = \begin{pmatrix} x_1 \vee \overline{y_1} \\ x_2 \wedge y_3 \\ x_3 \wedge y_2 \end{pmatrix} =$$

$$\begin{pmatrix} x_1 \\ x_2 \\ x_3 \end{pmatrix} \vee \wedge \wedge \begin{pmatrix} \overline{y_1} \\ y_3 \\ y_2 \end{pmatrix} = X \vee \wedge \wedge N_1 Y$$

The mixed expressions like $x_2 \wedge y_3$, etc. should be constructed via the transition isomorphisms ϕ_{23}, etc., but we omit this for short.

(b) Analogously (cf. $[PF3, \text{section } 6]$): $N_5((N_5 X) \vee \vee \vee N_5 Y) = X \wedge \wedge \wedge Y$.

6.3 Transjunctions

In a parallel system $\mathcal{L}^I$ the following situation arises naturally for bivariate operations: a local pair (x_i, y_i) in $L_i \times L_i$, $i \in I$, can be mapped into different subsystems $L_j, L_k, \ldots$.

With respect to the four possible local input pairs from $\Omega_i \times \Omega_i$ there are maximally four different subsystems for the images.

That means that such bivariate operations distribute images over different subsystems — *a new feature*.

Such bivariate operations are called *transjunctions*.

More details will be discussed in section 8 where we give a classification of transjunctions. It is helpful using evaluations to make things more transparent.

7 Evaluations, Semantical Aspects

7.1 Remark

As already remarked, starting from a free parallel system $\mathcal{L}^I$, there are many possibilities to find new logical spaces by introducing an equivalence relation $\equiv$ on the global value set Ω^I and then examining the passage $\mathcal{L}^I \rightarrow \mathcal{L}^{(I)}$.

Many multiple-valued logical spaces can be constructed systematically by this method.

Considering PCL systems, $\mathcal{L}^{(n)}$ is derived from $\mathcal{L}^n$ by the special $\equiv$-relation defined through the mediation scheme (cf. section 3).

From this point of view we obtain PCL systems as a particular class of certain logical fiberings.

To formalize the evaluation process we use an ad hoc notation which is useful for our purposes.

Again we point out that a rigorous formal treatment in categorial notions would be possible e.g. in the sense of Goldblatt $[GO, Ch.\ 6]$, cf. also $[LS]$.

7.2 Formalization of the Evaluation Process

In accordance with our notions of logical expressions and bivariate operations the (global) evaluation procedure w.r.t. a logical operation will be introduced componentwise hence being a family of local (classical) valuations.

We consider here only bivariate operations; evaluation of univariate operations can be introduced analogously.

Let Θ be a bivariate operation on $\mathcal{L}^I$. For every $i \in I$ let $w_i : L_i \to \Omega_i$ be a (classical) valuation and let $w := (w_i)_{i \in I}$ be the family of these valuations. This induces a global valuation $w : \mathcal{L}^I \to \Omega^I$.

Note, although we deal with families of expressions and truth values we do not express the domain and codomain of w as direct products explicitly (cf. similar remark at the end of section 4).

A global valuation $V(\Theta)$ of the operation Θ is defined componentwise by the following composition of maps (using functional notation).

$$V_i(\Theta) : \Omega_i \times \Omega_i \xrightarrow{J_i} L_i \times L_i \xrightarrow{\Theta} \mathcal{L}^I \xrightarrow{w} \Omega^I.$$

Where J_i denotes the local *input map* (substitution of pairs of truth values)

$$J_i : \Omega_i \times \Omega_i \to L_i \times L_i$$

which substitutes (x_i, y_i) by pairs of logical values in the expression $\Theta(x_i, y_i)$. The global valuation is then defined by the (family of the) $V_i(\Theta), i \in I$.

The evaluation procedure for a derived logical fibering $\mathcal{L}^{(I)}$ is induced by the previously described evaluation procedure in $\mathcal{L}^I$.

Let $p : \Omega^I \to \Omega^{(I)}$ be the canonical residue class map, where $\Omega^{(I)} = \Omega^I / \equiv$, for an equivalence relation $\equiv$ on Ω^I.

In this case, again, the evaluation is defined via the components, but we have to take into account the given $\equiv$ −relation and well-definedness properties (cf. the following example).

In formal notation, for $i \in I$, the *induced* valuation $V_{(i)}(\Theta)$ is defined by

$$V_{(i)}(\Theta) : \Omega_i \times \Omega_i \xrightarrow{V_i(\Theta)} \Omega^I \xrightarrow{p} \Omega^{(I)}$$

hence the local input is on pairs $\{T_i, F_i\}$ but respecting that these are representatives of equivalence classes and whenever two pairs (from different subsystems) belong to the same equivalence class the resulting value of $V_{(i)}(\Theta)$ has to be the same (this corresponds to well-definedness).

Thus we are led to certain constraints on the evaluation procedure.

REMARK. It is important to note again that the images of the four possible local input pairs of Θ can be distributed over maximally four different value sets $\Omega_\alpha, \Omega_\beta, \Omega_\gamma, \Omega_\delta$.

EXAMPLE. If we use 2×2 - matrix notation for the input pairs and corresponding output, the following is an example of a transjunction if $\{\alpha, \beta, \gamma, \delta\}$ contains at least 2 different indices.

$$\begin{array}{|cc|} \hline T_i T_i & T_i F_i \\ F_i T_i & F_i F_i \\ \hline \end{array} \xrightarrow{V_i} \begin{array}{|cc|} \hline T_\alpha & F_\beta \\ F_\gamma & F_\delta \\ \hline \end{array}$$

NOTATION. Subsequently, in all our considerations we express the value scheme (truth table) of a local bivariate operation by a 2×2 - matrix where only the output values (images of the operation) are represented (as in the right matrix above), their position (index pair) in the matrix is determined by the position of the corresponding input pair (cf. left matrix).

We use this convention analogously for 3-valued (multiple-valued) operations.

REMARK. We recall that we have chosen $(L_i \times L_i)_{i \in I}$ (the pullback) as domain of definition for a bivariate logical operation Θ on a free parallel system $\mathcal{L}^I$. This means that we do not consider input pairs $(x_i, y_j), i \neq j$, i.e. where x_i, y_j are from different subsystems (since we are mainly interested in forming logical connectives componentwise).

For a particular $\mathcal{L}^n$ the value set Ω^n totally contains $2n$ values.

A bivariate operation in the usual sense of multiple-valued logics would then be represented by a $(2n) \times (2n)-$ value matrix. In our definition of Θ on $\mathcal{L}^n$ the operation is represented by n $2 \times 2-$ submatrices which are arranged one after the other along the main diagonal of the whole $(2n) \times (2n)-$ scheme (in this sense it represents a restricted map). For further remarks on this we refer to $[PF3]$.

7.3 Two Examples

We briefly discuss the following evaluation problem. Let Z be the bivariate operation $Z = X \wedge \vee \to Y$.

No consistency problem arises, of course, when we form Z in the system $\mathcal{L}^3$.

We recall that the global value set $\Omega^{(3)}$ of $\mathcal{L}^{(3)}$ is given by $1 = [T_1] = [T_3], 2 = [F_1] = [T_2], 3 = [F_2] = [F_3]$ (corresponding to the mediation scheme $MS3$).

The corresponding local evaluations can be expressed by the following 2×2- matrices (on the right side we have inserted the corresponding global values):

$$x_1 \wedge y_1 : \qquad \begin{matrix} T_1 \ F_1 \\ F_1 \ F_1 \end{matrix} \equiv \begin{matrix} 1 \ 2 \\ 2 \ 2 \end{matrix}$$

$$x_2 \vee y_2 : \qquad \begin{matrix} T_2 \ T_2 \\ T_2 \ F_2 \end{matrix} \equiv \begin{matrix} 2 \ 2 \\ 2 \ 3 \end{matrix}$$

$$x_3 \to y_3 : \qquad \begin{matrix} T_3 \ F_3 \\ T_3 \ T_3 \end{matrix} \equiv \begin{matrix} 1 \ 3 \\ 1 \ 1 \end{matrix}$$

Applying the valuation $V_{(i)}(Z)$ - as defined above - we have to respect the relations between input pairs $(T_1, T_1) \equiv (T_3, T_3), (F_1, F_1) \equiv (T_2, T_2), (F_2, F_2) \equiv (F_3, F_3)$ according to the above mentioned identifications of local truth values.

In the evaluation procedure these identities have to be respected, i.e. equivalent pairs lead to equal images. (Alternatively, this can be expressed by the condition that the above three 2×2-value matrices are composable into one 3×3-scheme ('morphogram'), cf. next example).

Evaluating $X \wedge \vee \to Y$ leads to inconsistencies, since

$$(F_2, F_2) \overset{\vee}{\longmapsto} [F_2] = 3$$

$$\equiv \qquad\qquad \neq$$

$$(F_3, F_3) \overset{\to}{\longmapsto} [T_3] = 1$$

(Locally consistent, but not globally).

The pair (F_2, F_2) is local input in the second subsystem and (F_3, F_3) in the third, respectively. But globally, as input of the valuation of the bivariate operation Z on $\mathcal{L}^{(3)}$ both pairs are equal (since they belong to the same equivalence class). That means they have to produce the same image (output) of that operation (in the sense of a mapping).

APPLICATIONAL ASPECTS. It might be an interesting aspect whether such situations can be exploited to model specific applications where certain local operations are prohibited - from a global perspective (evaluation).

In the above case the implication in the third subsystem causes problems.

When we consider for example $X \wedge \vee \wedge Y$ these problems do not occur:

$$\begin{array}{|cc|}\hline T_1 & F_1 \\ F_1 & F_1 \\\hline\end{array} = \begin{array}{|cc|}\hline 1 & 2 \\ 2 & 2 \\\hline\end{array}$$

$$\begin{array}{|cc|}\hline T_2 & T_2 \\ T_2 & F_2 \\\hline\end{array} = \begin{array}{|cc|}\hline 2 & 2 \\ 2 & 3 \\\hline\end{array}$$

$$\begin{array}{|cc|}\hline T_3 & F_3 \\ F_3 & F_3 \\\hline\end{array} = \begin{array}{|cc|}\hline 1 & 3 \\ 3 & 3 \\\hline\end{array}$$

$$\begin{array}{|c|c|c|}\hline 1 & 2 & 3 \\\hline 2 & 2 & 2 \\\hline 3 & 2 & 3 \\\hline\end{array}$$

REMARK. The diagram on the right is an amalgamation of the three $2 \times 2-$ matrices on the left, suggesting that consistency is expressible in forming such a condensed form. The three $2 \times 2-$ submatrices along the diagonal have to be compatible in such a way that coinciding diagonal elements have to be equal (compatibility with the $\equiv -$ relation).

All the three 2×2-value schemes are represented (merged) in the $3 \times 3-$ value scheme as submatrices. In this form the operation $X \wedge \vee \wedge Y$ in $\mathcal{L}^{(3)}$ is represented by the complete $3 \times 3-$ matrix like a 3-valued logical connective, cf. [PF3].

Reversing this procedure leads to a method for decomposing a bivariate operation (given by a corresponding value matrix) in a multiple-valued logic into a system of 2-valued operations. In a certain sense this can be interpreted as a *parallelization* method for multiple-valued logics.

This will be subject of another work (forthcoming preprint in RISC-Linz publication series).

8 Classifying Transjunctions

For the classification of transjunctions it is convenient to consider the relevant evaluation procedures.

Let $\mathcal{L} := \mathcal{L}^I$, $\Omega := \Omega^I$ and Θ be a bivariate operation (actually we are interested in transjunctions).
For a local subsystem $L_i, i \in I$, we consider $\Theta : L_i \times L_i \to \mathcal{L}$ and w.r.t. $V_i : \Omega_i \times \Omega_i \to \Omega$ we can represent Θ locally by a 2×2 *pattern* (called morphogram in PCL notation), cf. the examples in section 7.
Suppressing the 4 indices $\alpha, \beta, \gamma, \delta$ in that T,F-*pattern* we obtain one of the sixteen $2 \times 2-$ value patterns corresponding to bivariate operations of classical (1st order) logic.

Using this, a transjunction can be described by such a $2 \times 2 - T, F - pattern$ followed by a *distribution* of the $T, F - values$ over (maximally four different) value sets $\Omega_\alpha, \Omega_\beta, \Omega_\gamma, \Omega_\delta$ corresponding to subsystems $L_\alpha, L_\beta, L_\gamma, L_\delta$.

More formally, let ϑ denote a classical bivariate operation $\vartheta : L_i \times L_i \to L_i$ and let $V = (w_i)_{i \in I}, w_i : L_i \to \Omega_i$ be valuations.

For $(T_i, F_i) \in \Omega_i \times \Omega_i$ let $\chi_{(T_i,F_i)} : \Omega_i \times \Omega_i \to \{0,1\}$ be the corresponding characteristic function. Then the local evaluation of the transjunction $\Theta : L_i \times L_i \to \mathcal{L}$ can be described by

$$\begin{aligned} w_{\alpha\beta\gamma\delta}(,) = \quad & \chi_{(T_i,T_i)}(,) \cdot w_\alpha \phi_{\alpha i} \vartheta(,) \;+ \\ & \chi_{(T_i,F_i)}(,) \cdot w_\beta \phi_{\beta i} \vartheta(,) \;+ \\ & \chi_{(F_i,T_i)}(,) \cdot w_\gamma \phi_{\gamma i} \vartheta(,) \;+ \\ & \chi_{(F_i,F_i)}(,) \cdot w_\delta \phi_{\delta i} \vartheta(,) \quad. \end{aligned}$$

We recall that the ϕ_{jk} denote the system changes (cf. section 4).

This local evaluation $w_{\alpha\beta\gamma\delta}$ could also be expressed by a map $D_{\alpha\beta\gamma\delta} \circ V_i$, with $V_i = w_i \circ \vartheta \circ J_i : \Omega_i \times \Omega_i \to \Omega_i$ and $D_{\alpha\beta\gamma\delta} : \Omega_i \to \Omega$ *distributes values* over different subsystems in the following way:
let for example F_i be the first value in the $2 \times 2-$ value matrix belonging to ϑ, then $D_{\alpha\beta\gamma\delta}(F_i) = F_\alpha$, analogously, if the third value would be T_i, then $D_{\alpha\beta\gamma\delta}(T_i) = T_\gamma$, etc..
In other words $D_{\alpha\beta\gamma\delta}$ transforms the $2 \times 2-$ value matrix of ϑ by substituting the indices $\alpha, \beta, \gamma, \delta$ (in this order) for the corresponding $T, F-$ values (cf. e.g. the following figure).

$$\begin{array}{|cc|}\hline T_i & F_i \\ F_i & F_i \\\hline\end{array} \rightarrow \begin{array}{|cc|}\hline T_\alpha & F_\beta \\ F_\gamma & F_\delta \\\hline\end{array}$$

This yields a systematic way to classify transjunctions in $\mathcal{L}^I$.

NOTATION. We can speak of conjunctional, disjunctional, implicational, ..., transjunctions corresponding to whether ϑ is a conjunction, disjunction, implication, etc., since the $T, F-$ value matrix of ϑ characterizes the transjunction type.

REMARK. Transjunctions extend the set of bivariate operations extensively.
Passing to $\mathcal{L}^{(I)}$ we have to respect well-definedness problems, similar to the example discussed in section 7. Therefore, the possibility for forming bivariate operations which involve transjunctions depends on the structure of the set of global values $\Omega^{(I)} = \Omega^I/\equiv$.
We are interested in possibilities to apply transjunctions in practical fields like robotic scenarios, for example.

9 Concluding Remarks

Concerning *labelled deductive systems (LDS)*, D.Gabbay pointed out that various structures for the label systems are of interest, e.g. semi groups, boolean algebras. Besides that it is a very interesting question whether there are possible links to LDS or possibilities to combine certain features.

We are in particular interested in logical fiberings which are deduced from free parallel systems by a *group action* on the value set.

Remarks on the *base space structure*:
Different structures for the base space (indexing set) may be of interest, e.g. totally ordered sets (in the case of certain PCLs); partially ordered sets (objects of the category POSET); semi groups; net structures; and others.
Of particular interest can be an *ultra metric base space*, these spaces appear naturally in the study of hierarchical structures (cf. [E]: 1-1 correspondence between indexed hierarchies and ultrametrics).

REMARK. We pointed out that we do not use here the categorical language systematically as, e.g., in dealing with $Bn(I)$ and $Sh(I)$, the category of bundles and sheaves, respectively (cf. [GO], [LS], [LNM1,2], [RB]), although this possibility exists for our approach.
We prefer a less abstract formulation here for a first attempt to present the main notions.
We adopt a more *engineering point of view* in the sense that we suppose that such logical fiberings might be suitable tools in situations where indexed systems play a role and this arises frequently.
Therefore we are motivated by *practical reasons* rather than by philosophical or purely theoretical ones.
    In general, we can say that the fibering approach allows many constructions since it is a very general 'formal mathematical language', in particular we think of topological and differentiable manifolds as base spaces - this might also be of interest from a physical point of view.
    The most general categorial framework seems to be "Indexed Categories" (cf. [LNM1]). We refer also to the corresponding remarks in [PF1], especially on the impact of the base space structure on the whole system.

REMARK. The generality of the fibering concept allows, in principle, to put different logics in the fibers over a common base space, that means to mix different logics. One has to describe carefully the transition ('communication') between different fibers.

APPLICATIONAL ASPECTS. Possible applications of the concept of logical fiberings can be seen from different perspectives. We find it interesting to try to apply it in the field of robotics, especially robot multitasking scenarios as they are discussed e.g. in the MEDLAR project.
For example, a space where actions are performed which are to be modeled formally (e.g. robots cooperating in a robot cell) may be covered by regions where each region has its own logic ('typical fiber'). This refers

to local triviality of an abstract bundle. Passing from one region to another causes a change of the logics applied; this is a certain local global interaction principle (typically included in the concept of fiber bundles and sheaves).

In particular, it is very interesting to find out possibilities how to apply transjunctions (some first ideas are in discussion).

10 References

Selected Literature on Polycontextural Logic

[G1] G. Günther.
Beiträge zur Grundlegung einer operationsfähigen Dialektik, 3 volumes. Felix Meiner Verlag, Hamburg 1980

[G2] G.Günther.
Cybernetic Ontology and Transjunctional Operations. Biological Computer Lab. Publ. vol.68 (Urbana, Ill.), published in 'Self-organizing Systems 1962', Spartan Books, Washington, D.C., pp..313-392 (1962).

[K1] R.Kaehr.
Materialien zur Formalisierung der dialektischen Logik und der Morphogrammatik 1973-1975. In: Idee und Grundriß einer nicht-aristotelischen Logik, 2.Auflage, Hamburg 1978.

[K2] R.Kaehr.
Excurs zu Logica. In: Die Logik des Wissens und das Problem der Erziehung, Hamburg 1981.

[R1] READER 1.
Texte G.Günther's zur Polykontexturalen Logik und Arithmetik, Morphogrammatik und Kenogrammatik, Proemial-Relation, Strukturtypentheorie, Cybernetic Ontology. Collected by R.Kaehr, University Witten/Herdecke 1988.

[R2] READER 2.
Texte R.Kaehr's zur Formalisierung der Polykontexturalen Logik. University Witten/Herdecke 1988.

Literature on Categories, Logics, Fiberings

[vD] D.van Dalen.
Logic and Stucture, 2nd ed. Springer Universitext, 1983.

[DO] C.T.J.Dodson.
Categories, Bundles and Spacetime Topology. Kluwer Academic Publ. 1988.

[E] P.Érdi.
On the Ultrametric Structure of Semantic Memory: Scope and Limits. In: R.Trappl (ed.), Cybernetics and Systems '88, pp.329-336.

[GA] D.Gabbay.
LDS-Labelled Deductive Systems. MEDLAR Milestone 1 Deliverables, Oct 1990. (Chapter 1 of the draft of a book). To be published.

[GO] R.Goldblatt.
Topoi. Studies in Logic and the Foundations of Mathematics, vol.98, North-Holland 1986.

[LNCS] D.Pitt, S.Abramsky, A.Poigné and D.Rydeheard (eds.).
Category Theory and Computer Programming. Springer Lecture Notes in Computer Science, vol.240, 1986.

[LNM1] P.T.Johnstone, R.Paré (eds.).
Indexed Categories and Their Applications. Springer Lecture Notes in Mathematics, vol.661, 1978.

[LNM2] M.P.Fourman, C.J.Mulvey and D.S.Scott.
Applications of Sheaves. Springer Lecture Notes in Mathematics, vol.753, 1979.

[LS] J.Lambek and P.J.Scott.
Introduction to Higher Order Categorical Logic. Cambridge Studies in Advanced Mathematics, vol.7, Cambridge University Press 1986.

[ML] S.MacLane.
Categories for the Working Mathematician. Springer Graduate Text in Mathematics, vol.5, 1971.

[PF1] J.Pfalzgraf.
Reasoning on a Möbius Strip. MEDLAR Newsletter No.1, Sept-Nov 1990, J.Cunningham,D.Gabbay, R.de Queiroz (eds.), Imperial College London.

[*PF2*] J.Pfalzgraf.
Representation of geometric spaces as Fibered Structures. Results in Math. Vol.12 (1987), 172-190 (in German).

[*PF3*] J.Pfalzgraf.
On Logical Fiberings and Polycontextural Systems. A First Approach. RISC-Linz Publ. Series No. 91-13.0 (1991).

[*RB*] D.E.Rydeheard and R.M.Burstall.
Computational Category Theory. Prentice Hall 1988.

[*S*] N.Steenrod.
The Topology of Fibre Bundles. Princeton University Press 1951.

Automated Deduction with Associative Commutative Operators

M. Rusinowitch and L. Vigneron
CRIN and INRIA-Lorraine
BP239, 54506 Vandoeuvre-les-Nancy Cedex, France
email : rusi@loria.crin.fr, vigneron@loria.crin.fr

Abstract

We propose a new inference system for automated deduction with equality and associative commutative operators. This system is an extension of the ordered paramodulation strategy. However, rather than using associativity and commutativity as the other axioms, they are handled by the AC unification algorithm. Moreover, we prove the refutational completeness of this system without needing the functional reflexive axioms or AC axioms. Such a result is obtained by semantic tree techniques, assuming that terms are compared with an AC-compatible and complete simplification ordering.

1 Introduction

Automated deduction with equality and associative commutative (AC) operators (i.e. binary operators f satisfying the following axioms: $f(f(x,y),z) = f(x,f(y,z))$ and $f(x,y) = f(y,x)$) has been considered as a difficult problem. The reason is that the presence of AC axioms increases dramatically the number of possible deductions. For instance, there are 1680 ways to write the following term $f(t_1, f(t_2, f(t_3, f(t_4, t_5))))$, where f is an AC operator and t_1, t_2, t_3, t_4, t_5 are different constants.

The approach we propose for dealing with AC-axioms is to work in the AC equivalence classes and to employ associative commutative identity checking, pattern matching and unification. This idea of building axioms within the unification procedures has been first initiated by Plotkin [Plo72]. In the context of automated deduction, it has been investigated too by Stickel [Sti84], Lankford [Lan79], Anantharaman et al. [AHM89]. However these works essentially refers to practical experimentations, and do not account for completeness results of the inference systems that they study. In the following, we focus on giving a complete set of inference rules for first order logic with equality and built-in AC unification. The recent development of efficient AC unification algorithms [BHK+88, Sti81] strongly argues in favour of the effectiveness of our method.

Our inference system includes resolution and paramodulation to deal with equality. Paramodulation performs substitution directly by replacing one argument of an equality atom by the other one, when the former occurs in some clause. Paramodulation has been introduced by [RW69]. Some important refinements have been proposed by introducing a simplification ordering on terms and forbidding the replacement of a term by a bigger one. These aspects are fully developed in [Pet83, HR86]. Here, we also confine the term replacement step of our paramodulation rule by a simplification ordering. However, for sake of completeness we also impose this ordering to be total on the set of AC congruence classes and AC-compatible.

The refutational completeness of our set of rules is derived by the transfinite semantic tree method of [Rus89]. Semantic trees represent the set of Herbrand interpretations for formulas in clausal form. Failure nodes are distinguished interpretations which falsify a consequence of the formula. When a tree associated to a formula is empty then we can be sure that the empty clause can be derived, and therefore the initial formula is unsatisfiable.

Associative commutative theories have been thoroughly studied in the context of term rewriting systems. We will not review here the Knuth-Bendix method [KB70]. Let us just mention that it has been extended to incorporate associativity and commutativity by [LB77, PS81, JK86].

The layout of this paper is as follows: Section 2 presents an overview of our approach on a simple example. Section 3 summarizes the basic material which is relevant to this work. In particular we give some details on the construction of orderings which are AC-compatible and total on Herbrand Universe. These orderings are fundamental for defining the refutationally complete inference system that we describe in Section 4. Finally the proof of refutational completeness is given in section 5, and the proof of a lifting lemma in section 6.

2   A simple example

Here is a simple example to show the problems due to the presence of the AC axioms and to present our approach for solving them. We consider the following system of equations S:

$$\begin{cases} a + b = d & (1) \\ c + b = e & (2) \\ g(c + d) = h(a) & (3) \\ g(e + a) = h(b) & (4) \end{cases}$$

assuming that $+$ is an AC-operator, a, b, c, d and e are constants, and g and h are unary operators.

Now, let us prove the following theorem: $h(a) = h(b)$.

The first step is to add to S the inequation $h(a) \neq h(b)$ (Th) and let us try to find a contradiction.

With the classical paramodulation method [HR86] we just add the $AC(+)$ axioms to the other ones and perform inferences in the empty theory, i.e. with syntactic unification.

$$\begin{cases} (x + y) + z = x + (y + z) & (A) \\ x + y = y + x & (C) \end{cases}$$

the refutation of the new system is performed as follows:

para(2,A)	$c + (b + z) = e + z$	(5)
para(C,5)	$c + (z + b) = e + z$	(6)
para(1,6)	$c + d = e + a$	(7)
para(7,3)	$g(e + a) = h(a)$	(8)
para(4,8)	$h(b) = h(a)$	(9)
res(11,Th)	$\square$	(10)

where para(i,j) means a paramodulation of the clause (i) in the clause (j), and res(i,j) means a resolution between the clauses (i) and (j).

We notice that the third of the steps needed for the refutation of the system use the AC axioms. However, when dropping these axioms we must replace them by other mechanisms. For instance using unification modulo AC allows to suppress the step using the commutativity.

However modifying the unification algorithm is not sufficient. The last problem is to avoid the creation of extended equations, which results from a paramodulation in the axiom of associativity. Extended equations (or clauses) dramatically increase the number of possible deductions. Therefore, instead of introducing a special control (as the *protection* of extended rules in AC-completion procedures [PS81, JK86]) we rather build these extended equations on the flight that is when they are immediately followed by a paramodulation step. Hence, we have designed an inference rule, named *extended paramodulation*, which, given two clauses $f(A, B) = C \vee P$ and $f(D, E) = F \vee Q$ generates an instance of the clause $f(C, G) = f(F, H) \vee P \vee Q$ whenever f is an AC operator and $f(f(A, B), G)$ and $f(f(D, E), H)$ are unifiable modulo AC. This rule may be viewed as a generalization of the superposition rule of Buchberger algorithm for computing Gröbner bases.

Now the refutation of the previous system is carried out as follows:

$$\begin{array}{lll}
\text{para-ext(2,1)} & a + e = d + c & (5) \\
\text{para(5,3)} & g(a + e) = h(a) & (6) \\
\text{para(4,6)} & h(b) = h(a) & (7) \\
\text{res(7,}Th) & \square & (8)
\end{array}$$

Hence we have reached our goal: all the deduction steps involving AC axioms have now disappeared. This very last refutation is the one that can be obtained when applying our inference rules to be introduced in Section 4.

3 Terms and orderings

3.1 Notations and preliminary notions

Let F be a finite set of functions with arities, and let X be a countably infinite set of variables. The algebra of terms composed from F and X is denoted by $T(F, X)$. We use $T(F)$ for the set of ground terms (the *Herbrand universe*).

Let P be a finite set of predicate symbols including the equality predicate $=$. The set of *atoms* $A(P, F, X)$ is $\{p(t_1, \ldots, t_n) : p \in P \text{ and } t_i \in T(F, X)\}$. We denote the set of ground atoms (the *Herbrand base*) by $A(P, F)$. An *equality atom* is an atom whose predicate symbol is $=$. Throughout this paper, we assume that $=$ is commutative in the sense that we do not distinguish between the atoms $s = t$ and $t = s$. A *literal* is either an atom or the negation of an atom, and a *clause* is a set of literals. In general we use the term *object* to indicate a term, an atom, a literal, or a clause, and the term *ground object* to indicate a ground term or a ground atom.

Let $V(t)$ denote the set of variables appearing in an object t. A *substitution* is a mapping σ from X to $T(F, X)$ such that $\sigma(x) \neq x$ for only finitely many variables. We use $Dom(\sigma)$ to denote the set $\{x : \sigma(x) \neq x\}$. We further assume that for every $x \in Dom(\sigma)$, $V(\sigma(x)) \cap Dom(\sigma) = \phi$. The substitution σ is *applied* to an object t if all variables x in t are replaced by $\sigma(x)$. The result is denoted as $t\sigma$. A substitution σ is a *unifier* of two objects s and t if $s\sigma = t\sigma$. A unifier σ of s and t is the *most general unifier* (*mgu*) if for every unifier θ of s and t, there exists a substitution ρ such that $\theta = \sigma\rho$. The mgu is unique upon renaming of variables.

To express subterms and substitutions more effectively, we sometimes use *positions*. Envision a term represented as a tree; a *position* (or *occurrence*) in a term indicates a node in the tree. Positions are usually represented as sequences of integers. Let u be an position, we use $t_{|u}$ for the subterm of t at u. More precisely, $t_{|\varepsilon} = t$ where ε is the empty position, and $g(t_1, \ldots, t_n)_{|i.u} = t_{i|u}$. We also use $s = s[u \leftarrow t]$ to denote that s is a term whose subterm at position u is t. For convenience, we sometimes express it by $s = s[t]$ if the particular position is not important. A subterm of t is *proper* if it is distinct from t. The function $top(t)$ returns the operator at the top of the term t. The function $args(t)$ returns the list of the arguments of the operator $top(t)$ in t. $Occ(t)$ denotes the set of all occurrences in the term t. We write $Occ\_max(t, f)$ the set of all maximal occurrences of the operator f in the termn t. if $f \in F_{AC}$; else, it is the set of all occurrences of the operator f in t.

We assume that the operators from a given subset F_{AC} of F are associative commutative, which means that for $f \in F_{AC}$ the following axioms are implicit in the theory to be considered:

$$AC(f) \begin{cases} f(x,y),z) & = & f(x,f(y,z)) \\ f(x,y) & = & f(y,x) \end{cases}$$

The congruence on $T(F, X)$ generated by the associative commutative equations satisfied by the symbols in F_{AC} will be written $\equiv_{AC}$. An AC operator can be seen as an operator of arbitrary arity and we will sometimes consider flattened terms, i.e. terms such that no argument of an AC operator has this operator at the root. More formally, we introduce the function fl for flattening terms, at the level of the AC operators. Let $t = f(t_1, \ldots, t_n)$ be a ground term. Then:

$$fl(t) = \begin{cases} t & \text{if } t \text{ is a constant} \\ f(fl(t_1), \ldots, fl(t_n)) & \text{if } f \notin F_{AC} \\ t' & \text{else} \end{cases}$$

with t' resulting from t by replacing t_i by $fl(t_i)$ if $top(t_i) \neq f$, and by replacing t_i by $s_1, \ldots, s_m$ if $fl(t_i) = f(s_1, \ldots, s_m)$.

3.2 Compatible orderings for associative-commutative theories

Orderings are used to define restricted versions of resolution and paramodulation. Firstly, resolution and paramodulation need only be performed on the maximal literals. Secondly, when using an equality in a clause to paramodulate, only the larger of the two terms in the equality needs to be considered for paramodulation. Our inference rules set to be introduced in the next section can be proved refutationally complete if it is defined with respect to a *complete simplification ordering*. Let us recall the definition of these orderings:

Definition 1 *A transitive irreflexive relation $>$ on the set of terms is a complete simplification ordering (CSO, in short) if*

1. $>$ *is total on the set of ground terms.*
2. $>$ *is well-founded.*
3. *(monotonicity) $s > t$ implies $w[s] > w[t]$.*
4. *(subterm) for any proper subterm s of t, we have $s < t$.*
5. *(stability) for any substitution σ, $s > t$ implies $s\sigma > t\sigma$.*

Now, since we want to perform inferences on literals representing AC congruence classes rather than terms, these inferences should be somewhat independant of the chosen congruence classes representatives. That is why we also require our ordering to have the *AC-compatibility* property:

Definition 2 *An ordering $>$ on $T(F)$ is AC-compatible iff whenever we have $s > t$, $s \equiv_{AC} s'$ and $t \equiv_{AC} t'$ we also have $s' > t'$.*

The design of AC-compatible orderings for proving termination of rewrite systems modulo AC has been considered as a hard task. In fact, to our knowledge, very few constructions are available in the literature. Perhaps the best known among them is the *associative path ordering* scheme [BP85], which extends the *recursive path ordering* (see also [Der82]). However, this ordering puts serious limitations on the precedence of AC symbols. In fact two AC symbols cannot be compared in the precedence unless they are related by a distributivity law. That explains why it seems difficult to extend the *associative path ordering* to get a total ordering when there are several AC symbols.

Up to now only one AC-compatible complete simplification ordering for a signature wich contains any number of AC-symbols has been found. It is described in [NR91] and is based on the method of polynomial interpretations [Lan79, CL87]. This ordering could be used for our purpose. In particular, inference rules built on it are refutationally complete.

In the case of **one** single AC-operator, k much more constructions of total AC-compatible ordering are available. In [Vig90] we present a construction which is based on an extension of Knuth-Bendix ordering proposed by Steinbach [Ste90] to deal with AC operators.

Given an AC-compatible CSO on terms, it can be extended to atoms in the following way (see [HR86]):

Definition 3 *Let $>_P$ be a total precedence on the predicates, "$=$" being the smaller. Let $\succ_{ORD}$ be a complete simplification ordering on the terms. The ordering $\succ_{A\_ORD}$ is defined on the set of the atoms by:*

$P(s_1, \ldots, s_n) \succ_{A\_ORD} Q(t_1, \ldots, t_m)$ *if*

- *either* $P >_P Q$,
- *or* $P = Q$, P *is the equality predicate, and* $\{s_1, s_2\} \gg_{ORD} \{t_1, t_2\}$,
  where $\gg_{ORD}$ *is the multiset extension of* $\succ_{ORD}$,
- *or* $P = Q$, P *is not the equality predicate, and* $(s_1, \ldots, s_n) \succ_{ORD}^{lex} (t_1, \ldots, t_m)$,
  where $\succ_{ORD}^{lex}$ *is the lexicographic extension of* $\succ_{ORD}$.

4 Inference rules

The inference rules, that we shall define in this section are compatible with the strategy of ordered clauses presented in [Pet83, Rus89]. This strategy permits to apply the paramodulation inference rule under a refined form: a term cannot be replaced by a more complex one, during a paramodulation step; in particular, we never paramodulate into a variable.

From now we assume that we are given an AC-compatible CSO on terms $\succ_{ORD}$ and that this ordering has been extended to atoms as in definition 3. Let us denote by $\succ_{A\_ORD}$ this extension to the atoms set.

The notation $s \not\preceq t$ means that either s is greater than t, or s and t are uncomparable for the ordering $\prec$.

Note that AC unification is used in the following inference rules.

$$(\text{OAC-fact}) \quad \frac{L_1 \lor L_2 \lor D}{(L_1 \lor D)\sigma}$$

$$\text{if} \quad \begin{cases} L_1\sigma \equiv_{AC} L_2\sigma \\ \forall A \in D, \ L_1\sigma \not\preceq_{A\_ORD} A\sigma \end{cases}$$

So, this ordered AC-factorisation rule is applied if σ is a most general unifier of L_1 and L_2, and if there does not exist an atom A of D such that $A\sigma$ would be greater or equal than $L_1\sigma$, for the ordering $\succ_{A\_ORD}$.

(OAC-res) $\dfrac{L_1 \vee D_1 \quad \neg L_2 \vee D_2}{(D_1 \vee D_2)\sigma}$

if $\begin{cases} L_1\sigma \equiv_{AC} L_2\sigma \\ \forall A \in D_1, \ L_1\sigma \not\preceq_{A\_ORD} A\sigma \\ \forall A \in D_2, \ L_2\sigma \not\preceq_{A\_ORD} A\sigma \end{cases}$

So, this ordered AC-resolution rule is applied if σ is a most general unifier of L_1 and L_2, and if $L_1\sigma$, respectively $L_2\sigma$, is not less or equal than any atom of $D_1\sigma$, resp. $D_2\sigma$.

(OAC-para) $\dfrac{(s = t) \vee D_1 \quad L \vee D_2}{(L' \vee D_1 \vee D_2)\sigma}$

if $\begin{cases} (L_{|_p})\sigma \equiv_{AC} M[s]\sigma \\ \text{where } p \text{ is a non variable occurrence of } L \\ \forall A \in D_1, \ (M[s] = M[t])\sigma \not\preceq_{A\_ORD} A\sigma \\ \forall A \in D_2, \ L\sigma \not\preceq_{A\_ORD} A\sigma \\ M[s]\sigma \not\preceq_{ORD} M[t]\sigma \\ L' = L[p \leftarrow M[t]\sigma] \\ M[.] \text{ is either empty (then } (L_{|_p})\sigma \equiv_{AC} s\sigma) \\ \qquad \text{or a term } f(., x) , \text{ where } f \text{ is an AC operator } (= top(s)). \\ \qquad\qquad\qquad\qquad\qquad\qquad \text{and } x \text{ is a new variable} \end{cases}$

So, this ordered AC-paramodulation rule applies if there is a non variable occurrence p of the literal L and a most general unifier σ, such that $(L_{|_p})\sigma$ holds $s\sigma$, that is either $(L_{|_p})\sigma \equiv_{AC} s\sigma$, or $(L_{|_p})\sigma \equiv_{AC} f(s,x)\sigma$, where f is the AC operator at the top of s. Moreover, $L\sigma$, resp. $(M[s] = M[t])\sigma$, must not be less or equal than any atom of $D_2\sigma$, resp. $D_1\sigma$. Another condition is that $L'\sigma$ must not be greater than $L\sigma$, this means that $M[t]\sigma$ must not be greater or equal than $M[s]\sigma$.

The use of context in L and of AC-unification holds us to avoid many operations with the AC axioms to obtain a term AC-equivalent to L and containing a subterm syntaxically equivalent to s.

(OAC-para-ext) $\dfrac{(s = t) \vee D_1 \quad (l = r) \vee D_2}{((f(t,x) = f(r,y)) \vee D_1 \vee D_2)\sigma}$

if $\begin{cases} top(l), top(s) \in F_{AC} \cup X \quad, \text{ where } top(s) = top(l) \text{ if } top(l), top(s) \in F_{AC} \\ \exists \sigma, \ f(l,x)\sigma \equiv_{AC} M[s]\sigma \quad, \text{ where } x \text{ is a new variable and } f \in F_{AC} \\ \forall A \in D_1, \ (M[s] = M[t])\sigma \not\preceq_{A\_ORD} A\sigma \\ \forall A \in D_2, \ (f(l,x) = f(r,x))\sigma \not\preceq_{A\_ORD} A\sigma \\ M[s]\sigma \not\preceq_{ORD} M[t]\sigma \\ (f(l,y) = f(r,y))\sigma \not\preceq_{A\_ORD} (M[s] = M[t])\sigma \\ M[.] \equiv f(.,y) \quad, \text{ where } y \text{ is a new variable} \end{cases}$

So, this ordered extended AC-paramodulation rule is applied if we can apply an OAC-paramodulation of $(s = t)$ in $(f(l,x) = f(r,x))$, at the top of $f(l,x)$ with σ as most general unifier. Moreover, $M[t]\sigma$ should not be greater or equal than $M[s]\sigma$.

5 Refutational completeness of AC paramodulation

The problem of proving the completeness of theorem proving strategies involving equality has been prominent in automated theorem proving since its first conception. A notorious instance is the question of whether the inference system consisting of resolution and paramodulation is complete without the functionally reflexive axioms and without paramodulating into variables. Brand [Bra75] an indirect proof (as a corollary of the completeness of the modification method) was described. A direct proof by semantic trees was given in [Pet83]. However, Peterson's proof requires the use of a simplification ordering which is also order isomorphic to ω on ground atoms. Hsiang and Rusinowitch [HR86] have developed a new method based on transfinite semantic trees for relaxing this condition and permitting a larger class of orderings.

The method has also been applied to Knuth-Bendix completion procedure [HR87] and to conditional completion [KR87].

This is the method that we shall use here for proving the refutational completeness of the inference rules which have been introduced in the previous section. Many additional subcases are introduced by the associative commutative axioms when considering failure nodes. We need to show that this additional failure nodes can be handled by our set of inference rules.

5.1 Semantic trees

The Herbrand base $A(P, F)$ can be ordered as an increasing sequence $\{A_i\}_{i<\lambda}$ by $\succ_{A\_ORD}$ (λ being its ordinal). Given an atom A_α, we write W_α the initial segment $\{A_i \ / \ i < \alpha\}$. The successor of the ordinal α is denoted by $\alpha + 1$.

Given the ordinal α, an *E-interpretation* is a mapping I from W_α to $\{T, F\}$ satisfying:

- if $(s = s) \in W_\alpha$, then $I(s = s) = T$

- if $(s = t)$, $B[s]$, $B[t] \in W_\alpha$ and $I(s = t) = T$, then $I(B[s]) = I(B[t])$

We extend E-interpretations to the set of ground clauses in the usual way as follows: let I be an E-interpretation on W_α, A an element in W_α, and $C = L_1 \vee \ldots \vee L_n$ a ground clause whose atoms are all in W_α. Then, $I(\neg A) = \neg I(A)$ and $I(C) = I(L_1) \vee \ldots \vee I(L_n)$. We sometimes write $I \models C$ for $I(C) = T$, and $I \not\models C$ for $I(C) = F$. Given an E-interpretation I and a clause C, I *E-satisfies* C (or C is *valid* in I) if for every ground instance C' of C, $I \models C'$. Otherwise, we say that I *falsifies* C. C is *E-satisfiable* if C is valid in some E-interpretation. Otherwise it is *E-unsatisfiable*. Given a set of clauses S, S is *E-satisfiable* if for every instance S' of S, there is an E-interpretation I such that I satisfies every clause in S'. Otherwise S is *E-unsatisfiable*.

Let v and w be two ground atoms and let I be an E-interpretation on W_α. We say that w is *I-reducible* in v by $(s = t)$ if there is an atom $(s = t) \in W_\alpha$ such that:

$$w \equiv w[s] , \ s \succ_{ORD} t , \ w \succ_{A\_ORD} (s = t) , \ I(s = t) = T \ and \ v \equiv w[t]$$

An atom which is not *I-reducible* is said to be *I-irreducible*. By the next theorem, it is possible to build inductively all the E-interpretations in a manner which is similar to that in [Pet83].

Theorem 1 [Pet83] *Let* $I : W_{\alpha+1} \rightarrow \{T, F\}$ *such that* I *is an E-interpretation on* W_α. *Let* J *be the restriction of* I *to* W_α. *Then,* I *is an E-interpretation on* $W_{\alpha+1}$ *iff:*

(1) A_α *is* J-reducible in one B and $I(A_\alpha) = I(B)$, *or*

(2) A_α *is* J-irreducible, of the form $(t = t)$, and $I(A_\alpha) = T$, *or*

(3) A_α *is* J-irreducible and not of the form $(t = t)$.

The collection of all E-interpretations is called an *E-semantic tree*. Given an E-semantic tree, a *node* is an element of the tree.

Let ET be an E-semantic tree. We call *coherent semantic tree* of a set of clauses S, and we write $MCT(S)$, the maximal subtree of ET such that:

For every node I of $MCT(S)$, every clause C of S, and every ground substitution θ, if the atoms of $C\theta$ belong to the domain of I, then $I(C\theta) = T$.

We call *failure node* a node I which falsifies a clause $C\theta$, but whose ancestors are in $MCT(S)$. In particular, if J is the last node of MCT(S) then every extension of J is a failure node.

5.2 Completeness

Let INF be a set of inference rules and S a set of clauses. Let $INF(S)$ denote the set of clauses obtained by adding to S all clauses generated by applying some rule in INF to S. Let $INF^0(S) = S$, $INF^{n+1}(S) = INF(INF^n(S))$, and $INF^*(S) = \bigcup_n INF^n(S)$.

Theorem 2 *Let S be an AC-unsatisfiable set of clauses, including the axiom $(x = x)$. Let INF be the following set of inference rules $\{$ OAC-factorisation, OAC-resolution, OAC-paramodulation, OAC-extended paramodulation$\}$. Then $INF^*(S)$ contains the empty clause.*

We assume that S only contains ground clauses. For the general case, lifting lemmas can be applied since we can always label a failure node K by a clause of S whose variables are instantiated by K-irreducible terms.

Proof :

We reason by contradiction. We assume that S is an AC-unsatisfiable set and that $INF^*(S)$ does not contain the empty clause. Therefore $MCT(INF^*(S))$ is not empty. Let us define the sets

$$AC = \{(s = t); \ s \equiv_{AC} t\}$$
$$\mathcal{ES} = \{(f(a, l) = f(a, r)) \vee C; \ top(l) = f \in F_{AC} \text{ and } a \in T(F) \text{ and } (l = r) \vee C \in INF^*(S)\}$$
$$S^* = \{C'; \ \exists C \in \mathcal{ES} \cup INF^*(S) \text{ such that } C \equiv_{AC} C'\} \cup AC$$

Let us notice that if two terms u and v are equivalent modulo AC, then the terms $ext(u)$ and $ext(v)$ are AC-equivalent too, $ext()$ being an extension.

Note that $MCT(S^*)$ is also non empty. We shall show that the rightmost branch of $MCT(S^*)$ is empty, in contradiction with the fact that $MCT(S^*)$ is not empty; that ends the proof.

Let $K = I_\alpha$ be the last node of the rightmost branch of $MCT(S^*)$. Assuming that $MCT(S^*)$ is not empty, such a node exists by the same arguments as in [HR86]. Hence, K is defined on all the atoms A_i where $i < \alpha$, K belongs to $MCT(S^*)$, and every extension of K is a failure node. Let us write B the atom A_α.

Three cases are possible. Since Case 1 and Case 2 are simple and very similar to the corresponding ones in [HR86], we omit them here. Let us just mention that they can be solved by

Ordered AC-factorisation and Ordered AC-resolution.

Case 1 : K admits an extension I, and B is K-irreducible.

$$(a = a) \qquad T \qquad \overset{\bullet\, K}{\underset{\bullet\, I}{\Big|}}$$

$$\boxed{\neg(a = a) \vee D}$$

Case 2 : K admits two extensions L and R, and B is K-irreducible.

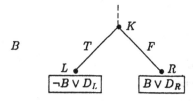

$$B \qquad T \qquad F$$

$$L \quad \boxed{\neg B \vee D_L} \qquad\qquad R \quad \boxed{B \vee D_R}$$

Case 3 : K admits an extension I, and B is K-reducible.

Let $(s = t)$ the smallest equation such that $s \succ_{ORD} t$, $K(s = t) = T$ and $B_{|p} \equiv s$. We verify that $(s = t)$ is K-irreducible (so it cannot be an element of $\mathcal{ES}$). Let β be the index of the atom $(s = t)$ and J the restriction of K to W_β (defined on the atoms smaller than $(s = t)$). Let M be the right successor of J. By the construction of K (right branch of $MCT(S^*)$), M is a failure node.

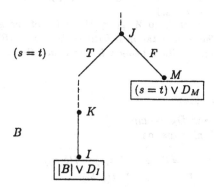

$$(s = t) \qquad T \qquad F$$

$$M \quad \boxed{(s = t) \vee D_M}$$

$$\bullet\, K$$

$$B$$

$$\bullet\, I$$

$$\boxed{|B| \vee D_I}$$

with : $|B| \equiv$ if $I(B) = F$ then B, else $\neg B$.
$\exists p \in Occ(B)$, $B_{|_p} \equiv s$.

We can establish the following facts :

(1) M falsifies the clause $C_M \sigma = (s = t) \vee D_M$, with $C_M \in S^*$ and σ a ground irreductible substitution. Moreover, each atom of D_M is smaller than $(s = t)$.

(2) I falsifies the clause $C_I\tau = |B| \vee D_I$, with $C_I \in S^*$ and τ a ground irreductible substitution. Moreover, every atom of D_I is smaller than B .

(3) $J(D_M) = K(D_M) = K(D_I) = F$.

(4) $I(|B|) = I(|B[p \leftarrow t]|) = K(|B[p \leftarrow t]|) = F$.

The facts (1), (2) and (3) are proved just as in [HR86]. Let us prove the fact (4).

By definition of an E-interpretation, as $I(s = t) = T$, it implies that B and $B[p \leftarrow t]$ (written $B[t]$) have the same truth value for I. Since $B[t] \prec_{A\_ORD} B$, we can deduce that $K(B[t]) = I(B[t])$. The fact (4) is so obtained.

We detail in the following the different possible sub-cases, and we illustrate them by examples. In those examples, we use the predicates P and $= (P >_P =)$, and the operators ordered by the precedence $f >_F g >_F a >_F b >_F c$, assuming that the operators f and g are AC.

Case 3.1 : If $B \notin \mathcal{AC}$

Case 3.1.1 : If $(s = t) \notin \mathcal{AC}$

Here is an example of this situation :

$$(s = t) \;\equiv\; (f(a, b) = c)$$
$$B \;\equiv\; P(f(a, b))$$

- If $|B| \vee D_I \in INF^*(S)$ then we can apply an OAC-paramodulation step from clause $(s = t) \vee D_M$ into clause $|B| \vee D_I$.
- If $|B| \vee D_I \equiv (f(m, l) = f(m, r)) \vee D_I$ with $(l = r) \vee D_I \in INF^*(S)$ then we can apply an extended OAC-paramodulation step from clause $(s = t) \vee D_M$ into clause $(l = r) \vee D_I \in INF^*(S)$.

In both cases we can derive the following clause $|B[t]| \vee D_I \vee D_M$.

But, $I(|B[t]|) = I(D_I) = I(D_M) = F$.

Let L be the restriction of K to $W_{\gamma+1}$, γ the index of the greatest atom in $|B[t]| \vee D_I \vee D_M$. The interpretation L falsifies the clause $|B[t]| \vee D_I \vee D_M$. This gives a contradiction with I being a failure node. Hence K cannot be an element of $MCT(S^*)$.

Case 3.1.2 : If $(s = t) \in \mathcal{AC}$

In this case $s \equiv_{AC} t$ and D_M is empty.

Here is an example of this situation :

$$(s = t) \;\equiv\; (f(a, b) = f(b, a))$$
$$B \;\equiv\; P(f(a, b))$$

Hence, $B \equiv_{AC} B[t]$. So, the clause $|B[t]| \vee D_I$ is equivalent modulo AC to the clause $|B| \vee D_I$.

Let L be the restriction of K to $W_{\gamma+1}$, γ the index of the greatest atom in $|B[t]| \vee D_I$. The interpretation L falsifies la clause $|B[t]| \vee D_I$. This gives a contradiction with I being a failure node, and, therefore, also K being an element of $MCT(S^*)$.

Case 3.2 : *If $B \in \mathcal{AC}$*

In this case D_I is empty and $B \equiv (l = r)$ with $l \equiv_{AC} r$. Moreover, $I(B) = F$.

Case 3.2.1 : *If $(s = t) \notin \mathcal{AC}$*

We assume that s is a subterm of l. If s was a subterm of r then we could reason in the same way.

Case 3.2.1.1 : *If $s \equiv l$*

Here is an example of this situation :

$$(s = t) \equiv (f(a, b) = c)$$
$$(l = r) \equiv (f(a, b) = f(b, a))$$

Then , $B[t] \equiv (r = t)$.
The AC-compatibility of the ordering implies that $r \succ_{ORD} t$.
The clause $(r = t) \vee D_M$ is equivalent modulo AC to the clause $(s = t) \vee D_M$.
Hence $(r = t) \vee D_M$ is falsified by the interpretation L, restriction of K to $W_{\gamma+1}$, γ index of the greatest atom of the clause $(r = t) \vee D_M$.
Therefore K cannot belong to $MCT(S^*)$.

Case 3.2.1.2 : *If $\exists q_l \in Occ(l)$, $l_{|q_l} \equiv s$ $(q_l \neq \varepsilon)$, and $q_l \in Occ\_max(l, top(s))$*

It means that, if $top(s)$ is an AC-operator, then q_l is a maximal occurrence of this operator in l.
Here is an example of this situation :

$$(s = t) \equiv (f(a, b) = c)$$
$$(l = r) \equiv (g(f(a, b), b) = g(b, f(b, a)))$$

Hence, $B[t] \equiv (r = l[q_l \leftarrow t])$ and there is an occurrence q_r of r such that $r_{|q_r} \equiv_{AC} s$. Let r' be the subterm $r_{|q_r}$. Since $(r' = s)$ is smaller then B, we have : $K(r' = s) = V$ *(subterm and AC-compatibility)*.

- If $r' \equiv s$, then $(l[t] = r)$ is K-reducible by $(s = t)$ into $(l[t] = r[t])$ which belongs to $\mathcal{AC}$. But $K(l[t] = r[t]) = F$. Hence, a failure node should occur at the level of $(l[t] = r[t])$ $(\prec_{A\_ORD} B)$.

- If $r' \succ_{ORD} s$, then $B[t]$ is K-reducible by $(r' = s)$ into $(r[q_r \leftarrow s] = l[t])$, and the latter atom is K-reducible into $(r[q_r \leftarrow t] = l[t])$ by $(s = t)$.
  The final atom denoted $(r[t] = l[t])$ belongs to $\mathcal{AC}$, since $r[t] \equiv_{AC} l[t]$. We have $K(r[t] = l[t]) = F$ by definition of an E-interpretation. Hence, a failure node should occur at the level of $(r[t] = l[t])$.

- If $s \succ_{ORD} r'$, then, by AC-compatibility we have : $r' \succ_{ORD} t$ and $(s = r') \succ_{A\_ORD} (s = t)$. Hence the atom $(s = r')$ is K-reducible by $(s = t)$ into $(r' = t)$. From that , $K(r' = t) = V$ and also $B[t]$ is K-reducible by $(r' = t)$ in $(r[q_r \leftarrow t] = l[t])$. We can conclude as in the previous case since we have an atom in $\mathcal{AC}$, which is smaller than B, and false in K.

Case 3.2.1.3 : If $\exists q \in Occ(l)$, $l_{|_q} \equiv s$ $(q \neq \varepsilon)$, $q \notin Occ\_max(l, f)$, and $\exists u \in T(F)$, $l \equiv_{AC} f(s, u)$ (with $f = top(s)$)

In this case f is an AC operator and there exists a term u such that : $r \equiv_{AC} f(s, u)$.

Here is an example of this situation :

$$(s = t) \equiv (f(a, b) = c)$$
$$(l = r) \equiv (f(f(a, b), b) = f(a, f(b, b)))$$

Consider the following clause $(f(s, u) = f(t, u)) \vee D_M$, which belongs to S^*, by definition.

Since $l[t] \equiv_{AC} f(t, u)$, the clause $(r = l[t]) \vee D_M$ is equivalent modulo AC to the clause $(f(s, u) = f(t, u)) \vee D_M \in S^*$.

The interpretation L, defined as the restriction of K to $W_{\gamma+1}$, with γ index of the atom $(l[t] = r)$, falsifies the clause $(r = l[t]) \vee D_M$.

Hence, K does not belong to $MCT(S^*)$.

Case 3.2.1.4 : If $\exists q \in Occ(l)$, $l_{|_q} \equiv s$ $(q \neq \varepsilon)$, $q \notin Occ\_max(l, f)$, and $\neg\exists u \in T(F)$, $l \equiv_{AC} f(s, u)$ (with $f = top(s)$)

In this case there is a non empty occurrence q_l of l, which belongs to the set $Occ\_max(l, top(s))$ and such that there is a term u verifying : $l_{|_{q_l}} \equiv_{AC} f(s, u)$.

Here is an example of this situation :

$$(s = t) \equiv (f(a, b) = c)$$
$$(l = r) \equiv (f(b, g(f(f(a, b), c))) = f(g(f(b, f(a, c))), b))$$

Then, there is a non empty occurrence q_r of $Occ\_max(r, f)$, such that : $r_{|_{q_r}} \equiv_{AC} f(s, u)$.

Let us denote $l' \equiv l_{|_{q_l}}$ and $r' \equiv r_{|_{q_r}}$.

- If $K(l' = r') = F$, then a failure node should occur at the level of $(l' = r')$, since this atom belongs to AC.

- If $K(l' = r') = V$:
  - If $l' \succ_{ORD} r'$, then the atom $(l = r)$ is K-reduced by $(l' = r')$ into $(l[q_l \leftarrow r'] = r)$, and we get : $K(l = r) = K(l[r'] = r) = F$. Note that $(l[r'] = r)$ belongs to AC. Hence, a failure node would occur at the level of this atom.

  - If $r' \succ_{ORD} l'$ then the atom $(l = r)$ is K-reduced by $(r' = l')$ into $(l = r[q_r \leftarrow l'])$, and we have : $K(l = r) = K(l = r[l']) = F$. But the atom $(l = r[l'])$ belongs to AC. We conclude as above.

Hence, in all cases we have a contradiction with $K \in MCT(S^*)$.

Case 3.2.2 : If $(s = t) \in AC$

Here is an example of this situation:

$$(s = t) \equiv (f(a, b) = f(b, a))$$
$$(l = r) \equiv (g(f(c, f(a, b)), c) = g(c, f(b, f(a, c))))$$

Then, $B[t] \equiv_{AC} B$. Hence, $B[t]$ also belongs to AC. We conclude as above.

Therefore case 3 always leads to a contradiction.

Since all cases are impossible, necessarily $MCT(S^*)$ is empty and therefore the empty clauses belongs to $INF^*(S)$.

□

6 Lifting lemmas

The lifting lemmas of the OAC-factorisation, OAC-resolution and OAC-paramodulation inference rules are developped and proved in [Vig90]. Let us detail the proof of the extended OAC-paramodulation lifting lemma.

Ordered Extended AC-paramodulation Lifting Lemma *Let $C_1 = (s = t) \vee C_1'$ and $C_2 = (l = r) \vee C_2'$ be two clauses whithout any common variable, such that l and s have the same AC-operator at their top. Let D result of an extended OAC-paramodulation of $C_1\sigma$ in $C_2\sigma$, where σ is a ground irreducible substitution; then, there is a clause C, resulting of an extended OAC-paramodulation of C_1 in C_2, such that D is an instance of C.*

Proof :

Let $C_1 = (s = t) \vee C_1'$, $C_2 = (l = r) \vee C_2'$ and σ a ground irreducible substitution, verifying

$$\begin{cases} top(l), top(s) \in F_{AC} \cup X \quad , \text{ where } top(s) = top(l) \text{ if } top(l), top(s) \in F_{AC} \\ f(l, x)\sigma \equiv_{AC} M[s]\sigma \quad , \text{ where } x \text{ is a new variable} \\ \forall A \in C_1', (M[s] = M[t])\sigma \not\succ_{A\_ORD} A\sigma \\ \forall A \in C_2', (f(l, x) = f(r, x))\sigma \not\succ_{A\_ORD} A\sigma \\ M[s]\sigma \not\succ_{ORD} M[t]\sigma \\ (f(l, x) = f(r, x))\sigma \not\succ_{A\_ORD} (M[s] = M[t])\sigma \\ M[.] \equiv f(., y) \quad , \text{ where } y \text{ is a new variable} \end{cases}$$

Let D be the extended OAC-paramodulant $(M[t] = f(r, x))\sigma \vee C_2'\sigma \vee C_1'\sigma$ of $C_1\sigma$ in $C_2\sigma$. As the terms $M[s]$ and $f(l, x)$ are AC-unifiable, let τ be one of their most general unifier; then there is a substitution ρ such that : $\tau\rho = \sigma$.

The orderings $\succ_{ORD}$ and $\succ_{A\_ORD}$ being stable by instanciation, we can deduce the following facts :

$$\begin{cases} \forall A \in C_1', (M[s] = M[t])\tau \not\succ_{A\_ORD} A\tau \\ \forall A \in C_2', (f(l, x) = f(r, x))\tau \not\succ_{A\_ORD} A\tau \\ M[s]\tau \not\succ_{ORD} M[t]\tau \\ (f(l, x) = f(r, x))\tau \not\succ_{A\_ORD} (M[s] = M[t])\tau \end{cases}$$

Indeed, for instance, for any terms u and v, using the previous substitutions σ, τ and ρ, if we had $u\sigma \not\preceq v\sigma$ and $u\tau \preceq v\tau$, then, by stability, we could deduce that $u\tau\rho \preceq v\tau\rho$; but, as $\tau\rho = \sigma$, we could write it $u\sigma \preceq v\sigma$, which is in contradiction with the hypotheses. So, if $u\sigma \not\preceq v\sigma$, then $u\tau \not\preceq v\tau$.

As all the conditions are now verified, using an extended OAC-paramodulation of C_1 in C_2 we can deduce the following clause : $C = (M[t] = f(r, x))\tau \vee C_2'\tau \vee C_1'\tau$.

Moreover, $\begin{aligned} C\rho &= (M[t] = f(r, x))\tau\rho \vee C_2'\tau\rho \vee C_1'\tau\rho \\ &= (M[t] = f(r, x))\sigma \vee C_2'\sigma \vee C_1'\sigma = D \end{aligned}$

This proves the lemma.

□

7 Further works and conclusion

We have designed a refutationally complete paramodulation-based strategy for associative-commutative deduction. It has been adapted from the ordered paramodulation strategy of [Pet83, HR86]. We can remark immediately that a similar treatment could be applied to other theorem-proving strategies such as the superposition strategy or the positive strategy of [Rus89]. More important, it can be shown that ordered AC-paramodulation is compatible with associative-commutative rewriting. Rewriting is a fundamental tool for the efficiency of theorem-proving. The idea of replacing axioms by ad-hoc mechanisms such as unification algorithm or inference rules can be further extended to other equational or non equational theories. In general the efficiency gain is noticeable, but this still needs to be carefully studied by experimentations.

References

[AHM89] Siva Anantharaman, Jieh Hsiang, and Jalel Mzali. Sbreve2: A term rewriting laboratory with (AC-)unfailing completion. In N. Dershowitz, editor, *Proceedings 3rd Conference on Rewriting Techniques and Applications, Chapel Hill, (North Carolina, USA)*, volume 355 of *Lecture Notes in Computer Science*, pages 533–537. Springer-Verlag, April 1989.

[BHK+88] H.-J. Bürckert, A. Herold, D. Kapur, J. Siekmann, M. Stickel, M. Tepp, and H. Zhang. Opening the AC-unification race. *Journal of Automated Reasoning*, 4(1):465–474, 1988.

[BP85] L. Bachmair and D. Plaisted. Associative path orderings. In *Proceedings 1st Conference on Rewriting Techniques and Applications, Dijon (France)*, volume 202 of *Lecture Notes in Computer Science*. Springer-Verlag, 1985.

[Bra75] D. Brand. Proving theorems with the modification method. *SIAM J. of Computing*, 4:412–430, 1975.

[CL87] A. Ben Cherifa and P. Lescanne. Termination of rewriting systems by polynomial interpretations and its implementation. *Science of Computer Programming*, 9(2):137–160, October 1987.

[Der82] N. Dershowitz. Orderings for term-rewriting systems. *Theoretical Computer Science*, 17:279–301, 1982.

[HR86] J. Hsiang and M. Rusinowitch. A new method for establishing refutational completeness in theorem proving. In J. Siekmann, editor, *Proceedings 8th International Conference on Automated Deduction, Oxford (UK)*, volume 230 of *Lecture Notes in Computer Science*, pages 141–152. Springer-Verlag, 1986.

[HR87] J. Hsiang and M. Rusinowitch. On word problem in equational theories. In Th. Ottmann, editor, *Proceedings of 14th International Colloquium on Automata, Languages and Programming, Karlsruhe (Germany)*, volume 267 of *Lecture Notes in Computer Science*, pages 54–71. Springer-Verlag, 1987.

[JK86] J.-P. Jouannaud and H. Kirchner. Completion of a set of rules modulo a set of equations. *SIAM Journal of Computing*, 15(4):1155–1194, 1986. Preliminary version in Proceedings 11th ACM Symposium on Principles of Programming Languages, Salt Lake City, 1984.

[KB70]   D.E. Knuth and P.B. Bendix. Simple word problems in universal algebras. In J. Leech, editor, *Computational Problems in Abstract Algebra*, pages 263–297. Pergamon Press, Oxford, 1970.

[KR87]   E. Kounalis and M. Rusinowitch. On word problem in Horn logic. In J.-P. Jouannaud and S. Kaplan, editors, *Proceedings 1st International Workshop on Conditional Term Rewriting Systems, Orsay (France)*, volume 308 of *Lecture Notes in Computer Science*, pages 144–160. Springer-Verlag, 1987. See also the extended version published in Journal of Symbolic Computation, number 1 & 2, 1991.

[Lan79]   D.S. Lankford. A unification algorithm for abelian group theory. Technical report, Louisiana Tech. University, 1979.

[LB77]   D.S. Lankford and A. Ballantyne. Decision procedures for simple equational theories with associative commutative axioms: complete sets of associative commutative reductions. Technical report, Univ. of Texas at Austin, Dept. of Mathematics and Computer Science, 1977.

[NR91]   P. Narendran and M. Rusinowitch. Any ground associative-commutative theory has a finite canonical system. In R. Book, editor, *Proceedings 4th Conference on Rewriting Techniques and Applications, Como, (Italy)*. Springer-Verlag, 1991.

[Pet83]   G. Peterson. A technique for establishing completeness results in theorem proving with equality. *SIAM Journal of Computing*, 12(1):82–100, 1983.

[Plo72]   G. Plotkin. Building-in equational theories. *Machine Intelligence*, 7:73–90, 1972.

[PS81]   G. Peterson and M. Stickel. Complete sets of reductions for some equational theories. *Journal of the Association for Computing Machinery*, 28:233–264, 1981.

[Rus89]   M. Rusinowitch. *Démonstration automatique-Techniques de réécriture*. InterEditions, 1989.

[RW69]   G. Robinson and L.T. Wos. Paramodulation and first-order theorem proving. In B. Meltzer and D. Mitchie, editors, *Machine Intelligence 4*, pages 135–150. Edinburgh University Press, 1969.

[Ste90]   J. Steinbach. Ac-termination of rewrite systems - A modified Knuth-Bendix ordering. In H. Kirchner and W. Wechler, editors, *Proceedings 2nd International Conference on Algebraic and Logic Programming, Nancy (France)*, volume 463 of *Lecture Notes in Computer Science*, pages 372–386. Springer-Verlag, 1990.

[Sti81]   M. Stickel. A unification algorithm for associative-commutative functions. *Journal of the Association for Computing Machinery*, 28:423–434, 1981.

[Sti84]   M. Stickel. A case study of theorem proving by the knuth-bendix method : Discovering that $x^3 = x$ implies ring commutativity. In *Proceedings 7th International Conference on Automated Deduction*, volume 170 of *Lecture Notes in Computer Science*. Springer-Verlag, 1984.

[Vig90]   L. Vigneron. Deduction automatique dans des théories associatives-commutatives. Rapport interne, Centre de Recherche en Informatique de Nancy, Vandœuvre-lès-Nancy, 1990.

Towards a Lattice of Knowledge Representation Systems

Andreas Strasser
Technische Universität München
Institut für Informatik
Forschungsgruppe Künstliche Intelligenz
zugeordnet dem Lehrstuhl für Rechnerarchitektur
Arcisstr. 21
D-8000 München 2
Tel. 089/52 10 97
email: strasser@lan.informatik.tu-muenchen.dbp.de

Abstract: Existing knowledge representation systems (KR-systems) are either very general and can, therefore, give no assistance or are very restricted and can become an obstacle for the knowledge engineer. In this paper, KR-systems are formalized using predicate logic. The semantics is given on different levels. A new approach is presented in which knowledge representation systems are linked together in order to make knowledge bases re-usable in different systems and to give the possibility to the knowledge engineer to generate his/her own specialized KR-system. A number of operators are presented which are useful for this purpose.
In this way, a lattice of KR-systems is generated. Along the links of this lattice knowledge bases can be transformed and can be, therefore, re-used within other systems.

1 Introduction

The developement of Expert Systems is usually an iterative process (rapid prototyping) rather than a straightforward one. Buchanan ([Buchanan, 1983]) identifies five steps for developing such a system: (1) identification of the task, (2) conceptualizing the knowledge,[1] (3) developing a formal representation, (4) implementing the prototype and (5) testing.

Step (1) and (2) mainly concern studies of the domain, proposals of possible formalizations and adequate problem solving techniques.

Based on the outcome of steps (1) and (2), the first action of the step (3) is to select an adequate *knowledge representation system* (KRS).[2] Unfortunately, this decision depends not only on the requirements of step (1) and (2) but also on additional matters, e.g. on hardware or financial limitations etc. Therefore, the decision to choose a knowledge representation system is mainly a compromise. Based on the selected system, methods to represent the required knowledge in terms of the language provided by the shell are explored. In step (4) the system is implemented. If in step (5) errors are found, which should be solved by jumping back to the responsible steps and making changes. As Buchman and other people (e.g. [Musen, 1989]) pointed out: *„frequent versions of the expert system commonly are dictated by errors that may have been introduced during any*

---

[1]Step (2) is called *knowledge level analysis* by Newell ([Newell, 1982]).
[2]Most times it is called a *shell* or a (knowledge-representation) *tool*.

of the first four stages of knowledge acquisition". This means that e.g. during the test phase it may become obvious that conceptualization of the domain knowledge is not adequate for solving the intended task. Sloman remarked (in [Sloman, 1985]): *"how ... to construe the world may depend on how much he* (the knowledge enigneer) *has already learnt, and what tasks or problems he has."* Therefore, it may be necessery to re-enter phase (2) and to make proper modifications.

It may also become obvious in the later steps that the selected knowledge representation system is not suitable to solve the given task. Errors in step (1) and (2) may also indirectly lead to a demand to revise the choice of a knowledge representation system. Buchanan's iterative model of expert system developement consequently requires, therefore, revisions *on all steps* of the model. But in practice it is hardly possible to discard the selected knowledge representation system because of the following problems:

- When reaching the test-phase, a substantial amount of knowledge has to be implemented. All this knowledge would have to be re-formalized and to be re-implemented according to the new system. This would be a very costly process.

- Knowledge representation system like KEE, EPITOOL or Knowledge Craft are expensive software products and, therefore, it can hardly be justified to pay for a new system after just buying the present one.

As a consequence, Buchman's model can be only partly realized in practice, because one of the most important decisions – the selection of a proper knowledge representation system – can not be revised. Therefore, revisions are limited to the possibilities given by the knowledge representation system. This could be a very serious restriction, because *"the knowledge representation scheme often makes the expression of knowledge awkward if not impossible"* ([Reichgelt and Harmelen, 1985, S. 21]).

An experiment done by Waterman and Hages-Roth ([Waterman and Hayes-Roth, 1983]) shows, that different representations used to formalize the same subject would lead to different results in essential aspects. Knowledge engineers tend to model all these aspects of the domain which can be easily expressed using the selected formalism; other aspects will be ignored ([Musen, 1989]). Therefore, it would be a great advantage if a knowledge engineer is able to make improvements on the selected system *without loosing the knowledge base* implemented so far. If you interchange an existing system with another system this is almost impossible, esp. because of the lack of a formal definition of semantics.

In this situation everyone tends to choose a system which includes as less restrictions as possible in order to keep "all doors open". But such systems can't be good assistants to the knowledge engineer.[3] If you choose a dedicated and strongly restricted system, it gives more support in complex situations. But if the choice is not adequate, you will meet big problems.

---

[3]See e.g. [Reichgelt and Harmelen, 1985]: *"they provide the knowledge engineer with a bewildering, array of possibilities and , if any guidance under what circumstance which of these possibilities should be used."*

2   Formalizing knowledge representation system

According to Delgrande and Mylopoulos ([Delgrande and Mylopoulos, 1986]) a knowledge representation system can be seen as a logic theory $\langle KB_0, \vdash_L \rangle$ based on a language L. KB_0 is a set of axioms forming the knowledge representation system combined with a particular knowledge base.

Brewka offers an example in [Brewka, 1987]. He shows a first order formalization of the BABYLON frame system ([Christaller et al., 1990]). In a similiar way Reimer described a formalization of his knowledge representation system called FRM ([Reimer, 1989]). But a knowledge representation system is not only characterized by its reasoning capabilities. It also guarantees that knowledge bases are consistent w.r.t. a set of constraints. Unlike Delgrande's and Brewka's approach, Reimer integrates consistency constraints into his formalization of FRM.

Horacek pointed out in [Horacek, 1989] that KL-ONE-like systems contain a number of such constraints. These constraints are - in his opinion - majorly due to guarantee a form of "syntactical consistence". Consequently, he defined a hierarchy of additional consistency requirements.

Above examples show[4] that predicate logic is a good candidate for formalizing knowledge representation systems. But the approach of Delgrande seems to combine different components into a single structure in KB_0. In my opinion they should be kept separately. On the other Hand, an important component –the consistency constraints– is missing in his schema.

I want to use predicate logic to describe knowledge representation systems in a different way: In my view a knowledge representation system consists of three components - *deduction rules* ($\mathcal{R}$), *consistency constraints* ($\mathcal{C}$) and a model-theoretic relation of deduceability $\vdash_L$. Formally, a knowledge representation system is given as:

$$\mathcal{KRS} = \langle \mathcal{C}, \mathcal{R}, \vdash_L \rangle$$

Unlike Delgrande's approach, the knowledge base KB is not included into the deduction rules and not in the set of constraints. In my approach, knowledge bases are separated from knowledge representation systems: a knowledge representation system $\langle \mathcal{C}, \mathcal{R}, \vdash_L \rangle$ together with a knowledge base $\mathcal{KB}$ forms a *Knowledge Based System* (KBS)[5], formally:

$$\mathcal{KBS} = \langle \mathcal{C}, \mathcal{R}, \mathcal{KB}, \vdash_L \rangle$$

$\mathcal{C}$, $\mathcal{R}$ and $\mathcal{KB}$ can contain any first order formula. But in practice, $\mathcal{KB}$ is supposed to contain knowledge about the particular case under consideration mainly *in the form of facts* and $\mathcal{R}$ will contain formulas representing deduction rules to deduce new information out of the given case in $\mathcal{KB}$.

---

[4]There are a number of other works in this area based on predicate logic, e.g. [Console et al., 1989, Hayes, 1981].

[5]Note, knowledge based systems are in some sense similar to deductive databases ([Moerkotte, 1990]), where $\mathcal{KB}$ corresponds to the base-relations of the database and $\mathcal{C}$ and $\mathcal{R}$ to the sets of constraints and rules.

_____ $klone^0$ _____

Constraints C^0:

c_1^0 :   Every concept C has to have at least one role R
$$\forall\, C\, \exists R\ :\ concept\,(\,C\,)\ \to\ role\,(\,C\,,\,R\,).$$

c_2^0 :   A role R can only be defined for existing concepts C
$$\forall C\,,\,R\ :\ concept\,(\,C\,)\ \leftarrow\ role\,(\,C\,,\,R\,).$$

Deduction Rules $\mathcal{R}^0$:

r_1^0 :   A concept C_1 subsumes C_2, if C_2 has all the roles R of C_1
$$\forall\ C_1\,,\,C_2\ :\ subs\,(\,C_1\,,\,C_2\,)\ \leftarrow\ (\forall R\ :\ role\,(\,C_1\,,\,R\,)$$
$$\to\ role\,(\,C_2\,,\,R\,)\,).$$

Figure 1: The knowledge represenation system $klone^0$

Figure 1 shows an example knowledge representation system called $klone^0$. It is a very simple KL-ONE-like system for representing only terminological knowledge. The constraint c_1^0 forces the user to define at least one role *for every* concept. c_2^0 states that definitions of roles are only allowed for *existing* concepts.

The knowledge base $\mathcal{KB}_{\text{small}}$ shown in Fig. 2 formalizes some concepts in the field of vehicles, and shows the simple language $klone^0$: we use the term concept(C) to declare the existence of a concept C. The term role(C,R) assigns a role R to the concept C. The only deductive ability of $klone^0$ is to deduce the *subsumption-relation* (called $subs(C_1, C_2)$) between concepts C_1 and C_2 using the deduction rule r_1^0. It is easy to see that for $\mathcal{KB}_{\text{small}}$ both constraints c_1^0 and c_2^0 hold.

_____ *Example Knowledge Base* $\mathcal{KB}_{\text{small}}$: _____

concept(car)	concept(vehicle)	concept(taxi)
	concept(lorry)	
role(car,has_wheels)	role(vehicle,has_wheels)	role(taxi,has_wheels)
	role(lorry,has_wheels)	
	role(lorry,has_max_loading)	role(taxi,has_max_passengers)

Figure 2: Example: The Knowledge Base $\mathcal{KB}_{\text{small}}$

3 Semantics

The semantics of knowledge representation systems and knowledge based systems, respectively, is defined on two different levels. Firstly, the semantics of a knowledge based system is defined in terms of all valid formulas within a theory constructed out of the knowledge base. Secondly the semantics of a knowledge representation system is defined in terms of all knowledge bases, which are correct w.r.t. the knowledge representation system, i.e. which are not violating the constraints of the system.

3.1 Semantics of Knowledge Based Systems

At the first level the semantics of a $\mathcal{KBS} = \langle \mathcal{C}, \mathcal{R}, \mathcal{KB}, \vdash_L \rangle$ is defined as the set of all formulae, which can be deduced from $\mathcal{KB}$ using the rules of $\mathcal{R}$. This semantics-function $M[\![.]\!]$ can be defined as:

Definition 3.1 (Semantics $M[\![\mathcal{KBS}]\!]$ of Knowledge Based Systems)

$$
\begin{aligned}
M[\![\mathcal{KBS}]\!] &= M[\![\langle \mathcal{C}, \mathcal{R}, \mathcal{KB}, \vdash_L \rangle]\!] \\
&= \{ X \mid \mathcal{R}, \mathcal{KB} \vdash_L X \} \\
&= Th(\mathcal{R} \cup \mathcal{KB})
\end{aligned}
$$

Note, other approaches, such as in [Hayes, 1981] or [Vilain, 1985] which is a formalization of NIKL, define the semantics of a KBS by using axiom-schemata, which are properly instantiated by the content of the knowledge base. The set of all such instances defines the semantics of the knowledge based system. The semantics of the underlaying knowledge representation system is not explicitly defined. In my approach a knowledge represenation system has a (logic-based) semantics without keeping a special knowledge base in mind. This semantics will be defined in the next section.

3.2 Semantics of Knowledge Representation Systems

Consistency constraints reduce the set of knowledge bases which can be properly handled by the knowledge representation system. They detect inconsistent situations in knowledge bases and can, therefore, be powerful mechanisms to support knowledge engineers in building *correct knowledge bases*. Knowledge bases are in our view of knowledge based systems units of their own containing knowledge about a particular domain. Because of that, it should be possible to integrate knowledge bases into different knowledge representation systems, esp. after a change of the knowledge representation system in step (3) of Buchanan's model.

It makes sense to describe the semantics of a knowledge representation system $\mathcal{KRS}$ also in terms of the set of knowledge bases, which are correct w.r.t. $\mathcal{KRS}$. The term "correctness" of knowledge bases wrt. a knowledge representation system is defined as follows:

Definition 3.2 (Correctness of knowledge bases)
A knowledge base KB is correct w.r.t. *a knowledge representation system KRS =
$\langle C, R, \vdash_L \rangle$, if there is no constraint $c \in C$, so that c an be falsified, i.e.:*
$$\neg \exists c \in C : KB, R \vdash_L \neg c$$

The correctness of knowledge bases can be achieved by the following semantics-function
$M^c[\![KRS]\!]$, which defines the semantics of KRS w.r.t. to the qualifying knowledge
bases:[6]

Definition 3.3 (Semantics $M^c[\![KRS]\!]$ of Knowledge Representation Systems)
$$\begin{aligned} M^c[\![KRS]\!] &= M^c[\![\langle C, R, \vdash_L \rangle]\!] \\ &= \{KB \mid KB \text{ is correct w.r.t. } KRS\} \end{aligned}$$

Clearly, all knowledge bases KB which are consistent w.r.t a knowledge representation
system KRS can be integrated into this KRS. This fact means that every knowledge
base KB can be re-used in all knowledge representation systems KRS to which it is
consistent, i.e. KB is in $M^c[\![KRS]\!]$.

This definition of correctness is based on the assumption that both KRS and the
knowledge base KB are formalized using the same vocabulary. This is a very strong
assumption, which is not true in normal cases. In the following section I want to discuss
some operations for the *modification* of knowledge representation systems in a way, that
possible relationships between knowledge representation systems become clear. This will
lead to a lattice of knowledge representation systems. Using this lattice the re-usability of
knowledge bases in other systems can be decided in most cases without explicit checking
of correctness. Additionally, it will be possible to change the vocabulary of knowledge
representation systems. In such cases, it will be possible to find a proper term-rewriting
system for transforming knowledge bases in order to re-use them in other knowledge
representation systems.

In this way different KR-systems can be compared. One very important criterion for
comparing KR-systems is the so-called *restriction-factor*.[7] A KR-system KRS' is more
restricted than KRS, if
$$M^c[\![KRS']\!] \subseteq M^c[\![KRS]\!]$$

This is equivalent to the following definition:

Definition 3.4 *A KR-system KRS' is* more restricted than *KRS, if every constraint
$c \in C$ can be deduced from KRS'; formally:*
$$\forall c \in C : C', R' \models c$$

---

[6]Baader defines in [Baader, 1990] KR1-languages (KR-languages based on first-order predicate logic)
basically as *„subsets L of the power set of FOL, i.e. sets of sets of formulas"*. $M^c[\![KRS]\!]$ corresponds to
this definition. Additionally, the set of constraints C gives us a characterization of these sets of formulas.
This characterization can e.g. be used to implement a system accepting only qualifying sets of formulas.

[7]In a similiar way KR-systems can be compared w.r.t. their *deductive strength* represented in R. For
this aspect see [Strasser, 1991].

This is the tird way to describe the semantics of KR-systems, namely in showing the relations to other systems, esp. the relation of restriction.

4 Generation of Domain-specific Knowledge Representation Systems

4.1 Simple Operators

As pointed out in section 1, in later stages of the development of an expert system it may become obvious that the selected knowledge representation system is not suitable and should be modified. If you e.g. select $klone^0$, you might realize later that there is, unfortunately, no possibility to describe *instances* of concepts given in your $klone^0$-knowledge base.

One solution would be to add new constraints and new rules to $klone^0$, which might lead to the following knowledge representation system $klone^1$ given in Fig. 3. Within this KR-system it is possible to define instances I of concepts C (instance(C,I)) and to set role-fillers V to roles of instances (roleval(I,R,V)).

$$\underline{\hspace{3cm}\ klone^1\ \underline{\hspace{3cm}}}$$

c_1^0 : $\qquad \forall C \exists R \ : \ concept(C) \rightarrow role(C, R).$

c_2^0 : $\qquad \forall C, R \ : \ concept(C) \leftarrow role(C, R).$

c_1^1 : An instance I can only be defined for an existing concept C.
$$\forall C, I \ : \ instance(C, I) \rightarrow concept(C).$$

c_2^1 : A role filler V of a role R belonging to an instance I has to be an instance of some concept C.
$$\forall I, R, V \exists C \ : \ roleval(I, R, V)$$
$$\rightarrow instance(C, V).$$

r_1^0 : $\qquad \forall C_1, C_2 \ : \ subs(C_1, C_2) \leftarrow (\forall R \ : \ role(C_1, R) \rightarrow role(C_2, R)).$

r_1^1 : I is an instance of C, if I has for every role of C a corresponding role-filler V.
$$\forall C, I \ : \ (\forall R : role(C, R) \rightarrow \exists V \ : \ roleval(I, R, V)) \rightarrow instance(C, I).$$

Figure 3: The Knowledge Representation System $klone^1$

It is obvious, that the knowledge representation system $klone^1$ can be easily constructed from $klone^0$ using some simple set-theoretic operations:

$$klone^0 \ = \ (\mathcal{C}_0, \mathcal{R}_0) = (\{c_1^0, c_2^0\}, r_1^0)$$

$$klone^1 = (\{c_1^0, c_2^0\} \cup \{c_1^1, c_2^1\} , \{r_1^0\} \cup \{r_1^1\})$$
$$= (C_0 \cup \{c_1^1, c_2^1\} , R_0 \cup r_1^1\})$$

All these set-theoretic operators will be subsumed under the term *simple operators*. Fig. 4 shows all simple operators. It can be read as follows: row (1) gives the operator-name, whereas in row (2) a corresponding symbol for the operator is defined; row (3) gives a little description and row (4) the definition of the operator. It has to be read as follows: the transformation of a knowledge representation system $\mathcal{KRS}$ using the corresponding operator leads to a system $\mathcal{KRS}$' which is shown in row(4). Another important point is the transformation of corresponding knowledge bases. In which cases can we re-use a knowledge base KB, built on $\mathcal{KRS}$, in the system $\mathcal{KRS}$'? In row (5) you can see that simple operators provide this re-use in a simple way.

The operator *deductive strengthening* $(\alpha_{R'}^+)$ transforms a knowledge representation system $(\mathcal{C}, \mathcal{R})$ into a more powerfull one in the sense, that the resulting system $\alpha_{R'}^+(\mathcal{KRS})$ allows more deductions than the original system $\mathcal{KRS}$. These additional rules are given in $\mathcal{R}$'.

Our example $klone^1$ can be constructed out of $klone^0$ as follows:

$$klone^1 = \alpha_{\{r_1^1\}}^+ (\varrho_{\{c_1^1, c_2^1\}}^+ (klone^0))$$
$$= \varrho_{\{c_1^1, c_2^1\}}^+ (\alpha_{\{r_1^1\}}^+ (klone^0))$$

Note, that these operators can be seen as transactions in deductive databases manipulating some parts of the database. A difficult problem is to decide in which cases such an operation leads to an inconsistent system, i.e. a system where *false* can be deduced. These problems are not discussed here. We are mainly interested in a manipulation of the knowledge representation system and in the question, in which cases knowledge bases of the original system can be re-used within the new system. Row (5) of Fig. 4 shows under which conditions knowledge bases can be integrated into new systems.

Operator		Description	Transformations for	
(1)	(2)	(3)	(4) KR-systems	(5) KBes
deductive strengthening	$\alpha_{R'}^+$	Adding new deductive capabilities	$\mathcal{KRS}' = (\mathcal{C}, \mathcal{R} \cup \mathcal{R}')$	$KB' = KB$, if KB is correct w.r.t. $\mathcal{KRS}'$
deductive weakening	$\alpha_{R'}^-$	Deleting deductive capabilities	$\mathcal{KRS}' = (\mathcal{C}, \mathcal{R} \backslash \mathcal{R}')$	$KB' = KB$
restriction	$\varrho_{C'}^+$	Building a more restricted system	$\mathcal{KRS}' = (\mathcal{C} \cup \mathcal{C}', \mathcal{R})$	$KB' = KB$, if KB is correct w.r.t. $\mathcal{KRS}'$
liberalization	$\varrho_{C'}^-$	Building a more liberal system	$\mathcal{KRS}' = (\mathcal{C} \backslash \mathcal{C}', \mathcal{R})$	$KB' = KB$

Figure 4: Simple manipulation-operations

4.2 Complex Operators

Using simple operators you can only add or delete some features of the representation system, e.g. to add restrictions to the system, or to change the reasoning capabilities of the system. It is not possible to change e.g. the vocabulary required for the system or to change the level of abstraction. An example of this problem is given below:

In our examples $klone^0$ and $klone^1$ you can only declare one single sort of concepts. If you have to make a difference between various types of concepts, e.g. between *individual* and *generic* concepts, which is a common distinction in KL-ONE-systems, then other mechanisms have to be used.

Operator		class of association axioms	Transformations for	
			$\mathcal{KRS}$s	KBes
1 Syntactical transformation	$\delta_{TX,I\to II}$	$P^I(X) \leftrightarrow Q^{II}(\bar{X})$	$\mathcal{KRS}' = $ $\delta_{TX,I\to II}(\mathcal{WRS}) = $ $\chi_{II}^{TX}(\mathcal{KRS})$	$\mathcal{KB}' = \chi_{II}^{TX}(\mathcal{KB})$
2 and-abstraction	$\beta_{TX,I\to II}^{\wedge}$	$P_1^I(X) \wedge \ldots \wedge P_n^I(X)$ $\leftrightarrow Q^{II}(\bar{X})\ (n>1)$	only in special cases	only in special cases
3 and-refinement	$\gamma_{TX,I\to II}^{\wedge}$	$P^I(X) \leftrightarrow Q_1^{II}(\bar{X})$ $\wedge \ldots \wedge Q_n^{II}(\bar{X})$ $(n>1)$	$\mathcal{KRS}' = $ $\gamma_{TX,I\to II}^{\wedge}(\mathcal{WRS}) = $ $\chi_{II}^{TX}(\mathcal{KRS})$	$\mathcal{KB}' = \chi_{II}^{TX}(\mathcal{KB})$
4 or-abstraction	$\beta_{TX,I\to II}^{\vee}$	$P_1^I(X) \vee \ldots \vee P_n^I(X)$ $\leftrightarrow Q^{II}(\bar{X})\ (n>1)$	$\mathcal{KRS}' = $ $\beta_{TX,I\to II}^{\vee}(\mathcal{WRS}) = $ $\chi_{II}^{TX}(\mathcal{KRS})$,	$\mathcal{KB}' = \chi_{II}^{TX}(\mathcal{KB})$;
5 or-refinement	$\gamma_{TX,I\to II}^{\vee}$	$P^I(X) \leftrightarrow Q_1^{II}(\bar{X})$ $\vee \ldots \vee$ $Q_n^{II}(\bar{X})(n>1)$	$\mathcal{KRS}' = $ $\gamma_{TX,I\to II}^{\vee}\mathcal{WRS} = $ $\chi_{II}^{TX}(\mathcal{KRS})$	only in special cases

Figure 5: Complex Operators

To do this it is necessary to logically describe that individual and generic concepts in the new system play the same role as concepts in the old one. Suppose we mark the old system with *1* and the new system with *2*, then this relationship could be described as follows:

$$\forall X\ indconcept^2(X) \vee genconcept^2(X) \leftrightarrow concept^1(X) \tag{1}$$

If we add (1) e.g. to $klone^1$, then it becomes possible to detect knowledge bases containing indconcept and genconcept as being correct w.r.t. this new $klone^1$-system.

Axiom (1) can also be used to generate this new system $klone^1$ which only uses the new vocabulary *indconcept* and *genconcept* instead of concept. This can be seen as a refinement of the original knowledge representation system. Such axioms can, therefore, be used to describe the relationships between two systems. In principle, there is no restriction for such axioms – they will be called *association axioms* ($\mathcal{TX}$ for short)–, but there are some special cases, in which the generation of the new system and transformations of knowledge bases are relatively simple (cf. section 4.3).

Fig. 5 shows a number of operators corresponding to axioms like e.g. axiom (1). Row (3) shows the type of association axioms. Row (4) shows a way to transform a knowledge representation system $\mathcal{KRS}$ into a new system $\mathcal{KRS}$'. Row (5) shows the same for knowledge bases. Both transformations are based on a function χ_s^{TX}. In section 4.3 a way for implementing this function is shown.

Axiom (1) is a refinement seperating the set of concepts into two subsets, which are *disjoint*. For knowledge engineers (and even for a system) it is an important information to know, that this disjointness-condition holds. It can be formalized as the following constraint c_1^2:

$$c_1^2 : \forall X \neg (indconcept^2(X) \wedge genconcept^2(X)) \tag{2}$$

Suppose you prefer the term *individual* instead of *instance*. You can do this by using the following – very simple – association axiom, which is of type syntactical transformation:[8]

$$\mathcal{TX}_2 : \forall I, C \ individual^2(I, C) \leftrightarrow instance^1(C, I) \tag{3}$$

These modifications may lead to the knowledge representation system shown in fig. 6. $klone^2$ can be constructed out of $klone^1$ as follows:

$$klone^2 = \varrho_{c_1^2}^+(\delta_{TX_2,1\to2}(\beta_{TX_1,1\to2}^\vee klone^1)))$$

For two KR-systems $\mathcal{KRS}$ and $\mathcal{KRS}$', where $\mathcal{KRS}$' is constructed out of $\mathcal{KRS}$ using a complex operator based on $\mathcal{TX}$ an important property can be shown: both systems are logically equivalent w.r.t. $\mathcal{TX}$, i.e. they are equivalent in the property of *restriction*:[9] Formally this means:[10]

$$\forall c \in \mathcal{C}' : \mathcal{TX}, \mathcal{C}, \mathcal{R} \models c$$
$$\forall c \in \mathcal{C} : \mathcal{TX}, \mathcal{C}', \mathcal{R}' \models c$$

It is important to see that, therefore, every knowledge base corresponding to the system $\mathcal{KRS} = (\mathcal{C}, \mathcal{R})$ can be expressed in terms of $\mathcal{KRS}$' $= (\mathcal{C}', \mathcal{R}')$ and vice versa. This is not true for simple operators.

---

[8]Note, in Baader's approach of comparing the expressive power of knowledge representation systems it is in a similiar way possible to define a function called ψ mapping predicate symbols from one system to another, which exactly corresponds to syntactical transformations.

[9]A similiar result concerning deductive strength can be found in [Strasser, 1991].

[10]This is a more general version of definiton 3.4.

$$\underline{\hspace{4cm} klone^2 \hspace{4cm}}$$

$c_1^{0'}:$ $\forall C\ \exists R\ indconcept^2\,(C) \vee genconcept^2\,(C) \rightarrow role^2\,(C, R).$

$c_2^{0'}:$ $\forall C, R:\ indconcept^2\,(C) \vee genconcept^2\,(C) \leftarrow role^2\,(C, R).$

$c_1^{1'}:$ $\forall C, I:\ individual^2\,(I, C) \rightarrow indconcept^2\,(C) \vee genconcept^2\,(C).$

$c_2^{1'}:$ $\forall I, R, V:\ \exists C:\ roleval^2\,(I, R, V) \rightarrow individual^2\,(V, C).$

$c_1^{2'}:$ $\neg\,(indconcept^2\,(C) \wedge genconcept^2\,(C)).$

$r_1^{0'}:$ $\forall C_1, C_2:\ subs^2\,(C_1, C_2) \leftarrow$
        $(\forall R: role^2\,(C_1, R) \rightarrow role^2\,(C_2, R)\,).$

$r_1^{1'}:$ $\forall C, I, R:\ (role^2\,(C, R) \rightarrow \exists V: roleval^2\,(I, R, V)) \rightarrow$
        $individual^2\,(I, C).$

Figure 6: The Knowledge Representation System $klone^2$

4.3   Using Term-Rewriting Techniques for Modifications

If we want to automatically transform a knowledge representation system $\mathcal{KRS}$ into $\mathcal{KRS}$' - e.g. $klone^1$ to $klone^2$ - a number of requirements should be fulfilled. Firstly, the structure of $\mathcal{KRS}$' should correspond to the structure of $\mathcal{KRS}$. This leads to comprehensable transformations. Secondly, the transformation-function should work locally and not in a context-sensitive way and thirdly, this function should be efficient. Point two makes it more easy to implement the function. Let us call this function χ_s^{TX}, where TX refers to the corresponding association axioms and s is the index of the new system. χ_s^{TX} can be built in transforming the association axioms into a term rewriting system. Axioms (1) and (3) may lead to the following term rewriting system:[11]

$$concept^1(X) \hookrightarrow indconcept^2(X) \vee genconcept^2(X) \qquad (4)$$

$$instance^1(C, I) \hookrightarrow individual^2(I, C) \qquad (5)$$

Rewrite rules (4) and (5) are terminating because of the different indexing of left hand and right hand side. Rewriting can be done locally on every constraint and on every deduction rule belonging to the system.

---

[11]The sign $\hookrightarrow$ seperates left-hand and right-hand side of a term-rewriting rule.

4.4   Transformation of Knowledge Bases

The question now is, how can we transform a knowledge base KB, built w.r.t. a knowledge representation system $\mathcal{KRS}$ into a system $\mathcal{KRS}$'? If $\mathcal{KRS}$' is constructed by a simple operator α_R^- or ϱ_C^- from $\mathcal{KBS}$, this problem is rather simple: KB can be used in $\mathcal{KRS}$' without any modifications. If it is built by applying α_R^+ or ϱ_C^+ to $\mathcal{KRS}$, then you have to check if the new constraints are violated in KB.

In the case of complex operators, however, the problem is much more difficult. But it is possible to use – in most cases– the same term rewriting system as shown in section 4.3 for transforming knowledge bases. The only thing to do is to conceive a knowledge base as a conjunction of literals.[12] Row (5) of fig. 5 shows all the cases in which this transformation can be done. In all other cases (such as and-abstraction) special modifications have to be done (see [Strasser, 1991]).

5   A Lattice of Knowledge Representation Systems

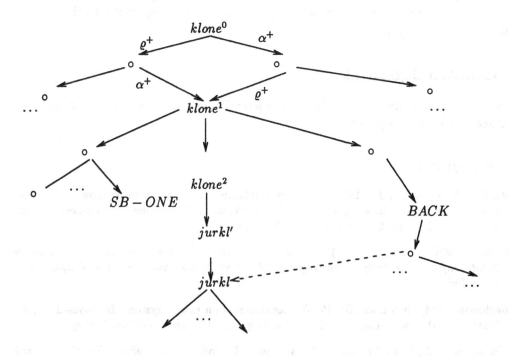

Figure 7: A Lattice of Knowledge Representation Systems

In section 4 a method for manipulating knowledge representation systems was presented. The use of this method makes it possible to generate a new knowledge representation

---

[12]Note, a knowledge base should contain only facts, otherwise the transformation would be much more difficult.

system $\mathcal{KRS}$' out of an existing system $\mathcal{KRS}$.

These two systems are, therefore, connected by the operation, which generates the new one out of the old one. Repeated manipulations of a knowledge representation system, possibly in different "directions", produces a *lattice of knowledge representation systems*, in which the arcs are labeled by the corresponding operators. Fig. 7 shows such a lattice built out of the previous examples. It becomes clear that knowledge representation systems are no longer "islands" where only dangerous ways lead frome one island to the other; the lattice shows exactly how to go from one system to another, eventually using islands in between.

Within this lattice existing knowledge representation systems can be integrated, e.g. KL-ONE-systems like BACK or SB-ONE. The only thing to do is to formalize these systems according to our methodology. The only difference between for example BACK and $klone^1$ is, therefore, the fact that BACK has an efficient implementation while $klone^1$ does not.

In the process of knowledge acquisition, the knowledge engineer can use both, the existing lattice and generate a new system, which can be integrated into the existing lattice in a simple way.

Acknowledgement

I would like to thank Zongyan Qiu and Dieter Dodenhöft for valuable comments on earlier versions of this paper.

References

[Baader, 1990] Baader, F. (1990). A formal definition for the expressive power of knowledge representation languages. In *ECAI-90European Conference on Artificial Intelligence,Stokholm*, page 53ff. Pitman Publishing, London.

[Brewka, 1987] Brewka, G. (1987). The logic of inheritance in frame systems. In *International Joint Conference on Artificial Intelligence, IJCAI-87, Milano*, pages 483–488. Morgan Kaufmann Publishers.

[Buchanan, 1983] Buchanan, B. (1983). Constructing an expert system. In Hayes-Roth, F., Waterman, D., and Lenat, D., editors, *Building Expert Systems*. Addison-Wesley.

[Christaller et al., 1990] Christaller, T., di Primio, F., and Voß, A. (1990). *Die KI-Werkbank BABYLON* . Addison-Wesley.

[Console et al., 1989] Console, L., Dupre, D. T., and Torasso, P. (1989). A theory of diagnosis for incomplete causal models. In *International Joint Conference on Artificial Intelligence, IJCAI-89, Detroit*, pages 1311–1317. .

[Delgrande and Mylopoulos, 1986] Delgrande, J. P. and Mylopoulos, J. (1986). Knowledge representation: Features of knowledge. In Bibel, W. and Jorrand, P., editors, *Fundamentals of Artificial Intelligence – An Advanced Course*, pages 3–38. Lecture Notes in Computer Science Nr. 232,Springer Verlag Berlin, Heidelberg, New York.

[Hayes, 1981] Hayes, P. (1981). The logic of frames. In Webber, B. L. and Nilsson, N. J., editors, *Readings in Artificial Intelligence*, pages 119–140. Tioga Publishing Company, Palo Alto, California.

[Horacek, 1989] Horacek, H. (1989). Towards principles of ontology. In *GWAI-89 German Workshop on Artificial Intelligence*, pages 323–330. Springer Verlag Berlin, Heidelberg, New York.

[Moerkotte, 1990] Moerkotte, G. (1990).   *Inkonsistenzen in deduktiven Datenbanken* . Informatik-Fachberichte 248Springer Verlag Berlin, Heidelberg, New York.

[Musen, 1989] Musen, M. A. (1989).   *Automated Generation of Model-based Knowledge-Acquisition Tools*. Morgan Kaufman, CA; Pitman,London.

[Newell, 1982] Newell (1982). The knowledge level. *Artificial Intelligence Journal*, 18:87–127.

[Reichgelt and Harmelen, 1985] Reichgelt, H. and Harmelen, V. F. (1985). Relevant criteria for choosing an inference engine in expert systems. In Merry, M., editor, *Expert Systems 85*, pages 21–31.

[Reimer, 1989] Reimer, U. (1989). *FRM: Ein Frame-Repräsentationsmodell und seine formale Semantik* . Informatik-Fachberichte 198Springer Verlag Berlin, Heidelberg, New York.

[Sloman, 1985] Sloman, A. (1985). Why we need many knowledge representation formalisms. In Bramer, M. A., editor, *Research and Development in Expert Systems*. Cambridge University Press.

[Strasser, 1991] Strasser, A. (1991).   *Generierung domänenspezifischer Wissensrepräsentationssysteme und Transformation von Wissensbasen mit einer Anwendung in der Rechtsinformatik* . PhD thesis, Technische Universität München.

[Vilain, 1985] Vilain, M. (1985). The restricted language architecture of a hybrid representation system. In *International Joint Conference on Artificial Intelligence, IJCAI-85, Los Angeles*, pages 547–551.

[Waterman and Hayes-Roth, 1983] Waterman, D. and Hayes-Roth, F. (1983). An investigation of tools for building expert systems. In Hayes-Roth, F., Waterman, D., and Lenat, D., editors, *Building Expert Systems*. Addison-Wesley.

Inconsistencies handling: nonmonotonic and paraconsistent reasoning

Ján Šefránek

Institute of Computer Science, Comenius University
Faculty of Mathematics and Physics, 842 43 Bratislava
Czechoslovakia

Abstract

A characterization of non-monotonic reasoning is given. Not-sound inference operator using, consistency maintaining and belief revision as key features of non-monotonic reasoning are discussed. A problem of inconsistencies handling leads to another approach interesting in our context, it is based on paraconsistent inference.

1 Introduction

The paper is aiming to state some characteristic features of non-monotonic reasoning. Its initial motivation was a deep insight into the nature of nonmonotonic reasoning given by Apt [1] (a formal proof that nonmonotony is unavoidable in some circumstances, see fact 1 in the following).

A characterization of non-monotonic inference is given in Sections 2 - 5. The main results of the paper are presented in sections 3 and 4. Some trivial facts (facts 2 - 5 in the following) and a discussion of them are used to show (in terms of formal arguments) that a reasonable understanding of non-monotonic inference is based on such concepts as not-sound inferencing, consistency maintaining and belief revision.

The approach used in the paper is based on a synthesis of two views: model theoretic and inference systems theoretic (in a Tarskian tradition). Classical model theory is assumed in sections 2 - 5: when a formula ϕ is not true in an interpretation I, then $\neg\phi$ is true in I and vice versa.

A comparison with the approach based on an axiomatic treatment of some reasonable properties of the non-monotonic inference operator Cn [3, 6, 8] is needed here. The approach of this paper is in some sense a complementary one. Some observations concerning a special and unavoidable non-monotonic Cn operator with a reasonable behaviour are given. An attempt to generalize the observations and to obtain a characterization of non-monotonic reasoning is undertaken. A better understanding of relations between both approaches will be the subject of a future paper. Some remarks concerning the problem are given in the Section 5 of this paper.

The problem of inconsistency seems to play a central role in our context. The remaining sections (6 to 8) of the paper are approaching to this problem from another point of view, it is the view of the so called paraconsistent logic. An introductory formulation of the problem is the only aim of this part of the paper. The perspective goal of an ongoing research is an understanding of connections between non-monotonic and paraconsistent reasoning using logic programming tools and concepts. A prerequisite of this intention is a reasonable introduction of an inconsistency concept into Horn logic.

The next two sections should answer the following question: What are the possibilities to introduce a monotonic negation into Horn logic? An interesting general characterization of non-monotonic reasoning is obtained from the answer.

2   Basic concepts and conventions

An (unspecified) language L is considered. S is the set of all formulae in L.

An (unspecified) inference is represented by an operator Cn. Let A is a set of formulae (a subset of S). It is supposed that $Cn(A)$ is the set of formulae inferable from A. (Cn is an operator from sets of formulae to sets of formulae.)

Cn is *finitary* iff for any set A and any formula $\phi \in Cn(A)$ there is X, a finite subset of A, such that $\phi \in Cn(X)$.

We assume only finitary Cn in the following. This assumption is not too restrictive in the context of computer science.

Cn is *non-monotonic* iff for some sets A, B, such that $A \subseteq B$, does not hold that $Cn(A) \subseteq Cn(B)$. (There is a consequence of the set A not inferable from its superset B.)

Let A is a consistent set of formulae, $\phi \in Cn(A)$. The operator Cn is a *sound* one iff every model of A is also model of ϕ. Cn is *weakly sound* iff $A \cup \{\phi\}$ is consistent. Cn *maintains consistency* iff $Cn(A)$ is consistent.

3   Basic facts

The following fact (Apt's lemma, [1]) shows that any introduction of a negation into Horn logic necessary results into nonmonotony (if the negation is understood in such a manner that $\neg\phi$ does not hold in a model, when ϕ holds in the model).

Fact 1 *Let P is a set of Horn clauses, ϕ is a negative ground literal. Let $\phi \in Cn(P)$. Then a) Cn is not sound. b) If Cn is weakly sound, then Cn is non-monotonic.*

A simple consequence (and a generalization) of the Apt's lemma provides an interesting characterization of non-monotonic reasoning.

Fact 2 *If Cn is not sound, but it maintains consistency, then Cn is non-monotonic.*

Proof: Let A is a consistent set of formulae. Cn is supposed to be not sound, so we can accept there is such $\phi \in Cn(A)$ that A is satisfied in an interpretation I but ϕ is not satisfied in I. It means that $\neg\phi$ is satisfied in I, i.e. $A \cup \{\neg\phi\}$ is consistent. Cn maintains consistency, i.e. ϕ cannot be in $Cn(A \cup \{\neg\phi\})$: Cn is non-monotonic.

4   Discussion

Basic facts mentioned in the preceding section provide a characterization of some key features of non-monotonic inference:

(i) **There is a close connection between** *not-soundness* **of an inference operator Cn and its non-monotonicity.** The following trivial fact completes the picture.

Fact 3 *If Cn is sound, it is monotonic.*

The proof is trivial, it suffices to realize that Cn is finitary.

A note: A not sound inference is of indispensable value in systems working in environments with incomplete knowledge. We have also seen it is unavoidable in Horn logic when a negation should be introduced in it. It is only a special case of incomplete knowledge - an implicit representation of negative facts is provided. In this case the limitation of a logical language to Horn clauses enables efficient implementation of inference. The limited expressive power of the language (and a work with incomplete knowledge) is a price we should pay for increased effectiveness of inference.

(ii) Non-monotony is not characterized by the not-soundness of a Cn operator sufficiently. *Consistency maintaining* **plays a key role in our characterization of non-monotonic reasoning.**

Definition of two concepts, at first: An operator Cn is *trivial* iff for every A, containing both some ϕ and $\neg\phi$, is $Cn(A) = S$. S is called the *trivial theory*.

Fact 4 *If Cn is trivial and does not maintain consistency, then there is a consistent set of formulae A such that $Cn(A)$ is the trivial theory.*

Proof: $\phi \wedge \neg\phi \in Cn(A)$, for some ϕ and A, because of non-maintenance of consistency. Hence, $Cn(A) = S$, according to the assumption that Cn is trivial.

The following notes comment the before-mentioned fact:

Note 1: The non-monotonic behaviour of Cn operator is excluded in situations when all sentences (the set S) are derived (and when there is no mechanism how to delete something from S). It means, in order to obtain an useful characterization of a non-monotonic Cn we should suppose Cn maintàining the consistency or a non-trivial Cn.

Example: Inductive inference, as studied traditionally, satisfies these assumptions. A falsification of consequences by a new observations does not lead to inference of all sentences but to a revision of the previous knowledge set.

Note 2: Of course, a non-monotonic inference is possible also without these two assumptions. As an example could serve the non-monotonic inference based on the *Closed World Assumption (CWA)* - it could be considered to be a rule of non-monotonic reasoning though it sometimes results in an inconsistency, and it enables (together with the classical logic in background) to derive the set of all formulae.

But an inconsistency leading to a trivial theory is an unpleasant property from the real systems development point of view. So, it is needed to maintain consistency or to require non triviality of inference, when designing and implementing systems with not sound inference.

(iii) Non-monotonic reasoning is in a close connection with *belief revision.*

Some definitions: Let $\mathcal{S}$ is the set of all subsets of S. Let a (partial) operator $R : \mathcal{S} \times \mathcal{S} \longrightarrow \mathcal{S}$ is defined for pairs of disjoint (and not empty) sets of formulae. It should be called a *revision operator* iff for every A, B, C such that $R(A, B) = C$ is $C \subseteq A \cup B$, and there is a $\phi \in A$ such that $\phi \notin C$.

Intuition: B, a set of formulae, is added to another set of formulae A in such a way that some formulae from A should be deleted.

A revision operator R is called *associated* with a consequence operator Cn iff for some subsets A, B of S (such that $B - Cn(A)$ is not empty) holds $Cn(B) = R(Cn(A), B - Cn(A))$.

Fact 5 *If Cn is non-monotonic and it maintains consistency, then there is a revision operator R associated with Cn.*

Proof: Let us consider such pairs A, B, that $A \subseteq B$, both are consistent, and $(Cn(A) \nsubseteq Cn(B))$. It is sufficient to define $R(Cn(A), B - Cn(A))$ by $Cn(B)$ and for other operands let R is undefined.

Summary: Non-monotonic reasoning is characterized (in some sense reasonable) by

- not sound inference operator using,

- consistency maintaining

- and belief revision.

5 Cumulation and consistency maintainig

Cumulative inference operation is defined in [6] as any operation $Cn : S \longrightarrow S$ such that the following properties are satisfied:

$A \subseteq Cn(A)$

if $A \subseteq B$ and $B \subseteq Cn(A)$, then $Cn(A) = Cn(B)$.

Makinson claims that *Closed World Assumption* satisfies these two conditions. There is one exception when it is not true. It is the case when CWA leads to a contradiction. In this situation $Cn(A) = S$ and there is a B such that $A \subseteq B$, and $B \subseteq Cn(A)$, but $Cn(A) = Cn(B)$ does not hold. The following counterexample is a proof: $A = \{b \leftarrow \neg a\}, CWA(A) = \{\neg a, \neg b\}, Cn(A) = S$, for Cn based on CWA and standard deduction. If $B = A \cup \{a\}$, then $\neg a \notin Cn(B)$.

It seems that consistency maintaining is a condition of cumulativeness of Cn. There are at least two ways how to guarantee consistency maintaining. The first consists in some syntactic restrictions. As an example could be mentioned Jäger's $HCWA$. An explicit embedding of a revision mechanism into inference mechanism represents the second way (see Fact 5).

The following fact argues that in a sense (in a conceptual environment) consistency maintaining is the necessary condition of cumulativeness.

A definition, at first: An operator Cn is *regenerable* iff there is a set A such that $Cn(A) = S$ and a superset B of A such that $Cn(B) \subset S$.

Fact 6 *Let Cn is trivial and regenerable. If Cn is cumulative, then it maintains consistency.*

Proof: Let Cn does not maintain consistency, i.e. for a consistent set A is $Cn(A) = S$. Because of the assumption that Cn is regenerable, we can suppose there is B, a superset of A such that $Cn(B)$ is a proper subset of S. So, A is a subset of B, B is a subset of $Cn(A) = S$, but $Cn(A) = Cn(B)$ does not hold, i.e. Cn is *not* cumulative.

6 Paraconsistency

One of the basic assumptions of the preceding sections is the following: a formula ϕ is satisfied in an interpretation I iff $\neg\phi$ is not satisfied in I. The assumption will be dropped in the following.

We have observed a defective behavior of the Cn operator, which is a trivial one and does not maintain consistency (fact 4). It was shown that one possible way how to solve the problem is consistency maintaining, in most cases using belief revision.

Belief revision is in general an expensive operation. There is another possible approach - to use a not trivial inference operator. Such operator enables to continue inference in presence of a contradiction. From practical point of view we can conceive inference with this property as a cautious inference combined with postponing a necessary belief revision.

At first, a brief initial characterization of an approach to the so called paraconsistent logic [2, 7]: A subset of {*truth, false*} is assigned to each atomic formula. Members of the set of all these subsets could be denoted by symbols t, f, b *(both)*, n *(none)*. Appropriate rules for values assigning to composed formulae are chosen. A formula with assigned value b is understood as a contradictory one. Inference in the logic is not a trivial one. It is impossible to derive S from contradictory premises.

7 Reasoning with inconsistency

A logic, presented in [5] and called RI (Reasoning with inconsistency) is briefly described in this section.

A belief lattice BL is given, with a set of members such that for every s an ordering $n < s < b$ holds, where $<$ is a reflective and transitive relation. If ϕ is an atomic formula of predicate calculus, then $\phi : s$ is a literal of RI, where $s \in BL$. Composed formulae of RI are created in an obvious manner. BL is restricted to $\{t,f,b,n\}$ in the following.

Interpretation is a triple (D, F, P), where D is the domain of I, F assigns to each function symbol a mapping from $D \times \cdots \times D$ to D, P assigns to each predicate symbol a mapping from $D \times \cdots \times D \times BL$ to $\{1,0\}$, such that the following condition holds: for every predicate p and tuple t from $D \times \cdots \times D$ there is an $r \in BL$ such that $P(p(t) : s) = 1$ iff $s < r$. (If $p(t)$ is believed to a degree r, it is impossible that it is not believed to a smaller degree.)

Concepts of valuation, satisfaction, truth, model, and logical entailment are defined in an usual manner.

Two new logical connectives are introduced, an epistemic negation N and an epistemic implication $==>$.

N is defined on four-valued BL as $Nt = f$, $Nf = t$, $Nb = b$, $Nn = n$. It is extended to formulae as follows. $Np : s$ iff $p : Ns$, $N\neg p : s$ iff $\neg p : Ns$, the behaviour of epistemic negation with respect to conjunction, disjunction, general quantifier and existential quantifier is the same as for classical negation.

Epistemic implication defined as $Np \lor q$ is overly cautious: from $T = \{q : t ==> p : t, q : t\}$ does not follow $p : t$. An interpretation $(q : b, p : n)$ is a model of T, but it is not a model of $p : t$. (An obvious abusing of the interpretation concept is used here and in the following.)

From this point of view it is preferable to consider only models with the least amount of inconsistency.

A model m is *e-inconsistent* iff there is a literal $f : b$, which is true in m. (This concept represents an epistemic inconsistency which is tolerated in RI; there are models of e-inconsistent sets of formulae, as we have seen. On the other hand, ontological inconsistency, is not tolerated. T is o-inconsistent in RI iff it contains both $\phi : t$ and $\neg \phi : t$.)

An interpretation J is *more or equal e-consistent* as I iff for every atom $p(t1, ..., tn)$ holds: if $p(t1, ..., tn) : b$ is true in J, then it is true in I. (All inconsistencies from J occur in I.) I is *most e-consistent* in a class of interpretations iff there is no I' such that I' is strictly more e-consistent than I. *Epistemic entailment* is restricted to the most e-consistent models: a formula ϕ is epistemically entailed by a set T iff every most e-consistent model of T is a model of ϕ.

An inference operator Cn implementing the epistemic entailment of RI is not a trivial one: from e-inconsistent set $\{q : t ==> p : t, q : b\}$ does not epistemically follow $p : t$. On the other hand, $p : t$ epistemically follows from $\{q : t ==> p : t, q : t\}$.

8 Conclusion

The following application of RI is an interesting one from our point of view. Let R is an inconsistent set of predicate calculus formulae. A natural embedding *epi*, called epistemic embedding, is defined in [5], $epi(R)$ is a set of RI-formulae. In RI it is possible to isolate the source of inconsistency : it can be identified with the set of all atoms p such that $p : b$ is true in a most e-consistent models of R', where R' is $epi(R)$.

A part of information which can be viewed as undamaged by the inconsistency could be isolated in RI, too.

An introduction of inconsistencies into logic programming in the style of [4] is promising one from the paraconsistent logic programming point of view.

An appropriate embedding of such logic programming language with inconsistency into RI via its embedding into predicate calculus is straigtforward.

A more detailed discussion of this idea will be the subject of a future paper.

Acknowledgment: I wish to thank Juraj Steiner and Juraj Waczulík for helpful comments on an earlier version of this paper.

References

[1] Apt, K.R. *Introduction to Logic Programming.* Technical Report CS-R8826, Centre for Mathematics and Computer Science, Amsterdam, 1988.

[2] Belnap, N.D. *How a computer should think.* In Contemporary Aspects of Philosophy. Proceedings of the Oxford International Symposium, 1976.

[3] Gabbay, D.M. *Theoretical foundations for non-monotonic reasoning in expert systems.* In K.R.Apt, ed., Logics and Models of Concurrent Systems, Springer, 1985.

[4] Gabbay, D.M., Sergot, M.J. *Negation as Inconsistency.* The Journal of Logic Programming 1986.

[5] Kifer, M., Lozinskii, E.L. *RI: A logic for reasoning with inconsistency.* In Proceedings of Fourth Annual Symposium on Logic in Computer Science, 1989, IEEE Computer Society Press.

[6] Makinson, D.: *General theory of cumulative inference.* In M.Reinfrank et al. (eds.): Non-Monotonic Reasoning, Springer Verlag, 1989.

[7] Priest, G., Routley, R., Norman, J.: *Paraconsistent Logic.* München, 1989.

[8] Wójcicki, R. *An axiomatic treatment of non-monotonic arguments.* Bulletin of the Section of Logic, vol. 17, No.2, Polish Acad.Sci., Inst. of Philosophy and Sociology, Warsaw - Lódz, 1988.

An approach to
structural synthesis of data
processing programs

Gražina Taučaitė and Justinas Laurinskas

Institute of Mathematics and Informatics
of the Lithuanian Academy of Sciences
232600 Vilnius, K.Požėlos 54

Abstract

An approach to the integration of the structural synthesis of programs and the synthesis of relations is proposed. The approach is based on a computation model (a set of formulas), describing both computations and navigation paths. The sound and complete system of inference rules for the class of formulas, used in such computation models, is given.

1. Introduction.

The structural synthesis of programs (Tyugu and Harf, 1980; Dikovskij and Kanovich, 1985) is based on a specification of computations called a computation model. A computation model describes the relationships among the initial program modules (operations) and is a set of named initial dependencies among some variables (values). A name of an initial dependency is a name of the program module which implements that dependency. The required program is specified with the help of two sets: the set of given data (variables) and the set of desired data (variables), i.e. with the help of unnamed dependency. The structural synthesis consists in the search for the corresponding named dependency, which is logically implied by the initial dependencies. The name of this derived dependency represents the plan of a program (in the form of a composition of initial modules), which implements that dependency.

The synthesis of relations (Lozinskii, 1980; Honeyman, 1980) is based on a specification of a relational database called a database scheme. A database scheme describes base relations and is a pair of sets: the set of named dependencies, representing the schemes of base relations, and the set of (unnamed) dependencies, describing the properties of base relations. A relation scheme is a set of attributes that define the format of the relation. In addition, the relational operations enabling to obtain new relations from the base ones are defined. The required relation is specified with the help of its scheme, i.e. with the help of unnamed dependency. The synthesis of the relation consists in the search for the corresponding derived dependecy, which is logically implied by a database scheme and the name of which represents the relational algebra expression that defines the required relation in terms of base relations.

The structural synthesis of programs does not take into account the possibilities of implementation of some initial dependencies by means of relations and does not consider possibilities of computation of relations by means of relational operations. On the other hand, the synthesis of relations does not consider possibilities of computation of relations by means of arbitrary (unrelational) operations. In this paper an approach to the integration of the structural synthesis of programs and the synthesis of relations is proposed, i.e. an approach to the structural synthesis of data processing (DP) programs, which combine computations by means of arbitrary operations with the computations by means of relational ones.

2. Example.

Let us consider a fragment of the salary computation problem (the example is taken from (Dikovskij and Kanovich, 1985)). Suppose EMP stands for employee name, REC for record, SAL for month's salary, A_SAL for average month's salary, and T_SAL for "thirteenth" salary. The structural synthesis of programs assumes there are four initial modules described by the following dependencies:

$F_1 : EMP.YEAR \rightarrow REC,$ $\qquad F_2 : EMP.YEAR.MONTH \rightarrow SAL,$

$F_3 : REC.A\_SAL \rightarrow T\_SAL,$ $\qquad F_4 : (MONTH \rightarrow SAL) \rightarrow A\_SAL.$

The first dependency, for instance, expresses the computability of REC from EMP and $YEAR$ by means of module F_1. The last dependency expresses the computability of A_SAL by means of module F_4 in the presence of computability of SAL from $MONTH$ (that is, module F_4 has one parameter-subprogram). In this model, it is possible, for instance, to formulate the task $TASK1$: "for employee $EMP = e$ find the 13th salary obtained per year $YEAR = y$". This task is solvable: from the initial dependencies it follows the derived dependecy

$$(F_4(F_2); F_1; F_3) : EMP.YEAR \rightarrow T\_SAL$$

(";" denotes the composite sequential operator). However, in the given model, it is impossible, for instance, to formulate the task $TASK2$: "for all employees whose work record in the year $YEAR = y$ has not exceeded the value $REC = r$ find employee names and their 13th salaries obtained per year $YEAR = y$". A problem with the models of the given type is that they do not take into account that some dependencies (dependencies $F1$ and $F2$ of our example) are represented by their extentionals.

The relation synthesis assumes there are four base relation schemes (or entity dependencies),

$R_1 : EMP.YEAR.REC,$ $\qquad R_2 : EMP.YEAR.MONTH.SAL,$

$R_3 : EMP.YEAR.A\_SAL,$ and $\qquad R_4 : EMP.YEAR.T\_SAL,$

and four (functional) dependencies,

$EMP.YEAR \rightarrow REC,$ $\qquad EMP.YEAR.MONTH \rightarrow SAL,$

$EMP.YEAR \rightarrow A\_SAL,$ and $\qquad EMP.YEAR \rightarrow T\_SAL,$

which describe the properties of base relations. In this model there can be formulated and solved the task to construct the derived relation with the scheme $EMP.YEAR.REC.T\_SAL$: from the initial dependencies it follows the dependency

$$(R_1 * R_4) : EMP.YEAR.REC.T\_SAL$$

(" $*$ " denotes the join operator). It is obvious that the relation $R_1 * R_4$ may be used when solving the task $TASK2$. However, this model does not take into account that relations R_3 and R_4 are not independent and that they are computed, by means of unrelational operations, on the basis of relations R_1 and R_2.

Our approach assumes there are two base relation schemes (or entity dependencies),

$$R_1 : EMP.YEAR.REC \text{ and } R_2 : EMP.YEAR.MONTH.SAL,$$

two (functional) dependencies,

$$EMP.YEAR \rightarrow REC \text{ and } EMP.YEAR.MONTH \rightarrow SAL,$$

which describe the properties of base relations, and two (operational) dependencies,

$$F_1 : (\rightarrow REC)\&(\rightarrow A\_SAL) \rightarrow T\_SAL \text{ and } F_2 : (MONTH \rightarrow SAL) \rightarrow A\_SAL,$$

which express the computability of relations by means of unrelational operations F_1 and F_2. From these dependencies it follows the dependency

$$(R_1 * F_1(R_1, F_2(R_2))) : EMP.YEAR.REC.T\_SAL,$$

which expresses the computability of relation with the scheme $EMP.YEAR.REC.T\_SAL$ by means of relational and unrelational operations.

3. Background.

Attributes are symbols taken from a given finite set U called the universe. All sets of attributes are subsets of the universe. The union of sets of attributes X and Y is denoted by XY. Each attribute is associated with the set called its domain.

A tuple on a set of attributes X is a function t mapping each attribute A of X to an element in its domain. If t is a tuple on X and Y is a subset of X, then $t[Y]$ denotes a restriction of t on Y. A relation scheme is a set of attributes. A relation s with the scheme S, denoted $s(S)$, is a set of tuples on S.

Given a relation $s(S)$ and a subset X of S, the projection of $s(S)$ onto X, denoted $\pi_X(s)$, is a relation with the scheme X, containing all the tuples $t[X]$ such that t belongs to s. The join of relations $s_1(S_1)$ and $s_2(S_2)$, denoted $s_1 * s_2$, is a relation with the scheme $S_1 S_2$, containing all the tuples t on $S_1 S_2$ such that $t[S_1]$ belongs to s_1, and $t[S_2]$ to s_2.

Relations often satisfy certain constraints. One class of constraints is the varies (data) dependencies. A dependency d is said to be logically implied from a set of dependencies D if for all relations s, s satisfies d if s satisfies all the dependencies of D. The usual method to describe a set of logically implied dependencies is the method of formal theories. An inference rule of a theory indicates how to derive dependency from other dependencies. The inference rules are sound if every dependency that can be derived from a set of dependencies D is also logically implied from D. The inference rules are complete if every dependency that is logically implied from a set of dependencies D can also be derived from D.

An expression of the form $X \to Y$, where X and Y are sets of attributes, is called functional dependency (F−dependency). A relation $s(S)$ satisfies $X \to Y$ if XY is a subset of S and for all tuples t_1 and t_2 of s, $t_1[X] = t_2[X]$ implies $t_1[Y] = t_2[Y]$.

A database scheme R is a set of named schemes of base relations, $R = \{R_1 : Q_1, \ldots, R_n : Q_n\}$, such that $Q_1 \ldots Q_n = U$. A database state r is a set of base relations, $r = \{r_1(Q_1), \ldots, r_n(Q_n)\}$. Let R be a database scheme and D be a set of dependencies. A relation $u(U)$ is called a universal relation for the database state r under D if u satisfies all dependencies of D and each r_i is a projection of u onto Q_i.

A (restricted) relational expression over a database R is a relational algebra expression, the operands of which are names of base relation schemes and the operators are projection and join. If E is a relational expression over R, then $E(r)$ denotes the value of E when each operand R_i of E is replaced by r_i.

4. Structural synthesis of DP programs.

Let G be a set of functions (unrelational operations). We say an expression of the form $F : (X_1 \to Y_1)\& \ldots \&(X_k \to Y_k) \to Z$ is operational dependency (O−dependency) if X_i, Y_i, and Z are sets of attributes and F is a k-ary function of G. Let D be a set of F−and O−dependencies and $F : (X_1 \to Y_1)\& \ldots \&(X_k \to Y_k) \to Z$ be an O−dependency in D. Given a relation $s(S)$ such that $X_1 Y_1 \ldots X_k Y_k Z$ is a subset of S, we say $s(S)$ satisfies $F : (X_1 \to Y_1)\& \ldots \&(X_k \to Y_k) \to Z$ under D if it satisfies $W_1 \ldots W_k \to Z$ and

$$\pi_{W_1 \ldots W_k Z}(s) = F(\pi_{W_1 X_1 Y_1}(s), \ldots, \pi_{W_k X_k Y_k}(s))$$

whenever for each i, $1 \le i \le k$, there exists a subset W_i of S such that the intersection of W_i and Y_i is empty whenever X_i is not empty, $s(S)$ satisfies $W_i X_i \to Y_i$, and $W_i X_i \to Y_i$ is logically implied from D.

Let $R = \{R_1 : Q_1, \ldots, R_n : Q_n\}$ be a database scheme such that $Q_1 \ldots Q_n$ is a subset of U and D be a set of $F-$ and $O-$dependencies. Then we say (R, D) is a database scheme with computations. A relation $u(U)$ is called a universal relation for the database state r under D if u satisfies all dependencies of D and each r_i is a projection of u onto Q_i.

A plan of DP program over (R, G) is defined as follows: if E is a relational expression over R, then E is a plan of DP program over (R, G); if F is a k-ary function of G and $E_1, \ldots, E_k$ are plans of DP programs over (R, G), then $F(E_1, \ldots, E_k)$ is a plan of DP program over (R, G). If E is a plan of DP program over (R, G), then $E(r)$ denotes the value of E when each operand R_i of E is replaced by r_i.

We say expression of the form $E : S$ is an entity dependency ($E-$dependency) if E is a plan of DP program over (R, G) and S is the scheme of E (the scheme of the resulting relation). Let $E_1 : S_1$ and $E_2 : S_2$ be two $E-$dependencies. We say $E_1 : S_1$ is a subdependency of $E_2 : S_2$ if E_1 is a subplan of E_2. Let u be a universal relation for r under D. We say r satisfies $E : S$ if $E(r)$ is a projection of u onto S and r satisfies all subdependencies of $E : S$.

Let D be a set of $F-$, $O-$, and $E-$dependencies and d be an $O-$dependency of D. Then we say $u(U)$ satisfies d under D if $u(U)$ satisfies d under D', where D' is obtained from D by removing all $E-$dependencies.

A pair $(R \cup D, X)$ is a task of the structural synthesis of DP program, where (R, D) is the database scheme with computations and X is a set of attributes. A solution of the task $(R \cup D, X)$ is the plan E of DP program over (R, G) such that the dependency $E : X$ is logically implied from the set $R \cup D$.

The inference rules we present for derivation of $E-$dependencies from a set of $F-$, $O-$, and $E-$dependencies are as follows.

P1. $X \to X$.

P2. $X \to Y$ implies $X \to Z$ if Z is a subset of Y.

P3. $X \to Y$ and $Y \to Z$ imply $X \to YZ$.

P4. $W_1 X_1 \to Y_1, \ldots, W_k X_k \to Y_k$, and $F : (X_1 \to Y_1) \& \ldots \& (X_k \to Y_k) \to Z$ imply $W_1 \ldots W_k \to Z$ if for each $i, 1 \leq i \leq k$, the intersection of W_i and Y_i is empty whenever X_i is not empty.

P5. $E : S$ implies $\pi_X(E) : X$ if X is a subset of S.

P6. $E_1 : S_1$, $E_2 : S_2$, and $S_1 \cap S_2 \to S_1$ or $S_1 \cap S_2 \to S_2$ imply $(E_1 * E_2) : S_1 S_2$.

P7. $E_1 : S_1, \ldots, E_k : S_k$, $W_1 X_1 \to Y_1, \ldots, W_k X_k \to Y_k$, $W_1 X_1 Y_1 = S_1$, $\ldots, W_k X_k Y_k = S_k$, and $F : (X_1 \to Y_1) \& \ldots \& (X_k \to Y_k) \to Z$ imply $F(E_1, \ldots, E_k) : W_1 \ldots W_k Z$ if for each $i, 1 \leq i \leq k$, the intersection of W_i and Y_i is empty whenever X_i is not empty.

To illustrate the derivation of dependency, let us continue with our example. Assume $SALARY$ denotes the database scheme with computations described above. Then the solution of the task $(SALARY, EMP.YEAR.REC.T\_SAL)$ is as follows. R_2, F_2, and $EMP.YEAR.MONTH \to SAL$ imply $F_2(R_2) : EMP.YEAR.A\_SAL$ by P7 and $EMP.YEAR \to A\_SAL$ by P4. Similarly, F_1, $F_2(R_2)$, $R1$, $EMP.YEAR \to A\_SAL$, and $EMP.YEAR \to REC$ imply $F_1(R_1, F_2(R_2)) : EMP.YEAR.T\_SAL$ by P7 and $EMP.YEAR \to T\_SAL$ by P4. Finally, R_1, $F_1(R_1, F_2(R_2))$, and $EMP.YEAR \to T\_SAL$ imply $(R_1 * F_1(R_1, F_2(R_2))) : EMP.YEAR.REC.T\_SAL$ by P6.

5. Soundness and completeness of the inference rules

Lemma 1. The inference rules are sound.

Proof. The soundness of $P1 - P3$ was shown in (Armstrong, 1974) and the soundness of $P6$ in (Rissanen, 1977).

P4. Assume a relation u satisfies $F : (X_1 \to Y_1) \& \ldots \& (X_k \to Y_k) \to Z$ under D. Suppose that for each i, $1 \leq i \leq k$, there exists a set of attributes W_i such that the intersection of W_i and Y_i is empty whenever X_i is not empty, $W_iX_i \to Y_i$ is derivable from D, and u satisfies $W_iX_i \to Y_i$. By hypothesis, since $W_iX_i \to Y_i$ is derivable from D, $W_iX_i \to Y_i$ is logically implied from D. Then, since u satisfies $F : (X_1 \to Y_1)\& \ldots \&(X_k \to Y_k) \to Z$ under D, u satisfies $W_1 \ldots W_k \to Z$.

P5. Assume a relation u satisfies $E : S$, that is, $E(r) = \pi_S(u)$. Suppose X is a subset of S. Then

$$(\pi_X(E))(r) = \pi_X(E(r)) = (\pi_X(\pi_S(u)) = \pi_X(u)$$

Thus u satisfies $\pi_X(E) : X$.

P7. Assume a relation u satisfies $F : (X_1 \to Y_1)\& \ldots \&(X_k \to Y_k) \to Z$ under D. Suppose that u satisfies $E_1 : S_1, \ldots, E_k : S_k$, and for each i, $1 \le i \le k$, there exists a set of attributes W_i such that $S_i = W_iX_iY_i$, and the intersection of W_i and Y_i is empty whenever X_i is not empty, $W_iX_i \to Y_i$ is derivable from D, and u satisfies $W_iX_i \to Y_i$. Since the inference rules $P1 - P4$ are sound and $W_iX_i \to Y_i$ is derivable from D, $W_iX_i \to Y_i$ is logically implied from D. Since u satisfies $E_i : S_i$, $E_i(r) = \pi_{S_i}(u)$. Then, since u satisfies $F : (X_1 \to Y_1)\& \ldots \&(X_k \to Y_k) \to Z$ under D,

$$\pi_{W_1 \ldots W_k Z}(u) = F(\pi_{W_1 X_1 Y_1}(u), \ldots, \pi_{W_k X_k Y_k}(u)) = F(E_1(r), \ldots, E_k(r)).$$

Thus u satisfies $F(E_1, \ldots, E_k) : W_1 \ldots W_k Z$. **Q.E.D.**

It follows from $P1 - P3$ the following inference rule (Armstrong, 1974) :

P8. $X \to Y$ and $X \to Z$ imply $X \to YZ$. Let D be a set of $F-$ and $O-$dependencies. For a set of attributes X, we define a closure of X under D, denoted X^+, to be the set of attributes A such that the dependency $X \to A$ can be derived from D (by using inference rules $P1 - P4$). It follows from $P2$ and $P8$ that an $F-$dependency $X \to Y$ can be derived from D if and only if Y is a subset of X^+.

Lemma 2. Let D be a set of $F-$ and $O-$dependencies. If an $F-$dependency d is logically implied from D, then d can be derived from D.

Proof. Suppose that a dependency $V \to T$ cannot be derived from D. We need to show that $V \to T$ is not logically implied from D. Consider a relation u with only two tuples t_1 and t_2 such that $t_1[V^+] = t_2[V^+]$ and $t_1[A] \neq t_2[A]$ for every A from $(U - V^+)$. We shall show that u satisfies all dependencies d in D. There are two cases to be considered.

Case 1. A dependency d is an $F-$dependency of the form $X \to Y$. Assume X is a subset of V^+ (otherwise $t_1[X] \neq t_2[X]$ and so u satisfies $X \to Y$). Hence $V \to X$ can be derived from D. The dependencies $V \to X$ and $X \to Y$ imply by rules $P3$ and $P2$, $V \to Y$, that is Y is a subset of V^+. Thus $t_1[Y] = t_2[Y]$. Consequently, u satisfies $X \to Y$.

Case 2. A dependency d is an $O-$dependency of the form $F : (X_1 \to Y_1)\& \ldots \&(X_k \to Y_k) \to Z$. Assume that for each i, $1 \le i \le k$, there exists a set of attributes W_i such that the intersection of W_i and Y_i is empty whenever X_i is not empty, $W_iX_i \to Y_i$ can be derived from D, and u satisfies $W_iX_i \to Y_i$ (otherwise u satisfies $F : (X_1 \to Y_1)\& \ldots \&(X_k \to Y_k) \to Z$). Then, by rule $P4$, $W_1 \ldots W_k \to Z$ can be derived from D. Suppose that there exists i, $1 \le i \le k$, such that W_i is a subset of V^+. Then $t_1[W_1 \ldots W_k] \neq t_2[W_1 \ldots W_k]$. Hence u satisfies $W_1 \ldots W_k \to Z$. Now suppose that for each i, $1 \le i \le k$, W_i is a subset of V^+, that is , $V \to W_i$ can be derived from D. It follows from $P8$ that $V \to W_1 \ldots W_k$ can be derived from D. The dependencies $V \to W_1 \ldots W_k$ and $W_1 \ldots W_k \to Z$ imply $V \to Z$, that is, Z is a subset of V^+. Hence $t_1[Z] = t_2[Z]$. Thus u satisfies $W_1 \ldots W_k \to Z$. Finally, we can pick a function F such that

$$\pi_{W_1 \ldots W_k Z}(u) = F(\pi_{W_1 X_1 Y_1}(u), \ldots, \pi_{W_k X_k Y_k}(u)).$$

Thus u satisfies $F : (X_1 \to Y_1)\& \ldots \&(X_k \to Y_k) \to Z$.

To complete the proof we need to show that u does not satisfy $V \to T$. Assume u satisfies $V \to T$. Then, since $t_1[V] = t_2[V]$, $t_1[T] = t_2[T]$. Hence T is a subset of V^+. Thus $V \to T$ can be derived from D. This contradicts the assumption that $V \to T$ cannot be derived from D. **Q.E.D.**

Lemma 3. Let D be a set of $F-$, $O-$, and $E-$dependencies. If an $E-$dependency $E : S$ is logically implied from D, then $E : S$ can be derived from D.

Proof. The proof is by induction on the number of operators in an expression E.

Basis. If there are no operators in E, then $E : S$ is contained in D. Thus $E : S$ can be derived from D.

Induction step. There are three cases to be considered.

Case 1. An expression E is of the form $\pi_S(E_1)$ and the condition S is a subset of S_1 holds, where S_1 is the scheme of E_1. According to the definition of $E-$dependency, since $E : S$ is logically implied from D, $E_1 : S_1$ is logically implied from D. Hence by the inductive hypothesis $E_1 : S_1$ can be derived from D. $E_1 : S_1$, by rule $P5$, implies $E : S$. Thus $E : S$ can be derived from D.

Case 2. An expression E is of the form $E_1 * E_2$ and the condition $S_1 S_2 = S$ holds, where S_1 and S_2 are the schemes of E_1 and E_2 respectively. The condition

$$\pi_{S_1}(u) * \pi_{S_2}(u) = \pi_S(u)$$

holds only if $S_1 \cap S_2 \to S_1$ or $S_1 \cap S_2 \to S_2$ is logically implied from D (Rissanen,1977). Hence, by Lemma 2, $S_1 \cap S_2 \to S_1$ or $S_1 \cap S_2 \to S_2$ can be derived from D. According to the definition of $E-$dependency, since $E : S$ is logically implied from D, $E_1 : S_1$ and $E_2 : S_2$ are logically implied from D. Hence, by the inductive hypothesis, $E_1 : S_1$, $E_2 : S_2$ can be derived from D. $E_1 : S_1$, $E_2 : S_2$, and $S_1 \cap S_2 \to S_1$ or $S_1 \cap S_2 \to S_2$, by rule $P6$, imply $E : S$. Thus $E : S$ can be derived from D.

Case 3. An expression E is of the form $F(E_1, \ldots, E_k)$, where $F : (X_1 \to Y_1) \& \ldots \& (X_k \to Y_k) \to Z$ is an $O-$dependency from D, and $S_1, \ldots, S_k$ are the schemes of $E_1, \ldots, E_k$, respectively. According to the definition of $O-$dependency for each i, $1 \le i \le k$, $E_i : S_i$ is logically implied from D and there exists a set of attributes W_i such that $S_i = W_i X_i Y_i$, the intersection of W_i and Y_i is empty whenever X_i is not empty, $W_i X_i \to Y_i$ is logically implied from D, and $S = W_1 \ldots W_k Z$. For each i, $1 \le i \le k$, $E_i : S_i$ can be derived from D by the inductive hypothesis and $W_i X_i \to Y_i$ can be derived from D by the Lemma 2. $E_1 : S_1, \ldots, E_k : S_k$, $W_1 X_1 \to Y_1, \ldots, W_k X_k \to Y_k$, and $F : (X_1 \to Y_1) \& \ldots \& (X_k \to Y_k) \to Z$ by rule $P7$, imply $E : S$. Thus $E : S$ can be derived from D. **Q.E.D.**

Theorem. The inference rules $P1 - P7$ are sound and complete for the derivation of $E-$dependencies from a set of $F-$, $O-$ and $E-$dependencies.

Proof. Follows immediately from Lemmas 1,2, and 3. **Q.E.D.**

6. Conclusions.

1. The methods of structural synthesis of programs do not take into account the possibility of implementation of some initial dependencies by means of relations and do not consider the possibility of computation of relations by means of relational operations.

2. The methods of relation synthesis do not consider a possibility of computation of relations by means of arbitrary (unrelational) operations.

3. The proposed approach to the program synthesis is based on a computation model (a database scheme with computations), describing computations by means of both unrelational and relational operations. A sound and complete system of inference rules for the class of dependencies, used in such computation models, is given.

REFERENCES

Armstrong, W.W. (1974). Dependency structures of database relationships. In Proc. IFIP 74, North-Holland, Amsterdam, pp. 580-583.

Dikovskij, A.J., and M.J. Kanovich (1985). Computation models with separable subtasks. Proc. Acad. Sci. USSR, Techn. Cybernetics, 5, pp. 36-59 (in Russian).

Honeyman, P. (1980) Extension joins. In Proc. 6th Int. Conf. Very Large Data Bases. (Montreal,Canada, Oct., 1-3, 1980). ACM, New York, pp. 239-244.

Lozinskii, E.L. (1980). Construction of relations in relational databases. ACM Trans. Database Syst., 5, 2, pp. 208-224.

Rissanen, J. (1977). Independent components of relations. ACM Trans. Database Syst., 2, 4, pp. 317-325.

Tyugu, E.H., and M.J. Harf (1980). The algorithms of structural synthesis of programs. Programming, 4, pp. 3-13 (in Russian).

Negation as Failure
and Intuitionistic Three-Valued Logic

Jacqueline Vauzeilles
LIPN
Université Paris-Nord
Avenue Jean-Baptiste Clément
93430 Villetaneuse
FRANCE
e-mail: jv@lipn.univ-paris 13.fr

Abstract

In this paper, we present a three-valued intuitionistic version of Clark's completion, denoted by Comp3I(P) ; we prove the soundness of SLDNF-resolution with respect to Comp3I(P), and the completeness both for success and failure, as far as allowed programs are concerned. Then we compare our results to Kunen, Cerrito and Shepherdson's works which are based on classical three-valued logic, linear logic, and on a system of rules which are valid in both intuitionistic logic and three-valued logic.

1.Introduction

It is notorious that SLDNF-resolution is sound with respect to Clark's completion Comp(P) of a program P, but not complete ; the underlying logic is classical logic.

What about replacing classical logic with one of its subsystems : intuitionistic logic, three-valued logic, three-valued intuitionistic logic, linear logic ? numerous works using those systems have so far led to partial completeness results, either by restricting the class of programs and queries (hierarchical programs, strict programs, allowed programs (i.e. programs such that each variable occurring in the head or in a negative literal of the body of a clause also occurs in a positive literal of the body of the clause), ...)[12], [2], [3], [17] or by extending the notion of SLDNF-resolution [16].

In this paper, we present a three-valued intuitionistic version of Clark's completion (now on referred as Comp3I(P)) based on Girard's three-valued intuitionistic logic ; SLDNF-resolution is sound with respect to Comp3I(P) and complete both for success and failure, as far as allowed programs are concerned. The use of three-valued intuitionistic logic (see later), or three-valued classical logic ([11], [12] : we shall designate by Comp3(P) Kunen's version of Clark's completion), or of linear logic ([2], [3] : CompL(P) will denote the version of Clark's completion proposed by Cerrito) thus lead to the same results for the latter class of programs. If we do limit ourselves to propositional programs, it is easy to show that it is a consequence of the following remarks : first, a query Q (or its negation) is provable from Comp3(P) if and only if it (resp. its negation) is provable from Comp3I(P) (see [17]) ; secondly, if a query Q (or its negation) is provable from Comp3I(P) (or Comp3(P)), it (resp. its negation) is provable without using the contraction rule [17] (which is excluded from linear logic ; see [10]).

If we do not restrict ourselves to the class of allowed programs, the previous three versions of Clark's completion are no longer equivalent, since one can build a program P (see example 4) and a query Q such that : Q is provable from Comp3(P) or from CompL(P), whereas Q does not succeed under SLDNF-resolution with answer the identity, but Q is not provable from Comp3I(P). There nonetheless exist programs for which SLDNF-resolution is not complete, whatever version of Clark's completion one considers.

Kunen ([11], [12]) introduced, in order to handle three-valued equivalence (<->), a connective different from Kleene's three-valued connectives ; that prevented him from having a corresponding deductive system. Remark that the system Comp3K(P) defined in section 3.2. gives a deductive system for Kunen three-valued logic.

In [16], Shepherdson presented a deductive system containing rules valid both from the intuitionistic and the classical points of view ; however, it seems to be an *ad hoc* system, built in the mere purpose of yielding the soundness of SLDNF-resolution ; in particular, it requires two connectives for entailment : $\supset$ and ->. We show, in section 4.3, that the results of Shepherdson [16] still hold if one replaces his system with ours.

Our system was introduced by Girard in 1973 ([8], [9]) to prove Takeuti's conjecture (cut elimination for second order sequent calculus) and gives a semantics for cut-free sequent calculus. Girard's three-valued classical system is a Gentzen-like sequent calculus and three-valued intuitionistic logic is obtained by limiting to at most one the number of occurrences of formulae appearing on the right side of the sequents (thus following the same pattern as the bivalued case). The models are either topological models or three-valued Kripke models [8].

Girard's three-valued logic is not a modal logic : the operators ! and ? are not internal to the formulae (for example !A &?B is not a formula of Girard's logic) ; so, we cannot easily compare our results with those by Gabbay [6] and Balbiani [1] ; these works contain soundness results for SLDNF-resolution with respect to modal logical interpretations, and completeness as far as allowed programs are concerned.

Why intuitionistic three-valued logic

In the following examples, we denote by Comp(P) Clark's completion of program P, by $\models_K$ the classical consequence relation, by $\vdash_I$ the intuitionistic derivability and by $\models_3$ the three-valued consequence relation (if we interprete, in Comp(P), the equivalence "<->" as Kunen do, i.e. A <-> B is true iff A and B have the same truth value ; recall that we denote this version of Comp(P) by Comp3(P)) ; these examples show that, according to the case, three-valued or intuitionistic completion are more suitable to get the completeness of SLDNF-resolution.

• example 1 : let P be the program :

p :- q

p :- ¬q

q :- q

Comp(P) contains the formulae p <-> q v ¬q, q <-> q; hence, Comp(P) $\models_K$ p since q v ¬q is classically valid ; but the query <--p does not succeed under SLDNF-resolution ; remark that we have neither Comp(P) $\vdash_I$ p nor Comp3(P) $\models_3$ p, because q v¬q is no sound in intuitionistic three-valued logic (disjunction property) nor in three-valued logic (since q can be undetermined).

• example 2 : let P be the program :

p :- ¬p

Comp(P) contains the formula p <-> ¬p; hence, Comp(P) is classically and intuitionistically inconsistent; then, Comp(P) $\models_K$ p , Comp(P) $\models_K$ ¬p, Comp(P) $\vdash_I$ p and Comp(P) $\vdash_I$ ¬p but the query <--p does not succeed under SLDNF-resolution. On the other hand, p and ¬p are not three-valued consequences of Comp3(P).

• example 3 [15] : let P be the program :

p(a)

p(f(a)) :- p(f(a))

Comp(P) $\models_K \exists x$ (p(x) &¬p(f(x))) but the query <-- p(x) &¬p(f(x)) does not succeed under SLDNF-resolution ; remark that we have neither Comp(P) $\vdash_I \exists x$ (p(x) & ¬p(f(x))) (existence property of intuitionistic logic) nor Comp3(P) $\models_3 \exists x$ (p(x) & ¬p(f(x)))

• example 4 [11] : let P be the program :

p(x) :- isc(x)

p(x) :- nonc(x)

isc(c)

nonc(x) :- ¬ isc(x)

We have both Comp(P) $\models_K \forall x p(x)$ and Comp3(P) $\models_3 \forall x p(x)$ but the query <--p(x) does not succeed with answer the identity under SLDNF-resolution. Remark that $\forall x p(x)$ is not an intuitionistic consequence of Comp(P).

We can believe, in view of the above examples (since in example 2, three-valued logic is more adapted, and in example 4, intuitionistic logic is more suitable) that we will get closer to **SLDNF**-resolution, if we use three-valued intuitionistic logic.

2. Three-valued sequent calculus

Let L be a fixed first-order language. Following Girard [8], Fitting [5] and Kunen [11], we give Kleene's interpretation in a three-valued structure M of connectives and quantifiers $\neg$, &, v, $\supset$,$\forall$,$\exists$: a formula A takes the value t (resp. f) iff all possible ways of putting in t (resp. f) for the various occurrences of the third value u, lead to a value t (resp. f) in ordinary bivalued logic ; otherwise, A takes the value u. We denote this fact respectively by M(A) = t, M(A) = f, M(A) = u.

We define the **language 3L** as follows : the terms are the terms of L and the formulae are the formal expressions ?A and !A, where A is a formula of L. We shall represent an arbitrary formula of 3L by ξA, ηA, where ξ, η vary through the set $\{?,!\}$. We use the symbol ° as follows : $\xi° = !$ if $\xi = ?$, and $\xi° = ?$ if $\xi = !$.

A sequent in 3L is a formal expression $\Gamma \vdash \Delta$, where Γ and Δ are finite sequences (possibly empty) of formulae in **3L**.

The closed formula ξA of 3L[M] (that is the language obtained by adding to L a new constant $\underline{c}$ for each c $\in$ M) is satisfiable in the three-valued structure M iff $\xi = !$ and M(A) = t (we say that A is necessary in M) or $\xi = ?$ and M(A) $\neq$ f (we say that A is possible in M). We shall denote this fact by M $\models \xi A$ (and M $\not\models \xi A$ otherwise).

In the context of SLDNF, the operator ! is used to express *success* and the operator ? to express *not failure*.

The closed sequent $A_1,...,A_n \vdash B_1,...,B_m$ of 3L[M] is satisfiable in the three-valued structure M iff :
- if n $\neq$0 and m $\neq$0 : if M $\models$ A_1 and ... and M $\models$ A_n then M $\models$ B_1 or ... or M $\models$ B_m
- if n $\neq$0 and m = 0 then M $\not\models$ A_1 or ... or M $\not\models$ A_n
- if n = 0 and m $\neq$ 0 then M $\models$ B_1 or ... or M $\models$ B_m.
- if n = 0 and m = 0 the sequent $\vdash$ means absurdity.
—

Let S be a set of sequents of 3L. A **three-valued model** M of S is a three-valued structure where any closed instance of a sequent of S is satisfiable in S.

If all closed instances of a formula A of 3L are satisfiable in any three-valued model of a set S of sequents of 3L, we say that A is a (three-valued) logical consequence of S.

Girard has defined the sequent calculus 3LK (resp. 3LI) in the spirit of Gentzen's calculus **LK** (resp. **LI**) ; these systems enjoy completeness and soundness with respect to classical (resp. intuitionistic) three-valued models and can be extended to second order logic (see [8] and [9]). We present these systems in the appendix. The intuitionistic system is obtained, as in the binary case, by considering sequents with at most one formula on the right part of the sequent. For the definition of intuitionistic three-valued models (topological or Kripke three-valued models) see [8].

If a formula A or a sequent $\Gamma \vdash \Delta$ of 3L is provable in the theory S in 3LK (resp. 3LI), we write :
$S \vdash_{3K} \vdash A$ or $S \vdash_{3K} \Gamma \vdash \Delta$ (resp. $S \vdash_{3I} A$ or $S \vdash_{3I} \Gamma \vdash \Delta$).

3. Intuitionistic three-valued completion

Let P be a normal program, and let $C_1, \dots, C_m$ be the clauses with the predicate p occurring in their head ; we can describe C_i as the sequent $L_{i1}, \dots, L_{iq} \vdash p(t_{i1},\dots,t_{in})$ and let E_i be the formula $\exists y_1 \dots \exists y_k (x_1 = t_{i1} \& \dots \& x_n = t_{in} \& L_{i1} \& \dots \& L_{iq})$; the completed definition of p is the following set of sequents :
$!L_{i1}, \dots, !L_{iq} \vdash !p(t_{i1},\dots,t_{in})$ (pour $1 \leq i \leq m$)
$? p(x_1,\dots,x_n) \vdash ? E_1 \vee \dots \vee E_m$;

if the predicate p does not occur in the head of any program clause, then the completed definition of p is the set containing only the sequent $? p(x_1,\dots,x_n) \vdash$.

We define **CET** (Clark's equational theory) to be the (closure under substitution of the) set of sequents :

1) $\vdash !x = x$ for each variable x ;
2) $! t(x) = x \vdash$ for each term t(x) different from x in which x occurs ;
3) $! x_1 = y_1, \dots, ! x_n = y_n \vdash ! f(x_1,\dots,x_n) = f(y_1,\dots,y_n)$ for each function f ;
4) $! x_1 = y_1, \dots, ! x_n = y_n, \xi p(x_1,\dots,x_n) \vdash \xi p(y_1,\dots,y_n)$ for each predicate p ;
5) $! f(x_1,\dots,x_n) = f(y_1,\dots,y_n) \vdash ! x_i = y_i$ for each n-place function f and for each i ($1 \leq i \leq n$) ;
6) $! f(x_1,\dots,x_n) = g(y_1,\dots,y_m) \vdash$ for all pairs of distinct functions f and g ;
7) $? x = y \vdash ! x = y$ for all pairs of variables.
Axioms 1-6 are the usual ones (particularly, we can prove all usual properties of equality ; for example, we have CET $\vdash_{3I} ! x = y \vdash ! y = x,$) ; axioms 7 say that the equality relation "=" is bivalued.

Let P be a normal program. The **intuitionistic 3-valued completion** of P, denoted Comp3I(P) is the union of the completed definitions for each predicate p occurring in P, and of **CET**. Comp3I(P) is consistent, because as we shall prove later, it is a three-valued consequence of Clark's three-valued completion Comp3(P) used by Kunen ([11], [12]) .

• example :

Let P be the following program : a :- b ; a :- ¬c .

Then Comp3I(P) = { !b ⊢ !a ; !¬c ⊢ !a ; ?a ⊢ ? (b ∨ ¬c) } with the following very straightforward meaning : " if b succeeds then a succeeds ; if c fails then a succeeds ; if a does not fail then b does not fail or c does not succeed ".

Lemma

a) If $p(s_1,...,s_n)$ and $p(t_1,...,t_n)$ are not unifiable then

$$CET \vdash_{3I} ?s_1 = t_1 ,...,? s_n = t_n \vdash$$

b) If $p(s_1,...,s_n)$ and $p(t_1,...,t_n)$ are unifiable with mgu $\theta=(x_1/r_1,...,x_k/r_k)$

then $\quad CET \vdash_{3I} ! s_1 = t_1 ,...,! s_n = t_n \vdash ! x_i = r_i \quad$ (for $1 \leq i \leq k$) and

$\qquad CET \vdash_{3I} ! x_1 = r_1 ,...,! x_k = r_k \vdash ! s_i = t_i \quad$ (for $1 \leq i \leq n$)

proof : an adaptation of the proof of Lloyd [13] (lemma 15.1).

3.1.Soundness of SLDNF-resolution with respect to Comp3I(P)

Theorem [17].

SLDNF-resolution is sound with respect to Comp3I(P) in intuitionistic three-valued logic, i.e. suppose that Q is a conjunction of literals :

if Q succeeds with answer θ then Comp3I(P) $\vdash_{3I}$ ⊢ ! Qθ

if Q fails then Comp3I(P) $\vdash_{3I}$ ⊢ ! ¬ ∃ Q

proof :

we present the proof (first published in [17])of this theorem because we complete it, in a larger context in section 4.3.

basis

i) Q is the positive literal M , and M matches with a clause ⊢ A , i.e. there exists a mgu θ of A and M , then since ⊢ !A is a sequent of Comp3I(P), then Comp3I(P) $\vdash_{3I}$ ⊢ ! Mθ

ii) if $M = p(s_1,...,s_n)$ is the chosen positive literal of Q :

- if p does not appear in the head of any clause of program P :

then the completed definition of p is $\quad ? p(x_1,...,x_n) \vdash \quad$; thus obviously Comp3I(P) $\vdash_{3I}$? M ⊢

and Comp3I(P) $\vdash_{3I}$ ⊢ ! ¬ ∃ Q

- if M does not unify with the head of any clause of P, then suppose that the 3-valued translations of the clauses with head p are : $! L_{i1}, ... , ! L_{iq_i} \vdash \ ! p(t_{i1},...,t_{in})$; if the completed definition of p contains the sequent : $? p(x_1,...,x_n) \vdash ? E_1 v ... v E_m$ then, by the above lemma,

CET $\vdash_{3I} ?(s_1 = t_{i1} \& ... \& s_n = t_{in}) \vdash$ for each i $(1 \leq i \leq m)$;

hence, CET $\vdash_{3I} ?s_1 = t_{i1} \& ... \& s_n = t_{in} \& L_{i1} \& ... \& L_{iq} \vdash$ for each i $(1 \leq i \leq m)$; and hence, using the completed definition of p : Comp3I(P) $\vdash_{3I} ? p(s_1,...,s_n) \vdash$, and then Comp3I(P) $\vdash_{3I} \vdash ! \neg \exists Q$;

inductive step

- if M_j is the chosen positive literal of Q which matches one or more clauses of P ; $M_j = p(s_1,...,s_n)$:

i) let $Q_1,...,Q_p$ be the resulting derived queries and suppose that we have a success tree for one Q_i ; then if $Q = M_1 \& ... \& M_r$, if ρ is a mgu of $p(s_1,...,s_n)$ and of $p(t_{i1},...,t_{in})$ and if $Q_i = M_1 \& ... \& M_{j-1} \& L_{i1} \& ... \& L_{iq_i} \& M_{j+1} \& ... \& M_r$, by induction hypothesis, Comp3I(P)$\vdash_{3I} \vdash ! Q_i \sigma$ (if Q_i succeeds with answer σ) ; then, using the clause $! L_{i1}, ... , ! L_{iq_i} \vdash ! p(t_{i1},...,t_{in})$, we see that Comp3I(P) $\vdash_{3I} \vdash ! Q\theta$ (with $\theta = \rho \circ \sigma$);

ii) let $Q = M_1 \& ... \& M_r$ and $Q_1,...,Q_p$ be the resulting derived queries and suppose that we have failure trees for all Q_i ; then, by induction hypothesis,

Comp3I(P) $\vdash_{3I} \vdash ! \neg \exists Q_i$ (for each i $(1 \leq i \leq p)$),

and $Q_i = (M_1 \& ... \& M_{j-1} \& L_{i1} \& ... \& L_{iq_i} \& M_{j+1} \& ... \& M_r)\rho$

(if ρ is a mgu of $p(s_1,...,s_n)$ and of $p(t_{i1},...,t_{in})$); then using the above lemma and axioms 7 of CET,

CET$\vdash_{3I} ?s_1 = t_{i1} \& ... \& s_n = t_{in} \vdash ? x_1 = r_1 \& ... \& x_k = r_k$;

then, we prove that :

Comp3I(P)$\vdash_{3I} ?\exists M_1 \& ... \& M_{j1} \& s_1 = t_{i1} \& ... \& s_n = t_{in} \& L_{i1} \& ... \& L_{iq_i} \& M_{j+1} \& ... \& M_r$

$\qquad\qquad \vdash ? \exists (M_1 \& ... \& M_{j-1} \& L_{i1} \& ... \& L_{iq_i} \& M_{j+1} \& ... \& M_r)\rho$;

therefore, Comp3I(P) $\vdash_{3I} ! \neg \exists Q$

- if M_j is the chosen ground negative literal $\neg A$, and if A has a failure tree and Q' a success tree, Q' being the query obtained from Q by deleting $\neg A$, then by induction hypothesis,

Comp3I(P) $\vdash_{3I} \vdash ! Q'\theta$ and Comp3I(P) $\vdash_{3I} \vdash ! \neg A$ (since A is ground) ; therefore, Comp3I(P) $\vdash_{3I} \vdash ! Q\theta$

-if M_j is the chosen ground literal $\neg A$, and if A has a failure tree and Q' (defined as above) has a failure tree, then by induction hypothesis,

Comp3I(P) $\vdash_{3I} \vdash ! \neg \exists Q'$, then Comp3I(P) $\vdash_{3I} \vdash ! \neg \exists Q$

- if M_j is the chosen ground negative literal $\neg A$, and if A has a success tree, then by induction hypothesis, Comp3I(P) $\vdash_{3I} \vdash !A$ and therefore, Comp3I(P) $\vdash_{3I} \vdash ! \neg \exists Q$.

3.2. Completeness of SLDNF-resolution for Comp3I(P) (for allowed programs)

Let P be the following program :
p(x) :- ¬q(x)
r(a)

the query <--p(a) succeeds on P under SLDNF-resolution but the query <--p(x) (that is, in fact, the query ∃x p(x) ?) does not succeed, it flounders because ¬q(x) is a non-ground literal. To prevent such problems, we only study programs and queries which never flounder, that is allowed programs and allowed queries.

Definitions

Let P be a program and <--Q be a query ; we say that Q **flounders** iff there exists a SLDNF-derivation of Q with a derived query only containing non-ground negative literals.

A program P is **allowed** if and only if every variable occurring in the head or in a negative literal of the body of a clause of P, also occurs in a positive literal of the body of this clause.
A query <--Q is **allowed** if and only if every variable occurring in a negative literal of Q, also occurs in a positive literal of Q.

Remark that this definition is very restrictive since we cannot have in an allowed program P, clauses such as : equal(x,x).

Theorem [14]

Let P be an allowed program and <--Q an allowed query ; then
- <--Q does not flounder;
- if <-- Q succeeds with answer θ, then every variable of Q is ground in the substitution θ.

Theorem completeness for allowed programs

Let P be an allowed program and <--Q an allowed query ; then,
- if Comp3I(P) ⊢₃ᵢ ⊢ !Qθ then <--Q succeeds under SLDNF with answer θ ;
- if Comp3I(P) ⊢₃ᵢ ⊢ ¬∃ Q then <--Q fails under SLDNF

proof :

Kunen's completed definition in Comp3(P) of a predicate p being : p(x₁,...,xₙ)<->E₁v...v Eₘ, it is easy to see that a three-valued structure M is a model (Kunen sense) of p(x₁,...,xₙ) <-> E₁ v ... v Eₘ, iff M is a three-valued model (in the sense of section 2) of the following sequents :

! $p(x_1,...,x_n)$ |- ! E_1 v ... v E_m

?$p(x_1,...,x_n)$ |- ? E_1 v ... v E_m

! $(E_1$ v ... v $E_m)$ |- ! $p(x_1,...,x_n)$

?$(E_1$ v ... v $E_m)$ |- ? $p(x_1,...,x_n)$

We denote by Comp3K(P) the system obtained by adding to **CET** the above sequents for every predicate p occurring in the head of a clause of **P**, and the sequents !$p(x_1,...,x_n)$ |- and ?$p(x_1,...,x_n)$ |- if p does not occur in any head of a clause of **P**.

Then, by the completeness of Girard's system 3LK (section 2), a formula F of L is a Kunen three-valued consequence of Comp3(P) iff |- !F is provable in 3LK from Comp3K(P).

Since Comp3I(P) is a subsystem of Comp3K(P), we conclude that, if Comp3I(P) $\vdash_{3I}$ |- !F, then Comp3K(P) $\vdash_{3K}$ |- !F ; thus, our completeness result for allowed programs and queries is a consequence of Kunen's completeness result for allowed programs and queries.

If we prefer a direct proof of this theorem we can adapt the proof established by Cerrito ([2], [3]) for the linear completion; we set the following modifications to the definitions of [3] :if $p(t_1,...,t_n)$ is a ground atomic formula,! $p(t_1,...,t_n)$ is realizable with respect to P and E (E being a reference set : see [3]) iff $p(t_1,...,t_n)$ succeeds on P (modulo E) ; ?$p(t_1,...,t_n)$ is realizable with respect to P and E iff $p(t_1,...,t_n)$ does not fail on P (modulo E) ;

if F is a closed formula : $\xi \neg$F is realizable with respect to P and E iff $\xi°$F is not realizable with respect to P and E.

Others definitions of realizability for formulae (conjunction, disjunction, quantification, formulae with variables) are similar to the ones of [3].

A sequent $A_1,...,A_n$ |- B is realizable with respect to P and E, iff for every substitution θ of closed terms of the extended language $\mathcal{BT}$ang(P) (see [3]), if $A_1\theta$ and... and $A_n\theta$ are realizable with respect to P and to E, then Bθ is realizable with respect to P and to E.

The rest of the proof is very similar to the proof for linear completion.

Remark that in intuitionistic three-valued logic ξA and $\xi \neg\neg$A are not equivalent ; also, $\xi \neg$(A $\&$ B) is not equivalent to $\xi \neg$A v $\neg$B ; but we feel that it doesn't matter because a query is always a conjunction of literals and so, $\neg\neg$A or $\neg$(A $\&$B) are never goals.

4. Comparison with related works

4.1. Comparison with Kunen results

Kunen adopts a model-theoretical approach [11], [12] ; he does not define a deductive system for his three-valued logic with the connective <-> ; we have proved, in section 3.2. that our completeness result for allowed programs and queries is a consequence of Kunen's completeness result for allowed programs and queries, but Kunen's soundness result is a consequence of ours.

If we consider the program P in example 4, we see that the sequent ⊢ !p(x) is not provable from Comp3I(P) in intuitionistic three-valued logic, and also p(x) does not succeed with answer the identity under SLDNF-resolution ; ∀xp(x) is a (Kunen) three-valued consequence of Comp3(P), but the sequent ⊢ !∀x p(x) is not provable in 3LI from Comp3I(P). We have these results because equality is decidable in classical three-valued logic : the sequent ⊢ ! x = a ∨ x ≠ a is provable from Comp3K(P), and then, Comp3K(P) ⊢₃ₖ ⊢ ∀x p(x) ; on the other hand, equality is not decidable in intuitionistic logic ; the sequents ⊢ ! x = a ∨ x ≠ a and ⊢ ! ∀x p(x) are not provable from Comp3I(P).

4.2. Comparison with Cerrito's results

In [2], [3], Cerrito defines a completion in linear logic : CompL(P) ; she proves the soundness of SLDNF-resolution with respect to CompL(P), and completeness for allowed programs and queries ; she compares these results with Kunen's, and shows that the completeness for CompL(P) (for allowed programs) is a consequence of the completeness for Comp3(P), and that the soundness for Comp3(P) is a consequence of the soundness for CompL(P).

The equality relation, used in this work, is decidable, since in order to prove the soundness, she uses the fact that the sequent ⊢ x = y ⊕ x ≠ y is provable from CompL(P) ; then, in example 4, the sequent ⊢∀x p(x) is provable from CompL(P).

Therefore, we can conclude that our soundness result is not a consequence of his soundness theorem.

4.3. Comparison with Shepherdson results

In [16], Shepherdson uses Kleene's connectives &,∨,¬,⊃,∃,∀ ; he also uses the bivalued connective '->' in the following way : "if p is true then q is true" ; he defines many rules to obtain the soundness of an extension of SLDNF-resolution (SLDNFS-resolution) with respect to Comp3(P) : a non-ground negative literal ¬A can be selected, only to construct success branches, if there exists a substitution θ such that the query <-- Aθ fails (¬A succeeds with the answer θ) ; the soundness theorem can be extended to this notion of SLDNFS-resolution :

Theorem

SLDNFS-resolution is sound with respect to Comp3I(P) in intuitionistic three-valued logic, i.e. suppose that Q is a conjunction of literals :
• if Q succeeds under SLDNFS with answer θ, then Comp3I(P)⊢₃ᵢ⊢ ! Qθ
• if Q fails under SLDNFS then Comp3I(P) ⊢₃ᵢ ⊢ ! ¬ ∃ Q

proof :

we complete the proof of the theorem 3.1. for SLDNF-resolution, adding the following case :

-if M_j is the chosen negative literal $\neg$ A, and if Aρ has a failure tree and Q'ρ succeeds with answer σ (with $\theta = \rho$ o σ), Q' being the query obtained from Q by deleting $\neg$ A , then by induction hypothesis, Comp3I(P) $\vdash_{3I}$ $\vdash$! (Q'ρ)σ and Comp3I(P) $\vdash_{3I}$ $\vdash$! $\neg \exists$ A ; hence, Comp3I(P) $\vdash_{3I}$ $\vdash$! ($\neg$A ρ)σ and Comp3I(P) $\vdash_{3I}$ $\vdash$! Qθ

But, we do not get the completeness, even for this extension of **SLDNF**-resolution as the following example (given by Shepherdson) shows :

• example 5 :
p :- $\neg$q(x)
q(x)

Comp3I(P) $\vdash_{3I}$ $\vdash$!$\neg$p but <-- p does not fail under **SLDNF**-resolution, nor under **SLDNFS**-resolution.

Then Shepherdson suggests a further extension of **SLDNFS**-resolution, using the NF rule in the form :
If A succeeds with answer the identity then $\neg$A fails

The analogue of the above theorem remains true for this version of **SLDNFS**-resolution, which we call by **SLDNFS***-resolution :

proof :

we add the following case to the above proof :
- if M_j is the chosen negative literal $\neg$ A, and if A succeeds with answer the identity, then by induction hypothesis, Comp3I(P) $\vdash_{3I}$ $\vdash$!A and therefore, Comp3I(P) $\vdash_{3I}$ $\vdash$! $\neg \exists$ Q.

Then, in example 5, Comp3I(P) $\vdash_{3I}$ $\vdash$!$\neg$p and <-- p fails under **SLDNFS***-resolution.

But, like Shepherdson, we do not have the completeness of **SLDNFS***-resolution with respect to Comp3I(P) in intuitionistic three-valued logic :

• example 6 : let P be the program :
s :- $\neg$q(x)
q(x) :- $\neg$r(x)
r(a)
q(a)

Comp3I(P) $\vdash_{3\mathcal{I}}$ ⊢ ! ¬s but s does not fail under SLDNFS*-resolution ; indeed, CET $\vdash_{3\mathcal{I}}$?x = a ⊢ ! x = a (axiom 7, that is the equality relation is bivalued) and then, Comp3I(P) $\vdash_{3\mathcal{I}}$?¬q(x)⊢ and hence, Comp3I(P) $\vdash_{3\mathcal{I}}$ ⊢ ! ¬s.

We can prove the other results of Shepherdson for $T_\omega(P)$, replacing his consequence relation $\vdash_{3\mathcal{I}}$ with our derivability relation $\vdash_{3\mathcal{I}}$.

5.Conclusion

Although we have only proved the completeness for allowed programs and queries, we think that our version of Clark's completion is satisfactory enough : the three-valued logical system we use was built in a very natural way from Gentzen's sequent calculus ; we also feel that it enables a better understanding of the different connectives introduced by Shepherdson and Kunen.

A direction for further work will be to define a class of programs larger than the class of allowed programs, for which we would get the completeness of SLDNF-resolution with respect to Comp3I(P).

Aknowledgements

We are deeply indebted to the anonymous referees for their careful reading of our paper and their numerous comments and indications.

Appendix

Classical and intuitionistic three valued sequent calculus

1. Classical three valued sequent calculus

We suppose that the language 3L is defined as in section 2. We define the formal system 3LK (we use notations of section 2 and ξ, η vary through !,?) :

- **axioms :** $\quad \xi A \vdash \xi A \quad$ and
  $\quad\quad\quad\quad !A \vdash ?A \quad$ for each atomic formula A

- **logical rules :**

  - conjunction

$$\frac{\Gamma \vdash \Delta, \xi A \quad\quad \Lambda \vdash \Pi, \xi B}{\Gamma, \Lambda \vdash \Delta, \Pi, \xi\, A\, \&\, B}\, r\&$$

$$\frac{\xi A, \Gamma \vdash \Delta}{\xi\, A\, \&\, B, \Gamma \vdash \Delta}\, l1\&\quad\quad\quad\quad \frac{\xi B, \Gamma \vdash \Delta}{\xi\, A\, \&\, B, \Gamma \vdash \Delta}\, l2\&$$

- disjunction

$$\frac{\Gamma \vdash \Delta, \xi A}{\Gamma \vdash \Delta, \xi A \vee B} \; r1v \qquad\qquad \frac{\Gamma \vdash \Delta, \xi B}{\Gamma \vdash \Delta, \xi A \vee B} \; r2v$$

$$\frac{\xi A, \Gamma \vdash \Delta \qquad \xi B, \Lambda \vdash \Pi}{\xi A \vee B, \Gamma, \Lambda \vdash \Delta, \Pi} \; lv$$

- negation

$$\frac{\Gamma \vdash \Delta, \xi A}{\xi° \neg A, \Gamma \vdash \Delta} \; l\neg \qquad\qquad \frac{\xi A, \Gamma \vdash \Delta}{\Gamma \vdash \Delta, \xi° \neg A} \; r\neg$$

- implication

$$\frac{\Gamma \vdash \Delta, \xi° A \qquad \xi B, \Lambda \vdash \Pi}{\xi A \supset B, \Gamma, \Lambda \vdash \Delta, \Pi} \; l\supset \qquad\qquad \frac{\xi° A, \Gamma \vdash \Delta, \xi B}{\Gamma \vdash \Delta, \xi A \supset B} \; r\supset$$

- for all

$$\frac{\xi A(t), \Gamma \vdash \Delta}{\xi \forall x A(x), \Gamma \vdash \Delta} \; l\forall \; (**) \qquad\qquad \frac{\Gamma \vdash \Delta, \xi A(x)}{\Gamma \vdash \Delta, \xi \forall x A(x)} \; r\forall \; (*)$$

- there is

$$\frac{\xi A(x), \Gamma \vdash \Delta}{\xi \exists x A(x), \Gamma \vdash \Delta} \; l\exists \; (*) \qquad\qquad \frac{\Gamma \vdash \Delta, \xi A(t)}{\Gamma \vdash \Delta, \xi \exists x A(x)} \; r\exists \; (**)$$

(*) we have the following restriction on variables : x not free in Γ, Δ.
(**) t is an arbitrary term of **L**.

• **structural rules**

- weakening

$$\frac{\Gamma \vdash \Delta}{\xi A, \Gamma \vdash \Delta} \; lW \qquad\qquad \frac{\Gamma \vdash \Delta}{\Gamma \vdash \Delta, \xi A} \; rW$$

- contraction

$$\frac{\xi A, \xi A, \Gamma \vdash \Delta}{\xi A, \Gamma \vdash \Delta} \; lC \qquad\qquad \frac{\Gamma \vdash \Delta, \xi A, \xi A}{\Gamma \vdash \Delta, \xi A} \; rC$$

- exchange

$$\frac{\Pi, \xi A, \xi B, \Gamma \vdash \Delta}{\Pi, \xi B, \xi A, \Gamma \vdash \Delta} \; lE \qquad\qquad \frac{\Gamma \vdash \Delta, \xi A, \xi B, \Pi}{\Gamma \vdash \Delta, \xi B, \xi A, \Pi} \; rE$$

cut

$$\frac{\Gamma \vdash \Delta, \xi A \qquad \xi A, \Lambda \vdash \Pi}{\Gamma, \Lambda \vdash \Delta, \Pi} Cut$$

2. Intuitionistic three valued sequent calculus

We define **3LI** from **3LK** following the same pattern as in the bivalued case : intuitionistic three-valued sequents contain at most one formula on the right side ; we define the same rules (except for lv) for **3LI** as **3LK**, provided that the involved sequents are intuitionistic three-valued sequents. The lv-rule is the following :

$$\frac{\xi A, \Gamma \vdash \Delta \qquad \xi B, \Lambda \vdash \Delta}{\xi A \vee B, \Gamma, \Lambda \vdash \Delta} lv$$

(with Δ containing at most one formula).

Bibliography

1. Balbiani, P. (1991) Une caractérisation modale de la sémantique des programmes logiques avec négation, *Thèse de Docorat*, Université Paul Sabatier, Toulouse.

2. Cerrito, S. (1990) A Linear Semantics for Allowed Logic Programs, *Proceedings of Fifth Annual IEEE Symposium on Logic in Computer Science*, IEEE Computer Society Press, 1990, pp. 219-227.

3. Cerrito, S. (1990) Contribution de la logique linéaire au problème de la négation par l'échec, *Thèse de Doctorat*, Université de Paris-Sud Centre d'Orsay.

4. Clark,K.L. (1978) Negation as failure, in *Logic and Databases* (H. Gallaire and J.Minker, Eds.), Plenum Press, New York, pp. 293-322.

5. Fitting, M. (1985) A Kripke-Kleene Semantics for General Logic programs, *Logic Programming* 2, pp. 295-312.

6. Gabbay, D.M. (1990) Modal Provability, Foundations for Negation by Failure, *Extensions of Logic Programming*, (P.Schroder Heister Ed.),Lecture Notes in Artificial Intelligence, Springer.

7. Gentzen, G. (1969) Collected works,ed. Szabo, North Holland, Amsterdam.

8. Girard, J.Y. (1976) Three-valued logic and cut-elimination : The actual meaning of Takeuti's conjecture, *Dissertationes Mathematicae*, Warszawa.

9. Girard, J.Y. (1987) Proof theory and logical complexity, vol 1. Napoli : Bibliopolis.

10. Girard, J.Y. (1987) Linear Logic, *Theoretical Computer Science* , vol. 50.

11. Kunen, K (1987) Negation in logic programming *J. Logic programming* 4, pp. 289-308.

12 Kunen, K (1989) Signed Data Dependencies *J. Logic programming* 7, 3, pp.231-245.

13. Lloyd, J. W. (1987) *Foundations of Logic programming*, Second Edition, Springer Verlag, Berlin.

14. Shepherdson, J.C. (1985) Negation as failure II, *J. Logic programming* 3, pp. 185-202.

15. Shepherdson, J.C. (1987) Negation in Logic programming, in : J.Minker (ed.), *Foundations of Deductive Databases and Logic Programming,* Morgan Kauffman, Los Altos, 1988, pp. 19-88.

16. Shepherdson, J.C. (1989) A Sound and Complete Semantics for a version of Negation as Failure, *Theoretical Computer Science* ,vol. 65, n° 3, pp. 343-371.

17. Vauzeilles, J., **Strauss,** A.(1991) Intuitionistic Three-Valued Logic and Logic Programming, Theoretical Informatics and Applications (R.A.I.R.O), n° 6 (to appear).

Part III
Appendix

Symbolic Computation and Artificial Intelligence

Alfonso Miola

Dipartimento di Informatica e Sistemistica
Università degli Studi di Roma "La Sapienza"
Via Salaria 113, 00198 Roma, Italy.

Abstract

The paper presents an overview of the research achievements on issues of common interest for Symbolic Computation and Artificial Intelligence. Common methods and techniques of non-numerical information processing and of automated problem solving are underlined together with specific applications. A qualitative analysis of the symbolic computation systems currently available is presented in view of the design and implementation of a new system. This system allows both formal algebraic and analytical computations and automated deduction to prove properties of the computation.

1. Introduction.

Symbolic Computation refers to the algorithmic solutions of problems dealing with symbolic objects, suitably represented in a computer. Mathematical objects (e.g. numbers, relations, functions) and their representations can be considered as symbolic objects and computing with them requires algebraic, analytical and logical methods and techniques.

In Symbolic Computation three major aspects have been considered:
- mathematical foundations, design and analysis of (sequential and parallel) algorithms;
- design and implementation of software systems;
- applications of advanced problem solving techniques in science and engineering.

The pioneering work on differentiation by Kahrimanian [Kah53] and by Nolan [Nol53] in 1953 is often considered the first effort in the development of software for Symbolic Computation.

Research partially supported by MURST projects "Calcolo Algebrico", "Metodi e strumenti per l'elaborazione non numerica"; and by CNR project "Sistemi Informatici e Calcolo Parallelo".

In 1960, the introduction of Lisp stimulated and facilitated a wide programming activity in the area of Symbolic Computation. Indeed a program for formal differentiation is the second most commonly written Lisp program (the first obviously being the factorial function).

In the following years, J. Slagle studied how to invert the process of differentiation. His work [Sla63] represents a first example of a special purpose symbolic computation system for indefinite integration. Then, the results of Slagle's work have been assumed by J. Moses as a basic achievement in symbolic integration to design and implement one of the most relevant component of the MACSYMA system developed at Project MAC of MIT.

The design and development of general purpose symbolic computation systems started in the sixties. The first generation of such systems, including ALPAK, FORMAC, MATHLAB and PM, provided a large number of different computing methods (e.g. polynomial and rational function algebra, polynomial factorisation, and Laplace transformation). MATHLAB also included facilities for interactive use and bidimensional output. The seventies presented the development of a second generation of very significant and efficient systems, such as ALTRAN, CAMAL, MACSYMA, muMATH, REDUCE, SAC-1, and SCRATCHPAD I. Many of these systems, which are interactive, are still in use today.

The development of symbolic computation systems in the eighties has mainly followed two directions: the enlargement and enrichment of systems already available, as MACSYMA and REDUCE, and the design of new systems, as DERIVE, MAPLE and MATHEMATICA, expecially designed also to run on powerful workstations with graphic facilities. In this same period, the SCRATCHPAD II project started at IBM Research Center in Yorktown Heigths. This project now represents the most innovative achievement in symbolic computation system design and it has brought to a new generation system, particularly suitable to treat applications with a strong algebraic characterization. A detailed description of these two generations of systems can be found in [Mio91].

Several practical results were also obtained for the algorithms: i.e. solution of differential equations and of systems of differential equations in closed form. Algorithms have been defined for the ordinary, linear and non linear, cases. They are available in many systems and have allowed users to solve significant problems in real applications. It is worth mentioning a particular work in Celestial Mechanics as an example of the power of the methods and of the software systems available today. Deprit et al. [Dep70] have discovered, by using Symbolic Computation methods and systems, that the research effort completed by Delunay in 1867 after 20 years of work to manually derive a 2000-page symbolic expression of the moon orbit only contained one incorrect coefficient.

A first observation that emerges from this short description of the historical evolution of Symbolic Computation, is that most of the scientific results in the early years have originated from research achievements in Artificial Intelligence. Actually it is worth noticing how the major accomplishments in Symbolic Computation were obtained in scientific environments where research in Artificial Intelligence was very active and valuable. The conception and the development of the MACSYMA system at MIT is the best example of this cooperation between Symbolic Computation and Artificial Intelligence.

Clearly, Artificial Intelligence is here considered in its "weak thesis", i.e. as the research area concerned with the design and implementation of systems to solve problems which require intelligence. According to this interpretation, the MACSYMA system is an *expert system* in symbolic mathematics.

For several years the two fields of Symbolic Computation and of Artificial Intelligence have had a parallel development with clear overlapping. Actually, the overlap and differences have varied over time. To some extent, in its early stage, the area of Symbolic Computation was considered as part of Artificial Intelligence. Later on, Symbolic Computation has become a specific research area. The same as for other areas originated as parts of Artificial Intelligence (e.g. computer vision or logic programming).

Thenafter, in the seventies, the study of the algebraic basis of Symbolic Computation produced various classes of algebraic algorithms (e.g. polynomial algebra and polynomial factorisation) following formal algebraic specifications and the connection to the evolution of Artificial Intelligence was less evident. The history of the algorithms for indefinite and definite integration best underlines the step from the original Artificial Intelligence approach of the heuristic procedure by Slagle (i.e. assuming the problem of formal integration as a game to be played) to the algebraic algorithm by R.H. Risch [Ris69].

Both the methodological and technological accomplishments of Artificial Intelligence research in the last decade have now brought to a renewed cooperation between the two fields. In particular, many of the basic methods and software tools identified and developed in both Symbolic Computation and Artificial Intelligence are now interpreted in some uniform framework as to represent the methodology and the technology common to the two fields.

In the next sections the commonalities of Symbolic Computation and Artificial Intelligence will be briefly analyzed together with the proposal of an innovative methodological approach to the design and the implementation of Symbolic Computation systems.

2. Issues of Common Interest.

The exchanges between Symbolic Computation and Artificial Intelligence have been meaningful over time and the two fields concour to the solution of relevant applications (e.g. motion planning in robotics, computer aided design, intelligent tutoring systems for mathematics). They have common theoretical origins from mathematics and theory of computation and several methods and techniques are used interchangeably. In particular, Symbolic Computation provides formal methods for symbolic processing and for pure computer mathematics, while Artificial Intelligence offers heuristic methods, pattern matching techniques, knowledge representation methods based on logical formalisms, and the technology for building expert systems.

The kernel of the intersection of the two areas of Symbolic Computation and of Artificial Intelligence can be easily identified as *non-numerical information processing*, with particular reference to programming languages and software environments used as common tools in the two areas.

This intersection is also characterized by the topic of automated problem solving, together with methods and techniques for automated deduction as other basic issues of common interest for the two areas. Furthermore, the interpretation of automated deduction problems as algebraic problems, as proposed by several authors [JSC87], once again stresses the overlapping of Symbolic Computation and Artificial Intelligence.

2.1. Languages.

There are specific features that are important in a language to facilitate the writing of Symbolic Computation and Artificial Intelligence programs. They derive from the common need to guarantee efficient non-numerical information processing. Actually, both languages for Symbolic Computation and for Artificial Intelligence need features for:

• symbol manipulation, with particular reference to list manipulation, since lists are such a widely used data structure in almost all programs in the two areas;

• flexible control structures, to allow both recursion and parallel decomposition of the system;

• high level of abstraction for data and procedures, to guarantee a high software quality, with high level of correctness and software reusability;

• modularity, to facilitate the decomposition of a large system in small parts so that it is easier to make changes to a part of the system without modifying the others;

• pattern matching facilities, to identify data in large knowledge bases and to determine control forms for the execution of production systems;

• automated deduction mechanisms, to obtain logical deductions on the basis of a stored database of assertions;

• uniform processing of procedures and data, to guarantee the ability to interchangeably process data as programs and viceversa;

• late binding, to allow the late definition of the size of a data structure or of the type of an object;

• incremental design, to offer adequate support to the entire process of program development;

• interactivity, both for program development and for maximally effective use of the final software system.

No existing language provides all these features. However the successive improvements in language design and implementation have brought to Lisp and Prolog systems which have represented the basic programming tools in Symbolic Computation and Artificial Intelligence for years.

2.2. Automated Deduction.

The strict relationship between logical deduction and (symbolic) computation revealed in the last two decades is one of the most outstanding scientific achievements of this century.

Foundamental steps in many symbolic computations are solution of (polynomial) equations and of systems of (polynomial) equations. At the same time, automated theorem proving represents a relevant part of Artificial Intelligence as to assume a role similar to that played by calculus in natural sciences; every expert system is based on a suitable deduction machinery.

The discovery of the similarity between equation solving and logical deduction (within particular theories) further stresses the overlap of Symbolic Computation and Artificial Intelligence. In particular, the methods of pattern matching, unification and completion can be viewed as common computational tools in both Symbolic Computation and Artificial Intelligence, if suitably interpreted and formulated.

2.2.1. Unification.

Pattern matching and unification play a very important role in Symbolic Computation. For instance, one can again refer to the formal integration problem where a major need is the verification of the possible matching between the integrand and a given pattern, as in Slagle's approach. Moreover, MACSYMA, as most other systems, extensively uses the pattern matching procedure. At the same time pattern matching and unification represent basic problem solving procedural steps also in Artificial Intelligence and specifically in automated deduction.

Actually, unification and, in general, equation solving have been studied since mathematics exists and, in the last decade, it has been recognized as a basic problem in Computer Science and then in Symbolic Computation and Artificial Intelligence.

The problem of unification can be formulated as follows. Let s and t be two words in a language L with variables, then, for a given binary relation $\sim$ in L find a substitution σ such that $\sigma s \sim \sigma t$.

The relation $\sim$ can be specified by a set E of equational axioms; if L is the language of first order terms then the unification of s and t in E turns into solving the equation $s = t$ in the variety defined by E. Then, particular selections of axioms for the set E define several unification problems as equation solving problems in different algebraic structures.

Various applications of the unification methods have been made in several fields of both Symbolic Computation and Artificial Intelligence (e.g. Databases and Information Retrieval, Computer Algebra, Computer Vision, Natural Language Processing, Pattern Directed and Logic Programming Languages, Knowledge Representation Languages, Deduction Systems). A comprehensive presentation of the Unification Theory is in [Sie89].

2.2.2. Completion.

The technique for completion and critical-pair formation is a problem solving strategy that has been applied in the following apparently independent areas: automated theorem proving [Rob65], polynomial ideal theory [Buc65], solution of word problems in universal algebra [Knu67].

Typical problems treated by this technique in the three above mentioned areas can be formulated in the following way. Given a set T of symbolic objects (i.e. a set logical clauses over an alphabet of functions and predicate symbols, a set of polynomials over a coefficient domain, a set of words over a finite alphabet, respectively) and a binary relation $\rightarrow$ on T, assumed as a *reduction relation* and generated by a finite set of *basic reductions F*, for $s, t \in T$ decide if (s, t) is in the reflexive, symmetric, transitive closure of the relation $\rightarrow$.

Since the early work in the mid sixties, this technique has been widely applied in several areas of algorithmic problem solving. Significant results have been obtained on the analysis and the improvement of algorithms for completion and critical-pair formation, by studying specific important aspects such as termination, definition of strategies for selecting and omitting critical pairs, complexity. Various implementations of the algorithmic formulation of this technique are now available.

A complete characterization of the technique for completion and critical-pair formation in different application areas can be found in [Buc87]. In that paper this technique is presented with a very general formulation that turns out to be uniform for different application areas, in order to show the equivalence of the related algorithmic problem solving strategies and to introduce the concept of *algorithm types* in analogy to that of *data types*.

3. Toward an Intelligent Symbolic Computation System.

Symbolic and algebraic computation systems (DERIVE, MACSYMA, MAPLE, MATHE-MATICA, REDUCE, to name a few) have been widely available for several years. These systems have supported mathematical problem solving in several application areas of sciences and engineering with significant achievements [Buc83, Cav86, Dav88, Mio90a, Mio90b, Yun80]. However these systems are still far from being completely correct and safe from the qualitative point of view of a computation.

Let us address some of the major problems which can rise in using the currently available systems throughout very simple examples. We do not refer here to any specific system. As a matter of fact, the problems we consider here are common in symbolic computation, when using a system designed and implemented with inadequate methodological approaches.

$$Example\ 1$$

1.1)
$$(\sqrt{x})^2 \longrightarrow x$$

1.2)
$$\sqrt{x^2} \longrightarrow x$$

$$Example\ 2$$

Step 1:
$$Let\ y\ be\ \frac{1}{x}$$

Step 2:
$$Let\ z\ be\ \frac{d}{dx} \log x$$

Step 3:
$$y - z \longrightarrow 0$$

$$Example\ 3$$

Step 1:
$$Let\ y\ be\ x^i$$

Step 2:
$$\sum_{0 \le i \le 10} y \longrightarrow 11x^i$$

$$Example\ 4$$

Step 1:
$$Let\ y\ be\ \sum_{i=1}^{n} x^i$$

Step 2:
$$\frac{\partial y}{\partial x_1} \longrightarrow 0$$

Example 5

5.1)

$$\infty - \infty \longrightarrow 0$$

5.2)

$$\infty \cdot 0 \longrightarrow 0$$

Examples 1 and 2 are mathematically incorrect. The functions $(\sqrt{x})^2$ and $\sqrt{x^2}$ are different, being defined in two different domains. The function $\frac{1}{x}$ and the function $\frac{d}{dx} \log x$, even with the same analytical expression $\frac{1}{x}$, are defined in two different domains. This is not taken into any consideration by a system designed simply on the basis of syntactic rewriting rules. Examples 3 and 4 are wrong for completely different reasons. The symbol i is used with different meanings in the two successive steps. A system designed without appropriate semantics can often bring to such an unpleasant situation. Example 5 shows the obvious need for a precise algebra and therefore for an adequate semantics to deal with infinity.

Example 6

$$\sum_{-1 \leq i \leq 10} \int x^i dx \longrightarrow ?$$

In general a message like "division by zero" occurs in this case.

This example again shows the case where the symbol i of the integrand function x^i is not bound by the constraint given in the summation. In order to get ride of this, the computation could proceed as follows, by interchanging the sum and the integral operators and splitting the sum into parts:

Step 1:

$$\longrightarrow \int \sum_{-1 \leq i \leq 10} x^i dx \qquad \longrightarrow \int (x^{-1} + \sum_{0 \leq i \leq 10} x^i) dx$$

Step 2:

$$\longrightarrow \log x + x + \frac{x^2}{2} + \cdots + \frac{x^{11}}{11}$$

Certainly some mathematical knowledge is needed to carry out this computation. But a powerful and flexible system is needed to support the computation.

From these motivations we started a research project to design and implement a new system TASSO for symbolic computation. This project is mainly based on specifications and programming methodologies for manipulating axiomatizable objects.

3.1. The TASSO Project.

The TASSO system deals with abstract entities as objects described by axiomatic specifications. It considers logic formulas and algebraic structures. Each object has a unique formal definition with the specification of its attributes and algebraic, analytical, logic computation are possible under fixed constraints.

On the basis of the instantiation mechanism, dynamically defined objects (e.g. matrices over polynomials over ... matrices of integers) can also be treated.

The specification of a computing method is given at the highest possible level of the object hierarchy, in an abstract way. For instance, the Euclidean Algorithm is specified as an attribute of the Euclidean Domain object. Then the instantiation of the Euclidean Domain will imply the instantiation of all the related computing methods, e.g. the Euclidean Algorithm.

This specification mechanism allows one to exploit the main results on the integration and the amalgamation of numerical and algebraic methods [Mio88]. A specification at very high level of abstraction can be given, for instance, for approximation methods in abstract domains, so to include both the classical Newton method as well as the Hensel construction for polynomial equations over the integers, as special cases.

The TASSO system includes two main modules: TASSO-L and TASSO-D. TASSO-L is an object oriented language which allows the user to define objects, to instantiate abstract structures into actual computing domains, to compute with available objects.

TASSO-D is the module for automated deduction. It can be used both by the user to check properties of given objects and by the system itself in order to generate and derive new properties of given objects from already known properties.

3.2. The language TASSO-L.

The TASSO language follows the object oriented approach. According to [Str87], this approach can be well characterized by the following equation

$$Object\text{-}Oriented = ADT + Inheritance$$

and it appears particularly suitable to our purpose. In fact, all the mathematical objects we like to consider, can be easily modelled by ADT, whose implementation is encapsulated and hidden to the user. The mathematical objects represented by ADT can be considered as generic structures in which data, methods and attributes are specified and localized. The strong type checking mechanism, acting on all the defined objects, is the key to support the correctness of the types and subtypes used at all the different steps of the computation. The available inheritance mechanism gives the possibility to specialize or extend an already defined ADT.

Moreover, a correct cooperation between abstraction and inheritance allows us also to obtain *parametric polymorphism*. Following this approach, the common properties of similar data structures are defined at the highest possible level of abstraction and the methods to perform specific operations are dynamically defined only upon the appropriate operands.

We have considered the most interesting, correct and flexible features of the existing object-oriented programming languages [Lim90, Reg90]. We have also been experimenting with Eiffel, Smalltalk and C++. On the basis of these experiments we have selected the language Loglan [Kre90] as implementation language, also because it refers to the theory of Algorithmic Logic [Mir87] as the main theoretical framework where the specification and the correctness verification can be carried out. In TASSO-L objects are connected in a tree structure by a multi-level *prefixing*. Attributes of the objects can be dynamically defined and redefined by *virtual* specifications, as supported by Loglan.

3.3. The automated deduction mechanisms TASSO-D.

TASSO-D incorporates different deduction mechanisms: *Resolution* with different kinds of *Backtracking* [Bon89], *Connection Method* [Bib87, For90] and *Sequent Calculus* [Gal86, Bon90].

These mechanisms can be activated either directly by the user to accomplish some steps of deduction, or by the system itself, as a tool to guarantee the correctness of the object definitions.

The extension to abduction has been proposed for the Connection Method [For89], as for the extension proposed for the *Resolution* [Kow83].

Furthermore, a *Sequent Calculus* has been defined, following Gallier's proposal, as a single method to support different kinds of deductions: verification, generation and abduction.

4. Conclusion.

The evolution of the research area of Symbolic Computation has been reviewed, first as subarea of Artificial Intelligence, then as autonomous specific area. The exchanges and commonalities of Symbolic Computation and Artificial Intelligence have been considered and the issue of non-numerical information processing has been identified as a basic common component of the two areas.

The design and implementation of a new system for symbolic computation is considered as a natural ground in which the commonalities of the two areas play a fundamental role, allowing one to enlarge the spectrum of symbolic computation systems with relevant processing facilities of automated deduction.

References

[Bib87] Bibel, W., Automated theorem proving, II Ed., Fried. Vieweg & Sohn, Braunschweig/Wiesbaden, (1987).

[Bon89] Bonamico, S., Cioni, G., Embedding flexible control strategies into object oriented languages, LNCS 357, Springer Verlag, (1989).

[Bon90] Bonamico, S., Cioni, G., Colagrossi, A., A Gentzen based deduction method for the propositional theory, Rapp. IASI 289, (1990).

[Buc83] Buchberger, B., Collins, G.E., Loos, R.G.K., (eds.), Computer Algebra - Symbolic and Algebraic Computation, 2nd ed., Springer-Verlag, (1983).

[Buc65] Buchberger, B., An Algorithm for Finding a Basis for the Residue Class Ring of a zero-dimensional polynomial Ideal, Ph.D. Thesis, Univ. Innsbruck (Austria), (1965).

[Buc87] Buchberger, B., History of Basic Features of the Critical-Pair/Completion Procedure, J. Symb. Comp., 3:3-38, (1987).

[Cav86] Caviness, B.F., Computer Algebra: Past and Future, J. Symb. Comp., 2:217-236, (1986).

[Dav88] Davenport, J.H., Siret, Y., Tournier, E., Computer Algebra - Systems and Algorithms for Algebraic Computation, Academic Press, (1988).

[Dep70] Deprit, A., et al., Lunar ephemeris: Delunay's theory revised, Science, 168:1569-1570, (1970).

[For89] Forcellese, G., Temperini, M., A system for automated deduction based on the Connection Method, Rapp. IASI 268, (1989).

[For90] Forcellese, G., Temperini, M., Towards a logic language: an object-oriented implementation of the Connection Method, LNCS 429, Springer Verlag, (1990).

[Gal86] Gallier, J.H., Logic for Computer Science, Harper & Row Publishers, (1986).

[JSC87] Rewriting Rules Techniques and Applications, Special Issue of the J. Symbolic Computation, 1/2, (1987).

[Kah53] Kahrimanian, H.G., Analytic Differentiation by a Digital Computer, M.A. Thesis , Temple Univ., Philadelphia, Pennsylvania, (1953).

[Knu67] Knuth, D.E., Bendix, P.B., Simple Word Problems in Universal Algebra, Proc. Conf. on Computational Problems in Abstract Algebra, Pergamon Press, (1970)

[Kow83] Kowalski, R., Logic for Problem Solving, The Computer Science Library, (1983).

[Kre90] Kreczmar, A., et al., LOGLAN '88 - Report on the Programming Language, LNCS 414, Springer Verlag, (1990).

[Lim90] Limongelli, C., et al., Abstract Specification of Mathematical Structures and Methods, LNCS, 429:61-70, Springer Verlag, (1990).

[Mio91] Miola, A., Symbolic Computation Systems, In: Encyclopedia of Computer Science and Technology, (Kent, A., Williams, J.G., eds.), Marcel Dekker, to appear, (1991).

[Mio88] Miola, A., Limongelli, C., The amalgamation of numeric and algebraic computations in a single integrated computing environment, Rapp. DIS 18.88, (1988).

[Mio90a] Miola, A., (ed.): *Computing tools for scientific problems solving*, Academic Press, (1990).

[Mio90b] Miola, A., (Ed.): *Design and Implementation of Symbolic Computation Systems*, LNCS 429, Springer Verlag, (1990).

[Mir87] Mirkowska, G., Salwicki, A., Algorithmic logic, PWN Warszawa & D. Reidel Publishing Company, (1987).

[Nol53] Nolan, J., Analytic Differentiation on a Digital Computer, M.A. Thesis, M.I.T., Math. Dept., Cambridge, Massachusetts, (1953).

[Reg90] Regio, M., Temperini, M., A short review on Object Oriented Programming: Methodology and language implementations, Proc. ACM Sympos. on Personal and Small Computers, (1990).

[Ris69] Risch, R.H., The problem of Integration in Finite Terms, Trans. A.M.S., 139:167-189, (1969).

[Rob65] Robinson, J.A., A machine Oriented Logic Based on resolution Principle, J. ACM, 12/1:23-41, (1965).

[Sie89] Siekmann, J.H., Unification Theory, J. Symb. Comp. 7:207-274, (1989).

[Sla63] Slagle, J.R., A Heuristic Program that Solves Symbolic Integration Problems in Freshman Calculus, J. ACM, 10:507-520, (1963).

[Str87] Stroustrup, B., What is Object-Oriented Programming, ECOOP 87, (1987).

[Yun80] Yun, D.Y.Y., Stoutemyer, R.D., Symbolic Mathematical Computation. In: Encyclopedia of Computer Science and Technology, Vol.15, (Belzer, J., Holzman, A.G., Kent, A., eds.), Marcel Dekker, 235-310, (1980).

Lecture Notes in Artificial Intelligence (LNAI)

Lecture Notes in Computer Science